Cell Calcium and the Control of Membrane Transport

Society of General Physiologists Series • Volume 42

Cell Calcium and the Control of Membrane Transport

Society of General Physiologists • 40th Annual Symposium

Edited by
Lazaro J. Mandel
Duke University
and
Douglas C. Eaton
Emory University

Marine Biological Laboratory
Woods Hole, Massachusetts

3–7 September, 1986

The Rockefeller University Press
New York

Library of Congress Catalog Card Number 87-050562
ISBN 0-87470-043-4
Printed in the United States of America

This volume is dedicated to
the memory of
Roger Eckert and
Peter F. Baker

Contents

Preface

This book is a compendium of the written contributions submitted by the invited speakers to the 40th Annual Meeting of the Society of General Physiologists, held in Woods Hole, Massachusetts, during September 4–7, 1986. There were also 118 abstracts of contributed papers submitted to this meeting, which have been published in the December 1986 issue of *The Journal of General Physiology.*

In selecting the topic for this meeting, we were influenced by the increasing attention that the role of calcium has been receiving as a possible modulator of a number of cellular processes. This continuing interest is probably due to, first, the enormous success in the near past at demonstrating specific cellular mechanisms that are influenced or modulated by calcium, and second, a recognition that, although the questions about the role of calcium are more specific, complete answers are still not available. The number of investigators in various disciplines who are interested in this topic, as well as the number of publications in this area, have been rapidly increasing in the past few years. Our goal was to attempt to achieve a synthesis by bringing together investigators who work in different systems and are interested in the regulation of intracellular calcium and the modulation of cellular events, especially at the plasma membrane, that are mediated by alterations in intracellular calcium and associated events. A more global view of the role of calcium emerged as investigators described their results using a variety of approaches in numerous tissues.

Besides being a time for an exchange of ideas, the symposium was also a time of particular sadness for some of us. Roger Eckert particularly loved Woods Hole and the fall meeting of the Society of General Physiologists. He so enjoyed the meeting in general, and the topic of the effects of cellular calcium in particular, that he had originally agreed to return early from a research period in Europe to be a speaker for us. Sadly, in the summer of 1986, Roger succumbed to the ravaging effects of a malignant melanoma. We had invited Roger to speak because of the major contributions he has made to our understanding of the control and regulation of calcium flux across cellular membranes. His scientific contributions speak for themselves: almost 20 years of work on various aspects of calcium channels and their control; numerous well-cited publications; and the training of several young investigators who have become distinguished in their own right and who have stepped in with sadness to fill Roger's shoes at this symposium. We respect and applaud Roger's scientific contributions and will miss them in the future. But, as significant as his scientific contributions were, we will miss more the energy, exuberance, and good humor that imbued both his work and his life. We feel that Roger would be pleased to be remembered in a book such as this, which contains material so close to his heart.

While Roger Eckert's death occurred prior to the symposium and was remembered at the time of the presentation of his work, another tragic event took place as we were preparing this volume for press. We were dismayed to hear that the keynote speaker for our symposium, Professor Peter Baker, suffered a fatal heart attack while driving home from his laboratory. At the age of 47, Professor Baker's death seemed particularly distressing; especially so considering his enthusiastic discussion and interaction with other symposium participants. Despite his relatively young age, Professor Baker's accomplishments were numerous. His contributions to the development of the perfused axon preparation were the basis for the work of an entire generation of neurophysiologists and biophysicists. More recently, he played a major role in an examination of the role that intracellular calcium plays in exocytosis. It was primarily on the basis of this recent work that we invited him to provide an overview of the field as the keynote speaker for our symposium. He also was gracious enough to provide an excellent introductory chapter for our symposium volume based on the material of his keynote address.

Because of the major contributions that both Dr. Eckert and Professor Baker have made to the role of calcium as a regulator of physiological activity and because of the personal drive and spirit that they brought to the field, we would like to dedicate this volume to their memory. As mentioned above, the volume begins with a chapter by the late Professor Baker on exocytosis. This is followed by chapters representing the four broad but overlapping areas to which the meeting was devoted. The first area is the regulation of cytosolic free calcium. Chapters by Carafoli and Longoni, Nicholls et al., Benham and Tsien, Mullins and Requena, and Somlyo et al. detail the properties of calcium pathways in the plasma membrane and intracellular organelles involved in the regulation of free calcium and the compartmentation of calcium within the cell. The second major topic concerns receptor-mediated changes in intracellular calcium and phosphatidylinositides, with chapters by Williamson et al., Schulz et al., and Vergara et al. These chapters explore the role of cytosolic calcium and other possible mediators, such as phosphatidylinositides, in receptor-mediated responses in liver, pancreas, and skeletal muscle. The third part of the book is devoted to the modulation of membrane transport by intracellular calcium and calcium channels. The chapter by Grinstein et al. discusses interactions between cytosolic calcium and proton transport; chapters by Snutch et al., Chad et al., Strong et al., and Cannell et al. cover voltage-gated calcium channels from their genesis to their modulation by cytosolic calcium. The fourth part of the book describes various examples of calcium involvement in intracellular events. Chapters by Hannun and Bell, Hansford and Staddon, Ausiello et al., Mandel et al., and Rasmussen and Means describe the activation of protein kinase by diacylglycerol and calcium, the control of mitochondrial pyruvate dehydrogenase by cytosolic calcium, the cytoskeleton and vasopressin action, calcium and anoxic cell injury, and calmodulin in relation to cell growth and gene expression.

We are ourselves very much interested in the role of calcium in regulating cellular events and are pleased and amazed at the significant recent progress that is reflected by the contents of this volume. Nonetheless, the complexity and interrelationship between cellular events provoked one of us (D.C.E.) to relate the following story to the others during the meeting.

> As a young and aspiring graduate student, I was investigating mechanisms of pacemaking in single gastropod neurons. Often I would have difficulty in interpreting the results of my experiments and would seek advice from my thesis adviser, Susumu Hagiwara. On one day in particular, I was especially confounded by results which

seemed more complex and baffling than usual. After confronting Susumu with the data, I was disappointed to find that this once he did not have an immediate explanation, or at least a suggestion for an additional experiment. As I was about to leave his office, he said he was sorry he couldn't help me, but remarked with a smile, "If the results are so complicated, then calcium must be involved somewhere!" Of course, he was right, although it required a number of more years and other investigators to demonstrate the various interactions of calcium ions with the spontaneous pacemaking activity of gastropod neurons. Nonetheless, as I listened to the invited speakers and examined the various contributed posters of this, the 40th Symposium of the Society of General Physiologists, I could not help but think of that conversation many years before.

Douglas C. Eaton
Lazaro J. Mandel

Keynote Presentation

Chapter 1

Theme and Variation in the Control of Exocytosis

P. F. Baker and D. E. Knight

Department of Physiology, MRC Secretory Mechanisms Group, King's College London, Strand, London, England

Introduction

In the early 1970's, I (P.F.B.) found myself on more than one occasion lecturing alongside George Palade, he on his elegant analysis of protein secretion (see Palade, 1975) and I on the newly discovered intricacies of Ca homeostasis in squid axons (Baker, 1972). It was apparent to me—and I assume to many others—that a major problem in the study of exocytosis was its inaccessibility to experimental manipulation. My first tentative attempts to learn about exocytosis were made in Cambridge with Andrew Crawford and Tim Rink. On moving to King's College in 1976, I was joined by Derek Knight and together we embarked on a more direct frontal attack on the problem.

The Problem Defined

Exocytosis is one of the better-defined components of membrane flow. It refers to the fusion of cytoplasmic vesicles with the inner surface of the plasma membrane and it serves two major physiological functions: to alter the composition of the cell surface, for instance, by the insertion of receptors, channels, and pumps, and to release into the external medium molecules previously trapped within the vesicle. Included in this latter category is the release of numerous nervous transmitter substances, hormones, and enzymes that are central to many systems of communication in multicellular organisms. Despite their general physiological importance, surprisingly little is known at the molecular level of the mechanisms that permit, direct, and control exocytosis and other components of membrane flow within cells.

Exocytosis can handle vesicles of markedly different sizes and, in the fastest-responding systems, can occur in <1 ms. It is normally monitored by the appearance in the extracellular fluid of some secretory product, but this does not necessarily provide much information about the underlying fusion process, as there may be a delay between fusion and secretion. More direct information can be obtained either by electron microscopy (Heuser et al., 1979; Chandler and Heuser, 1980; Torri-Tarelli et al., 1985), or, in some favorable cases, electrically, by monitoring fluctuations in membrane capacitance (Cole, 1935; Gillespie, 1979; Neher and Mary, 1982; Fernandez et al., 1984). Electron microscopy, especially of fast-frozen material, can give information about the time course and morphology of fusion, while capacitance, which for a particular membrane system is proportional to membrane area, permits individual fusion and fission events to be monitored directly.

The striking result from electron microscopy is that the initial fusion event is very highly localized (Chandler and Heuser, 1980; Ornberg and Reese, 1981; Schmidt et al., 1983) and may result from a single molecular interaction. The narrow pore that is generated widens rapidly into the equivalent of the wide-mouthed Ω-profile seen in transmission electron micrographs and may finally flatten out, becoming indistinguishable from the plasma membrane. In the particular case of the adrenal medulla, application to stimulated cells of membrane markers specific to secretory vesicle antigens reveals discrete patches of vesicle membrane incorporated into the plasma membrane (Lingg et al., 1983). These patches are slowly and selectively retrieved by endocytosis. The extent to which vesicles flatten out and the time course and route of membrane retrieval probably differ from cell to cell, but, in general, they are poorly understood.

There is good evidence that exocytosis, like motility, can be subject to a variety of different controls. Even within a single cell, it appears that exocytosis is not a homogeneous process, but rather a set of parallel processes, each with its characteristic features and controls. The two most obvious are the so-called constitutive, or nonregulated, and triggered, or regulated, forms of exocytosis in which vesicles either fuse readily with the plasma membrane (constitutive) or queue up, awaiting a suitable stimulus (triggered) (see Kelly, 1985). In many cells, triggered exocytosis can be further subdivided in terms of the stimulus required to effect secretion (see Lundberg and Hökfelt, 1983; Bartfai, 1985) and, by analogy, it seems possible that constitutive exocytosis will also prove not to be a single homogeneous process.

From the viewpoint of mechanism, it is necessary to seek molecular explanations for the following features of exocytosis: (*a*) How is membrane fusion brought about? (*b*) What determines which membranes fuse and which do not? (*c*) How are the various forms of exocytosis controlled?

Remarkably little is known to attempt even a rudimentary answer to the first two questions, but data are slowly accumulating to permit the third question, control, to be addressed more fully in at least a few selected tissues.

Mechanism and Specificity

Although the molecular mechanisms that underlie specificity and fusion remain a mystery, it is worth examining some general principles that may well apply. Most membrane systems are reluctant to fuse spontaneously and secretory vesicles can be subjected to very high pressures, for instance during centrifugation, without undergoing fusion. This observation suggests that membrane fusion is brought about by a specific set of biochemical events and the mere proximity of membranes is not enough. Virtually all key cellular processes involve proteins, and although localized alterations in lipid composition might serve to facilitate fusion under some circumstances, the involvement of proteins specialized for effecting fusion seems highly likely. This is certainly the case in the one biological fusion about which something is known at the molecular level. This is fusion of certain enveloped viruses with cells. The trick seems to be that the viral membrane has projecting from its surface a special pH-sensitive "spike" protein (White et al., 1981, 1983). Viruses attach to the cell surface, become internalized into endocytotic vacuoles, and, as the pH inside the vacuoles falls, the spike protein undergoes a conformational change, revealing a hydrophobic sequence that buries itself into the neighboring wall of the vacuole. It is this step that seems to lead to fusion of the viral and vesicular membranes, permitting escape of the viral contents into the cytosol. Clearly, the experimental evidence does not favor a pH-regulated process in exocytosis, but the dependence on a protein specialized for effecting fusion seems highly likely. Whatever mechanism is used, it must permit both of specificity—in that only certain membranes fuse with each other—and of control.

Two very general solutions can be envisaged (Fig. 1). By analogy with viral fusion, described above, the fusion mechanism could be totally built into one of the reacting membranes (one-sided fusion) or fusion could require a specific molecular contribution from both reacting membranes ("key-in-the-lock" fusion).

In its simplest form, one-sided fusion would not permit of any specificity or control, but by blocking the fusion reaction, both could be achieved, especially if removal of the block were dependent on interaction with a component of the target membrane. Key-in-the-lock fusion, on the other hand, has built-in specificity, but would occur spontaneously in the absence of a blocking reaction. As is clear from the very diagrammatic representation in Fig. 1, a single fusion process could be subject to more than one type of block, each of which could be controlled differently. Until more is known of the underlying molecular mechanisms of exocytosis, Fig. 1 is put forward in the hope that it may stimulate new experimental approaches to the problem.

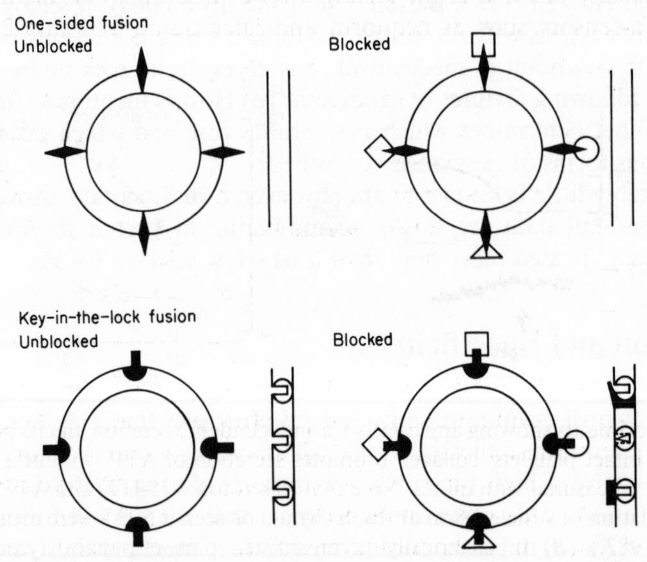

Figure 1. Membrane fusion may depend on specific proteins. The figure shows two possible scenarios for fusion of vesicles with the plasma membrane: one-sided fusion, in which the fusion mechanism is wholly built into one of the reacting membranes, and key-in-the-lock fusion, in which both membranes contribute essential components of the fusogenic process. The left-hand diagrams show the situation in which fusion will occur spontaneously and the right-hand diagram illustrates some ways in which spontaneous fusion can be blocked and thereby put under control. The location of the reacting molecules is only diagrammatic and their positions could be reversed.

Control of Exocytosis

General Features

As its name implies, continuous or nonregulated exocytosis has not so far been shown to be subject to specific control, although it is of some interest that fusion of microsomal membranes derived from the liver, a very active secretor, requires low levels of Ca (Judah and Quinn, 1978) and both continuous exocytosis and evoked exocytosis virtually cease during mitosis (Warren, 1985). In the terminology of Fig. 1, continuous secretion is an obvious candidate for either blocked, one-sided fusion with removal of the block at the reacting membrane or unblocked, key-in-the-lock fusion.

The rest of this discussion will focus on regulated or triggered exocytosis, which has, until recently, been viewed as a Ca-dependent process. This is particularly true in most nervous and endocrine tissues, although a role for cAMP is strongly implicated in salivary and pancreatic exocrine secretion. The classic view derives from studies of transmitter release at the neuromuscular junction, where the arrival of the nerve impulse can increase acetylcholine release transiently by 10,000-fold and the whole of this increase is dependent on external Ca (Katz, 1966; Ceccarelli and Hurlbut, 1980). Douglas (1968) subsequently widened the range of preparations examined and obtained evidence for a central role for Ca in stimulus-secretion coupling. The weight of experimental evidence was consistent with a triggering of exocytosis by a rise in free Ca inside the cell, although a slow rate of secretion is usually possible at the resting level of free Ca. With the development of suitable Ca-sensors such as aequorin and later quin2 and fura-2, it became

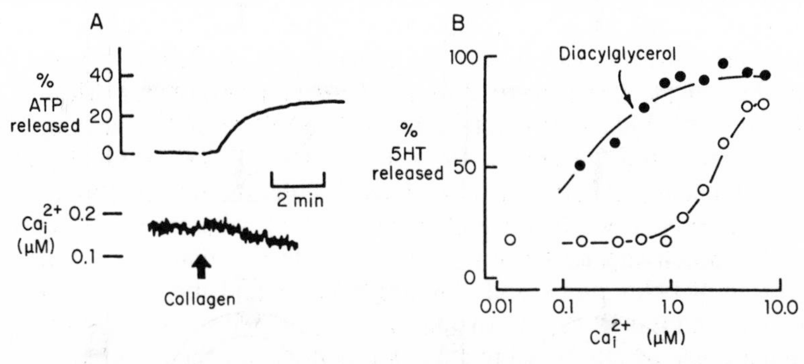

Figure 2. Experiments showing apparently Ca-independent secretion and its possible explanation. (*A*) In intact platelets, collagen promotes secretion of ATP without a concomitant rise in free Ca^{2+} measured with quin2. Note that as serotonin (5-HT) and ATP are stored in the same population of vesicles, similar results would be seen if 5-HT were measured. (From Rink et al., 1983.) (*B*) In electrically permeabilized platelets, diacylglycerol (which is produced after exposure to collagen) brings about a marked increase in the sensitivity of the exocytotic machinery for Ca. (From Knight and Scrutton, 1984*a*.)

possible to test this hypothesis directly. In general, it seems that a rise in free Ca from its resting level close to 100 nM into the micromolar range is a sufficient stimulus for most exocytosis systems (Baker, 1972, 1974; Llinas et al., 1981; Knight and Kesteven, 1983; Rink and Hallam, 1984). This view is supported by the finding that microinjection of Ca into secretory cells, or increasing their permeability to Ca by exposure to a Ca ionophore such as A23187 or ionomycin, also brings about a rise in intracellular free Ca and initiates secretion. However, close inspection shows that secretion evoked by exposure to an ionophore may need a somewhat higher free Ca than secretion evoked physiologically, which suggests that factors other than Ca may also be involved.

Of course, it is always possible that this discrepancy may arise either because a physiological stimulus brings about a highly localized rise in free Ca or because the ionophore is increasing Ca in the wrong part of the cell. However, such arguments are difficult to apply to experiments such as the one illustrated in Fig. 2*A*, where exposure of platelets to the physiological stimulus collagen triggers serotonin release with little or no detectable change in free Ca (Rink et al., 1983), and in some systems, such as the parathyroid gland, release is associated with a very clear *fall* in free Ca (Shoback et al., 1984). These experiments suggest that Ca may be only one of a number of factors involved in the control of exocytosis. Other factors probably include cyclic nucleotides, products of phospholipid metabolism, guanosine nucleotides, and messenger molecules yet to be discovered.

Studies on Permeabilized Cells

A major advance in elucidating the relative importance of these different systems has been the development of permeabilized preparations in which the intracellular environment of the exocytotic machinery is subject to experimental control (Baker and Knight, 1978, 1981, 1987; Knight and Baker, 1982; Baker et al., 1985; Gomperts and Fernandez, 1985; Knight and Scrutton, 1986). Permeabilization can be achieved by a variety of techniques, of which two, electro-permeabilization (Baker and Knight, 1978, 1981; Knight and Baker, 1982) and detergent permeabilization (Brooks and Treml, 1983; Dunn and Holz, 1983; Wilson and Kirshner, 1983), have been widely used. In suitable preparations, the cell membrane can also be permeabilized by exposure to certain toxins (Thelestam and Möllby, 1979), viruses (Gomperts et al., 1983), complement (Schweitzer and Blaustein, 1980), or ATP^{4-} (Cockcroft and Gomperts, 1980; Bennett et al., 1981). Of the two most common approaches, detergents make large lesions in the membrane, but also rapidly inhibit exocytosis (Baker and Knight, 1981), whereas electro-permeabilization makes much smaller lesions ("holes" of effective diameter 4 nm), but has the advantage of being quick and chemically clean and the "holes" can be very stable (Knight and Baker, 1982). An alternative approach suitable for some systems is to attach a single secretory cell to a patch pipette. Once the membrane under the pipette is broken, the cell interior can be perfused and exocytosis can be monitored as changes in capacitance (Neher and Marty, 1982; Fernandez et al., 1984). Cortical plaques generated by breaking sea urchin eggs attached to a suitable substrate provide an even more accessible preparation consisting of portions of plasma membrane with secretory vesicles attached (Vacquier, 1975; Whitaker and Baker, 1983).

A major conclusion from studies of these preparations is that Ca is required for the bulk of secretory systems (Baker et al., 1980; Knight and Scrutton, 1980; Bennett et al., 1981; Knight and Baker, 1982), even ones in which cAMP seems to play a prominent role (Knight and Koh, 1984). A major exception is the mast cell under whole-cell patch clamp, where exocytosis can be extremely difficult to trigger with anything (Neher and Almers, 1986). Other messenger systems seem, in the main, to modify the sensitivity of the exocytotic machinery to Ca. Thus, in the example shown in Fig. 2*A*, collagen triggers the hydrolysis of phosphatidylinositides, and the diaclyglycerol that is liberated increases the apparent affinity of exocytosis for Ca such that secretion can take place at the resting free Ca within the cell (Fig. 2*B*; Knight and Scrutton, 1984*a*, *b*). So far, it seems that the primary messengers acting rather directly in exocytosis are Ca, which is hydrophilic and presumably cytosolic, and diacyglycerol (DG), which is highly hydrophobic and almost certainly restricted to membranes. Both are under physiological control, and anything that alters either or both is likely to affect the rate of exocytosis. Other messenger systems such as cyclic nucleotides may only be secondary in that they serve to modulate the production of the primary messengers. For instance, cAMP can both increase the probability of opening of Ca channels in response to depolarization and also alter the rate of DG production, but it should be stressed that studies in this area are only in their infancy and much still remains to be discovered. Thus, GTP is required for DG production (Haslam and Davidson, 1984; Berridge, 1984), but it seems likely that its role is best described as permissive rather than as a messenger.

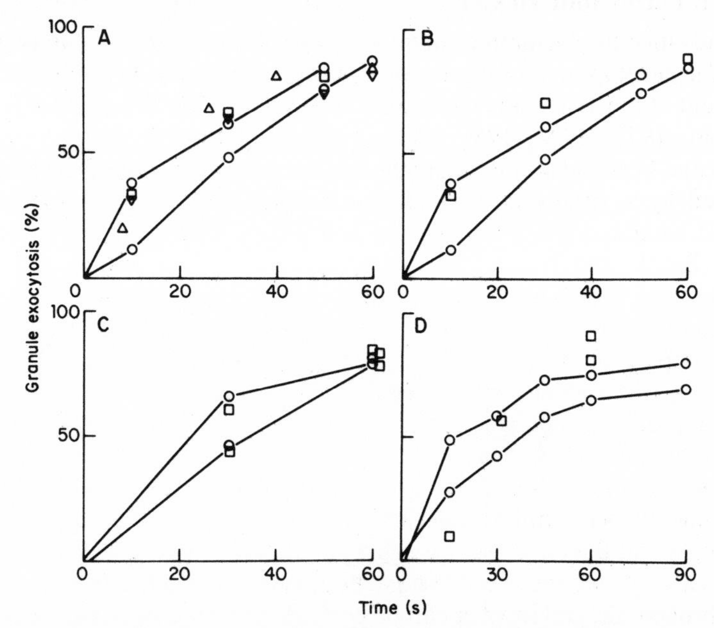

Figure 3. The effects of inhibitors of microtubule and microfilament function on the cortical granule exocytosis in portions of sea urchin egg cortex isolated by attachment to glass coverslips. (*A*) △, 100 μM colchicine; ▽, 100 μM vinblastine; □, 100 μM nocodazole; ○, control responses to test solution (~2 μM free Ca) obtained at the beginning and end of the experiment. (*B*) 100 μM phalloidin (added from stock in isolation medium). (*C*) 10 μM cytochalasin B. (*D*) 20 μM NEM-81 (added from 0.5 M KCl); □, experimental; ○, controls. The isolated cortical fragments were incubated for 10 min with each drug before the addition of test solution containing the same concentration of inhibitor. Drugs were added from dimethylsulfoxide (1%), which was applied at this concentration in controls. Cortical granule exocytosis was assessed photographically. (*A* and *C*) *Echinus esculentus*; (*B*) *Psammechinus miliaris*; (*D*) *Lytechinus pictus*. (From Whitaker and Baker, 1983.)

Responding to the Primary Message

How do Ca and DG alter the rate of exocytosis? It is still not possible to give even a remotely satisfactory answer to this crucial question, but in the terminology of Fig. 1, one possibility might be that they are involved in generating a fusogenic process. They might, for instance, be involved in the creation of a fusogenic molecule or in the removal of a block from an underlying fusogenic molecule.

Permeabilized cells have permitted some features of exocytosis to be elucidated under conditions where the normal physiological controls have been bypassed. Only a limited number of secretory systems have been investigated, but there are some common features. (*a*) Exocytosis is activated in the micromolar range of Ca concentrations. (*b*) It is largely unaffected by a wide variety of agents that interact specifically with the cytoskeleton (see Fig. 3). (*c*) It can usually be inhibited by agents that also inhibit both calmodulin-dependent processes and protein kinase C. (*d*) It is usually inhibited by raising the osmotic pressure of the medium by addition of sucrose.

Systems differ, however, in other features, especially their sensitivity to removal of ATP. The cortical reaction in the sea urchin egg persists for at least 1 h after complete removal of ATP from the superfusion fluid (Baker and Whitaker, 1978), whereas secretion from adrenal medullary cells, human platelets, mast cells, and exocrine and endocrine pancreas ceases within minutes of exposure to an ATP-free medium. Where it exists, this requirement for ATP is very specific and the preferred substrate is MgATP, which suggests that ATP may be involved in phosphorylation. The cells that require MgATP also usually display some sensitivity to phorbol esters and analogues of DG, as well as GTPγS, which probably acts, at

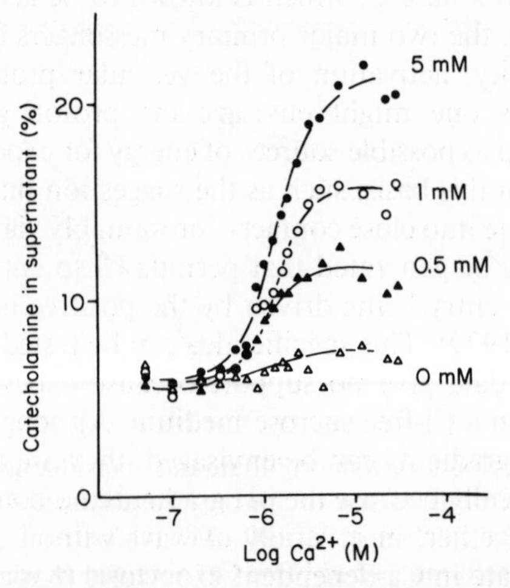

Figure 4. Dependence on MgATP of catecholamine release from permeable bovine adrenal medullary cells. Cells were suspended in K glutamate medium and rendered permeable by exposure to 10×2 kV pulses; $\tau = 200\ \mu s$ (see Knight and Baker, 1982). Note that lowering MgATP reduces the extent of secretion but not its apparent affinity for Ca.

least in part, by stimulating production of endogenous DG. Although the sea urchin egg may have a tightly bound store of ATP, the simplest hypothesis consistent with the data is that ATP is not essential for exocytosis in this tissue and that secretory control is exerted directly via a calmodulin-like molecule. This seems not to be the case in the various mammalian systems examined, where there is a rather specific requirement for millimolar amounts of MgATP (Fig. 4).

What Is the Role of MgATP?

Although a number of proteins are phosphorylated in a Ca-dependent fashion, no one has discovered a substrate that is phosphorylated in a manner uniquely associated with exocytosis in eukaryotes, but in view of the many possibilities inherent in Fig. 1, this may be too restrictive a criterion. The failure to find a protein that is phosphorylated (or dephosphorylated) in parallel with exocytosis may also have a purely technical explanation in that minor phosphorylated species are very hard to detect. For instance, if only one phosphorylation occurred per vesicle, the total amount of phosphate incorporated would be extremely difficult to detect in a permeabilized preparation, which is effectively a whole cell, and exocytosis has not so far been reconstituted reproducibly in simpler systems. One phosphoprotein, synapsin I, seems to be generally distributed in the mammalian nervous system, where it apparently coats the cytosolic faces of synaptic vesicles and may play some role in exocytosis (Navonne et al., 1984; Llinas et al., 1985). In certain protozoans, there is very clear evidence for dephosphorylation associated with exocytosis of trichocysts (Zeiseniss and Plattner, 1985).

Where MgATP is essential, one experimental approach is to examine possible ATP-dependent reactions that may be involved in secretion. Two stand out: (*a*) as substrate for the proton pump that generates the acid internal environment and positive internal membrane potential of the secretory granule, and (*b*) as substrate for the enzyme protein kinase C, which is known to be activated both by Ca and DG (Nishizuka, 1984), the two major primary messengers for exocytosis.

The first possibility, activation of the vesicular proton pump, is of some theoretical interest, as one might envisage the proton gradient and potential developed by the pump as possible sources of energy for exocytosis. Specific models have been proposed on this basis, such as the suggestion that when the vesicle and plasma membrane come into close contact—presumably via a Ca-dependent step—an anion channel may be generated that permits Cl to enter the vesicle from the extracellular fluid, the entry being driven by the positive internal potential of the vesicle (Pollard et al., 1979). This specific idea can be tested in permeabilized cells, but the experimental data give no support because exocytosis can be triggered, apparently normally, in a Cl-free sucrose medium. Although other possible mechanisms utilizing these gradients can be envisaged, they are rather unlikely to be of importance physiologically because the pH gradient and potential can be collapsed, either separately or together, in a variety of ways without affecting the ability of the vesicles to participate in Ca-dependent exocytosis (Knight and Baker, 1985*a*).

The second possibility seems much more interesting, especially in view of the finding that most exocytotic systems are affected by analogues of DG (Knight and Baker, 1983; Peterfreund and Vale, 1983; DiVirgilio et al., 1984; Jones et al., 1985) and can be inhibited by a variety of compounds that inhibit protein kinase C, albeit rather nonspecifically. Protein kinase C requires Ca, a phospholipid, and DG for maximal activity. It utilizes MgATP as substrate and is activated in the low millimolar range. DG increases the apparent affinity of the enzyme for Ca and can be replaced in this action by certain phorbol esters, including 12-*O*-tetradecanoyl phorbol 13-acetate (TPA). So far, protein kinase C is the only known substrate for these phorbol esters, and, from the viewpoint of exocytosis, it is particularly interesting that protein kinase C becomes strongly associated with the plasma membrane in the presence of TPA and DG (Kraft and Anderson, 1983). Secretory systems do not all respond to phorbol esters in the same way. Fig. 5 summarizes the three main types of response that have been observed in permeabilized cells. Type I exocytosis is characterized by a rather small leftward shift in the Ca-activation curve after addition of the phorbol ester TPA or analogues of DG. Type II has a very large leftward shift such that, at high concentrations of TPA or DG, secretion seems to become independent of Ca. In type III exocytosis, phorbol esters do not alter the affinity for Ca but increase the extent of secretion. Some examples are: type I: catecholamine release from the adrenal medulla (Knight and Baker, 1983); type II: serotonin release from human platelets (Knight and Scrutton, 1984*a*, *b*) and possibly histamine release from mast cells (White et al., 1984); type III: β-*N*-acetylglucosaminidase release from human platelets (Knight et al., 1984) and insulin release from pancreatic β cells (Jones et al., 1985).

In the platelet, there are at least two distinct populations of secretory granules, one of which exhibits type II and the other type III exocytosis. The coexistence of different types of exocytosis in the same cell may provide a mechanism for effecting differential release of secretory products. Thus, at low cytosolic free Ca, a rise in DG will provoke type II exocytosis but will have little or no effect on type III until the free Ca is also elevated.

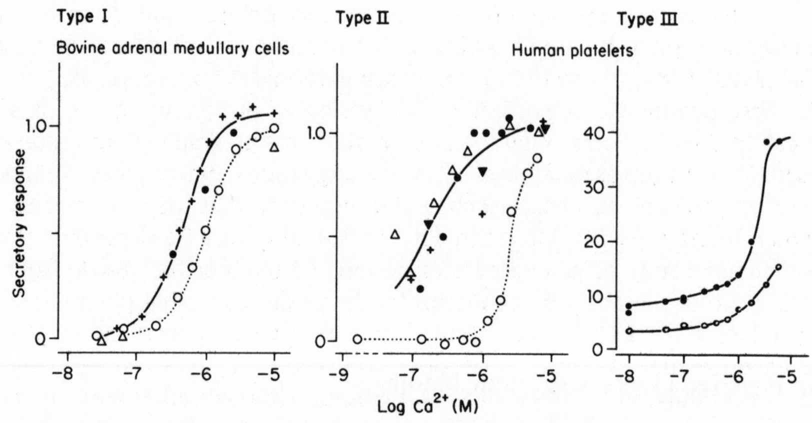

Figure 5. Modification of Ca-dependent secretion in "leaky" cells by exposure to the phorbol ester TPA and other agents: examples of the three classes of response that have been observed. Data were collected in absence of TPA (○). Responses in the presence of TPA: Type I: a small leftward shift seen in catecholamine secretion from bovine adrenal medullary cells in presence of 3 nM TPA (+), 100 μM dioctanoxyglycerol (●). (From Knight and Baker, 1983.) Type II: a marked leftward shift seen in 5-HT secretion from human platelets: 30 nM (TPA) (●); 20 μM DG (▼); 50 μM GTPγS (△). Type III: the lack of any shift, only an increase in the extent of exocytosis of *N*-acetylglucosaminidase also from platelets: 10 ng/ml TPA (●). (From Knight et al., 1984.)

Protein Kinase C and Exocytosis

Comparisons between the properties of protein kinase C determined in the test tube and exocytosis in "leaky" cells are fraught with problems, two of the most important being a lack of knowledge of the physiological substrates of the enzyme (it is conventionally assayed by phosphorylation of histone) and the very real possibility that its properties may change during isolation or when it no longer has membranes with which to associate. In addition, diacylglycerol is a generic term and it is possible that different forms of C-kinase prefer different members of the diacyglycerol family.

As the enzyme requires both Ca and DG for activation, it is of interest to consider what would happen if the enzyme had a preferred order of binding of these two substrates. If we neglect the binding of phospholipid and MgATP, there are three possibilities:

$$E \xrightleftharpoons{DG} E^{DG} \xrightleftharpoons{Ca} E^{DG}_{Ca} \quad (1)$$

$$E \xrightleftharpoons{Ca} E_{Ca} \xrightleftharpoons{DG} E^{DG}_{Ca} \quad (2)$$

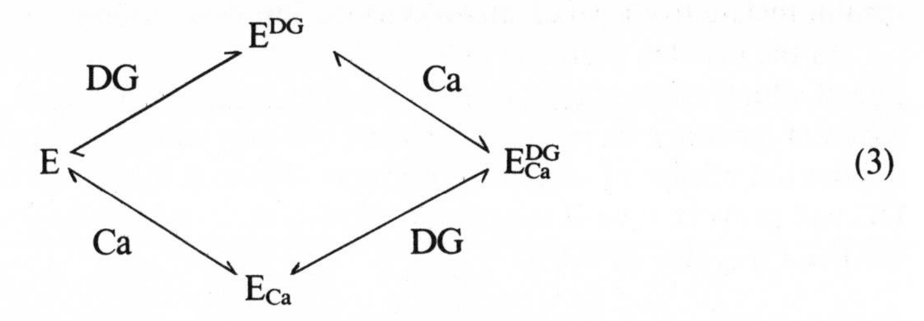

(3)

The Ca dependence of exocytosis predicted by these three schemes is shown in Fig. 6. They are quite different. Scheme 1 shows a small, but limited, leftward shift on adding DG; Scheme 2 shows a larger shift that continues to increase as the DG concentration is increased such that at high DG concentrations, secretion will seem to be Ca independent; and Scheme 3 shows no shift at all, only an increase in the extent of secretion.

The striking parallel between the Ca dependence of Fig. 6 and the three major patterns observed in permeable cells may be more than a coincidence (see Baker, 1986). Thus, Scheme 1 is remarkably similar to type I, Scheme 2 to type II, and Scheme 3 to type III. It is certainly a very interesting point that these apparently different kinetics can all be generated via the same enzyme simply by specifying different preferred orders of substrate binding.

Inhibitors of C-Kinase

In view of the data presented so far, it is important to establish whether protein kinase C is essential for exocytosis or whether it merely plays a modulatory role. This is a difficult question to anwer unequivocally because there are no really specific inhibitors of the kinase. The isolate kinase shares with calmodulin sensitivity to trifluoperazine and other so-called "calmodulin antagonists" and can also be inhibited by high concentrations of amiloride and polymixin B. Exocytosis is also sensitive to these agents, but none of the inhibitors is specific enough to clinch the argument.

It is particularly striking that no inhibitor has been found that can remove the TPA shift while leaving Ca-dependent exocytosis otherwise unaffected (Knight, 1986*a*). Agents that inhibit exocytosis all seem to leave a residual TPA shift, which is consistent with the view that the kinase is directly involved in the machinery of exocytosis.

In conclusion, for systems that are sensitive to both Ca and DG, the attraction of the protein kinase C hypothesis is that one single molecule serves as receptor for Ca, DG, and MgATP, but this economy should be viewed with caution. It does, however, seem a reasonable working hypothesis that at least one subset of exocytotic reactions may utilize protein kinase C as part of the control process. Not all exocytotic systems are sensitive to phorbol esters, however, and it seems quite possible that other subsets of exocytotic reactions may utilize other Ca receptors, such as calmodulin (Baker and Whitaker, 1980; Steinhardt and Alderton, 1982;

Trifaro and Konigsberg, 1983), calelectrin (Sudhof et al., 1985), synexin (Creutz et al., 1978), and the chromobindins (Creutz et al., 1983). Even where protein kinase C may be the main channel through which Ca acts, we do not know *how* it brings about exocytosis; in particular, we do not know the nature of the substrates, if any, that must be phosphorylated and how fusion itself is generated.

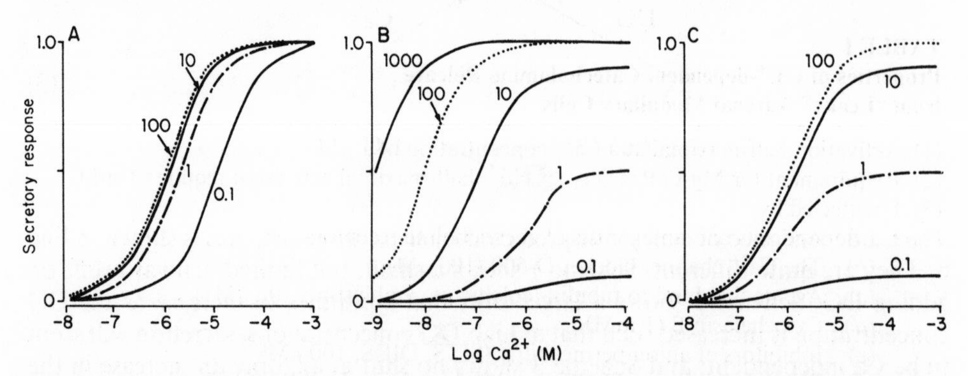

Figure 6. Ca dependence of secretion calculated on the assumption of ordered binding of DG and Ca to protein kinase C. The numbers by the curves refer to DG concentrations, given in units of micromolar.

$$(A)\quad E \underset{K_D}{\overset{DG}{\rightleftharpoons}} E^{DG} \underset{K_{Ca}}{\overset{Ca}{\rightleftharpoons}} E^{DG}_{Ca} \qquad F_{E^{DG}_{Ca}} = \frac{1}{1 + \frac{K_{Ca}}{Ca}\left(1 + \frac{K_D}{D}\right)}$$

$$(B)\quad E \underset{K_{Ca}}{\overset{Ca}{\rightleftharpoons}} E_{Ca} \underset{K_D}{\overset{DG}{\rightleftharpoons}} E^{DG}_{Ca} \qquad F_{E^{DG}_{Ca}} = \frac{1}{1 + \frac{K_D}{D}\left(1 + \frac{K_{Ca}}{Ca}\right)}$$

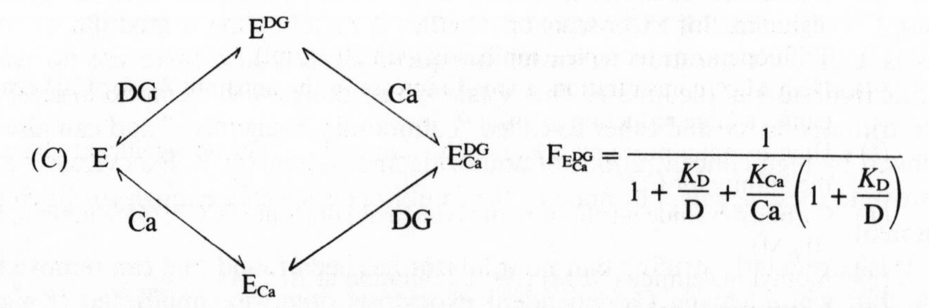

$$F_{E^{DG}_{Ca}} = \frac{1}{1 + \frac{K_D}{D} + \frac{K_{Ca}}{Ca}\left(1 + \frac{K_D}{D}\right)}$$

For ease of calculating the curves, the affinity constants K_{Ca} and K_D for Ca and DG (D), respectively, have been set at 1 μM. (From Baker, 1986.)

Other Control Factors in Exocytosis

Table I lists a number of agents that can inhibit Ca-dependent exocytosis in permeable adrenal medullary cells. They include chaotropic anions, high osmotic pressures, GTPγS, and botulinum toxins. In no instance is the mechanism of inhibition known, but each may provide important clues to the mechanism of exocytosis.

Thus, chaotropic anions may serve to dissociate some macromolecular assembly that is essential for exocytosis, and inhibition by high osmotic pressures may reflect the importance of osmotic factors at some stage in the exocytotic process (see reviews by Baker and Knight, 1984; Holz, 1986; Lucy and Ahkong, 1986). One possibility is that once a point fusion between vesicle and plasma membrane has been generated, entry of water through this microchannel may be essential to

TABLE I
Properties of Ca^{2+}-dependent Catecholamine Release from "Leaky" Adrenal Medullary Cells

(1) Activation half-maximal at a Ca^{2+} concentration of 1 μM
(2) Requirement for MgATP is very specific: half-maximal activation requires 1 mM
(3) Unaffected by
- (*a*) Agonists and antagonists of acetylcholine receptors
- (*b*) The Ca^{2+} channel blocker D-600 (100 μM)
- (*c*) Agents that bind to tubulin (colchicine, vinblastine, 100 μM)
- (*d*) Cytochalasin B (1 mM)
- (*e*) Inhibitors of anion permeability (SITS, DIDS, 100 μM)
- (*f*) Protease inhibitors TLCK (1 mM), leupeptin (1 mM)
- (*g*) Cyclic nucleotides (cAMP, cGMP, 1 mM)
- (*h*) *S*-adenosyl methionine (5 mM)
- (*i*) Phalloidin (1 mM)
- (*j*) Vanadate (10^{-4} M)
- (*k*) Leu- and Met-enkephalins, substance P (100 μM)
- (*l*) Somatostatin (1 μM)
- (*m*) NH_4Cl (30 mM)
- (*n*) Trimethyltin (0.2 mM)

(4) Inhibited by
- (*a*) Chaotropic anions: SCN > Br > Cl
- (*b*) Detergents (complete inhibition after 10 min incubation with 10 μg/ml of digitonin, Brij 58, or saponin)
- (*c*) Trifluoperazine (complete inhibition with 20 μg/ml)
- (*d*) High Mg^{2+} concentration: a small increase in the apparent K_m for Ca^{2+} accompanies a large reduction in V_{max}
- (*e*) High osmotic pressure: a large reduction in V_{max}, but no significant change in the affinity for Ca^{2+}
- (*f*) Carbonylcyanide *p*-trifluormethoxyphenylhydrazone (FCCP) (45% inhibition by 10 μM)
- (*g*) *N*-ethyl maleimide (NEM) (100% inhibition at 10^{-4} M)
- (*h*) Neomycin (1 mM)
- (*i*) Amiloride (1 mM)
- (*j*) GTPγS (50% inhibition at 5 μM)
- (*k*) Botulinum toxin types A, B, C, and D
- (*l*) Tetanus toxin

enlarge and complete the fusion process. Such water flow may be prevented or reversed by high external osmotic pressures, permitting the micropore to reseal.

Inhibition by GTPγS and botulinum toxin are quite different. That GTPγS should inhibit at all is surprising because it might be expected to stimulate phospholipase C and thereby raise the endogenous level of DG. Whether or not this occurs, the net effect of GTPγS is strongly inhibitory in bovine adrenal medullary cells, and inhibition persists in the presence of high concentrations of TPA that should saturate the TPA/DG-binding site on protein kinase C. The implication is that GTPγS is exerting a direct inhibitory effect on exocytosis quite distinct from its actions on phospholipase C and the endogenous level of DG. An analogous but stimulatory role of GTP analogues has been proposed by Barrowman et al. (1986) to explain their results on histamine secretion in permeabilized mast cells, and by Knight and Baker (1985*b*) to account for the response of permeabilized chicken adrenal medullary cells to GTPγS. Very recently, Oetting et al. (1986) have shown that 5′-guanylylimidodiphosphate (GPPNHP) and GTPγS stimulate a massive and apparently Ca-insensitive release of parathormone from permeabilized parathyroid cells. The nature of the GTPγS binding site is unknown, but one

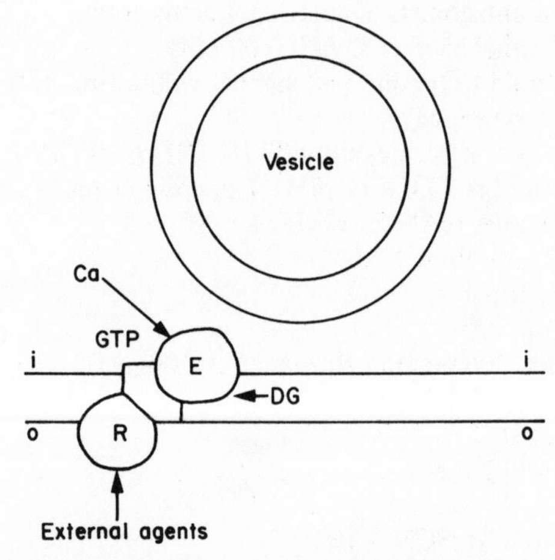

Figure 7. One possible scheme in which a GTP-binding protein is involved directly in the transmembrane control of exocytosis. GTP could be involved in either stimulatory or inhibitory control.

attractive possibility is that G-proteins may be involved directly in exocytosis. Speculating a little, it is possible that the membrane-associated form of protein kinase C, or some other molecule involved in exocytotic control, is also subject to modulation by inhibitory (or stimulatory) G-proteins, and, if this were true, it would not be a very large step to hypothesize that G-proteins might mediate transmembrane control of exocytosis (Fig. 7).

A direct effect of this kind might, for instance, underlie a number of presynaptic control mechanisms. One specific case where this might also apply is the inhibition of insulin release from pancreatic β cells after sympathetic stimulation. This effect persists after the application of noradrenaline to permeabilized β cells and a direct effect on exocytosis seems to be a real possibility (Jones et al., 1986).

Inhibition by botulinum toxin is also exciting since it persists for many days and may lead ultimately to the isolation of one of the components of the secretory machinery. Botulinum toxin is a potent inhibitor of neuromuscular transmission, where it appears to bind to specific acceptors on the outer surface of the nerve terminal, from where it is presumed to gain entry into the cytosol (Dolly et al., 1984; Simpson, 1985). Recently, an action of botulinum toxins has been demonstrated in cultured bovine adrenal medullary cells (Knight et al., 1985; Knight, 1986*b*). Inhibition of catecholamine release is not associated with block of Ca entry and persists in permeabilized cells that display a lack of response to Ca (Fig. 8). Very high doses are required, but this may reflect a virtual absence of acceptors on the adrenal cell surface. This interpretation has received dramatic confirmation in a very recent series of experiments in which various toxins were introduced directly into the interior of bovine adrenal medullary cells by perfusion through a patch pipette (Penner et al., 1986). In these experiments, a range of botulinum toxin types and also tetanus toxin were very effective blockers of the capacitance increase that follows the introduction of 10 μM free Ca into the cell interior. So far, the molecular target of these toxins has resisted isolation.

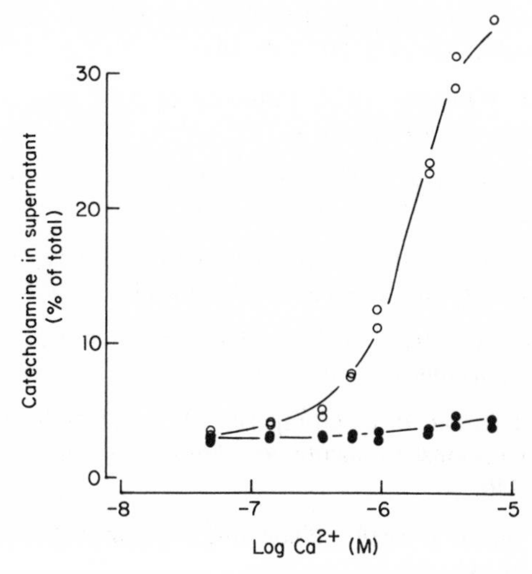

Figure 8. Inhibition of Ca-dependent exocytosis by pretreatment of cultured bovine adrenal medullary cells with botulinum toxin type D. Cells were subsequently rendered permeable and exposed to Ca: control (○); botulinum toxin–treated (●). (From Knight et al., 1985.)

A fascinating feature of botulinum action is that, after exposure to the toxin, adrenal medullary cells accumulate catecholamines and behave in all respects as if they have "forgotten" how to secrete, despite exposure to appropriate messages. Endogenous agents having actions similar to botulinum toxin could play a key role in the long-term alteration of synaptic efficacy, which opens up the possibility that direct effects on the machinery of exocytosis may contribute to both the short-term and the long-term regulation of synaptic function.

References

Baker, P. F. 1972. Transport and metabolism of calcium ions in nerve. *Progress in Biophysics and Molecular Biology.* 24:177–223.

Baker, P. F. 1974. Excitation-secretion coupling. *Recent Advances in Physiology.* 9:51–86.

Baker, P. F. 1986. Protein kinase C and exocytosis. *Progress in Zoology.* 33:265–274.

Baker, P. F., and D. E. Knight. 1978. Calcium-dependent exocytosis in bovine adrenal medullary cells with leaky plasma membranes. *Nature.* 276:620–622.

Baker, P. F., and D. E. Knight. 1981. Calcium control of exocytosis and endocytosis in bovine adrenal medullary cells. *Philosophical Transactions of the Royal Society of London, Series B.* 296:83–103.

Baker, P. F., and D. E. Knight. 1984. Chemiosmotic hypotheses of exocytosis: a critique. *Bioscience Reports.* 4:285–298.

Baker, P. F., and D. E. Knight. 1987. Experimental control of the internal environment of chromaffin cells. *In* In Vitro Methods for Studying Secretion. A. M. Poisner and J. M. Trifaro, editors. Elsevier/North-Holland, New York.

Baker, P. F., D. E. Knight, and J. A. Umbach. 1985. Calcium clamp of the intracellular environment. *Cell Calcium.* 6:5–14.

Baker, P. F., D. E. Knight, and M. J. Whitaker. 1980. The relation between ionized calcium and cortical granule exocytosis in eggs of the sea urchin *Echinus esculentus. Proceedings of the Royal Society of London, Series B.* 207:149–161.

Baker, P. F., and M. J. Whitaker. 1978. Influence of ATP and calcium in the cortical reaction in sea urchin eggs. *Nature.* 276:513–515.

Baker, P. F., and M. J. Whitaker. 1980. Trifluoperazine inhibits exocytosis in sea urchin eggs. *Journal of Physiology.* 298:55P. (Abstr.)

Barrowman, M. M., S. Cockcroft, and B. D. Gomperts. 1986. Two roles for guanine nucleotides in the stimulus-secretion sequence of neutrophils. *Nature.* 319:504–507.

Bartfai, T. 1985. Presynaptic aspects of the co-existence of classical neurotransmitters and peptides. *Trends in Pharmacological Sciences.* 6:331–334.

Bennett, J. A., S. Cockcroft, and B. D. Gomperts. 1981. Rat mast cells permeabilized with ATP secrete histamine in response to calcium ions buffered in the micromolar range. *Journal of Physiology.* 317:355–346.

Berridge, M. J. 1984. Inositol trisphosphate and diacylglycerol as second messengers. *Biochemical Journal.* 220:345–360.

Brooks, J. C., and S. Treml. 1983. Catecholamine secretion by chemically-skinned cultured chromaffin cells. *Journal of Neurochemistry.* 40:468–473.

Ceccarelli, B., and W. P. Hurlbut. 1980. Vesicle hypothesis of the release of quanta of acetylcholine. *Physiological Reviews.* 80:396–441.

Chandler, D. E., and J. E. Heuser. 1980. Arrest of membrane fusion events in mast cells by quick freezing. *Journal of Cell Biology.* 86:666–674.

Cockcroft, S., and B. D. Gomperts. 1980. The ATP^{4-} receptor of rat mast cells. *Biochemical Journal.* 188:789–798.

Cole, K. S. 1935. Electric impedance of *Hipponoë* eggs. *Journal of General Physiology.* 18:877–887.

Creutz, C. E., L. G. Dowling, J. J. Sando, C. Vilar-Palasi, J. H. Whipple, and W. J. Zaks. 1983. Characterization of the chromobindins: soluble proteins that bind to the chromaffin granule membrane in the presence of Ca^{2+}. *Journal of Biological Chemistry.* 258:14664–14674.

Creutz, C. E., C. J. Pazoles, and H. B. Pollard. 1978. Identification and purfication of an adrenal medullary protein (synexin) that causes calcium dependent aggregation of isolated chromaffin granules. *Journal of Biological Chemistry.* 253:2858–2866.

DiVirgillio, F., D. P. Lew, and T. Pozzan. 1984. Protein kinase C activation of physiological processes in human neutrophils at vanishingly small cytosolic Ca^{2+} levels. *Nature.* 310:691–693.

Dolly, J. O., J. Black, R. S. Williams, and J. Melling. 1984. Acceptors for botulinum neurotoxin reside on motor nerve terminals and mediate its internalization. *Nature.* 307:457–460.

Douglas, W. W. 1968. Stimulus-secretion coupling: the concept and clues from chromaffin and other cells. *British Journal of Pharmacology.* 34:451–474.

Dunn, L. A., and R. W. Holz. 1983. Catecholamine secretion from digitonin-treated adrenal medullary chromaffin cells. *Journal of Biological Chemistry.* 258:4989–4993.

Fernandez, J. M., E. Neher, and B. D. Gomperts. 1984. Capacitance measurements reveal stepwise fusion events in degranulating mast cells. *Nature.* 312:453–455.

Gillespie, J. I. 1979. The effect of repetitive stimulation on the passive electrical properties of the presynaptic terminals of the squid giant synapse. *Proceedings of the Royal Society of London, Series B.* 206:293–306.

Gomperts, B. D., J. M. Baldwin, and K. H. Micklem. 1983. Rat mast cells permeabilized with Sendai virus secrete histamine in response to Ca^{2+} buffered in the micromolar range. *Biochemical Journal.* 210:737–745.

Gomperts, B. D., and J. M. Fernandez. 1985. Techniques for membrane permeabilization. *Trends in Biochemical Sciences.* 10:414–417.

Haslam, R. J., and M. M. L. Davidson. 1984. Receptor induced diacylglycerol formation in permeabilized platelets; possible role for a GTP binding protein. *Journal of Receptor Research.* 4:605–629.

Heuser, J. E., T. S. Reese, M. J. Dennis, L. Yan, and L. Evans. 1979. Synaptic vesicle exocytosis explored by quick-freezing and correlated with quantal transmitter release. *Journal of Cell Biology.* 81:275–300.

Holz, R. W. 1986. The role of osmotic forces in exocytosis from adrenal chromaffin cells. *Annual Reviews of Physiology.* 48:175–189.

Jones, P. M., J. M. Fyles, and S. L. Howell. 1986. Regulation of insulin secretion by cAMP in rat islets of Langerhans permeabilized by high voltage discharge. *FEBS Letters.* 205:205–209.

Jones, P. M., J. Stutchfield, and S. L. Howell. 1985. Effects of Ca^{2+} and a phorbol ester on insulin secretion from islets of Langerhans permeabilized by high-voltage discharge. *FEBS Letters.* 191:102–106.

Judah, J. D. and P. S. Quinn. 1978. Calcium ion-dependent vesicle fusion in the conversion of proalbumin to albumin. *Nature.* 271:384–385.

Katz, B. 1966. Nerve, Muscle and Synapse. McGraw-Hill, New York.

Kelly, R. B. 1985. Pathways of protein secretion in eukaryotes. *Science.* 230:25–31.

Knight, D. E. 1986*a*. Calcium and exocytosis. CIBA Foundation Symposium 122: Calcium and the Cell. D. Everard and J. Whelan, editors. John Wiley & Sons, London. 250–270.

Knight, D. E. 1986*b*. Botulinum toxin types A, B, and D inhibit catecholamine secretion from bovine adrenal medullary cells. *FEBS Letters.* 207:222–226.

Knight, D. E., and P. F. Baker. 1982. Calcium-dependence of catecholamine release from bovine adrenal medullary cells after exposure to intense electric fields. *Journal of Membrane Biology.* 68:107–140.

Knight, D. E., and P. F. Baker. 1983. The phorbol ester TPA increases the affinity of exocytosis for calcium in 'leaky' adrenal medullary cells. *FEBS Letters.* 160:98–100.

Knight, D. E., and P. F. Baker. 1985*a*. The chromaffin granule proton pump and calcium dependent exocytosis in bovine adrenal medullary cells. *Journal of Membrane Biology.* 83:147–156.

Knight, D. E., and P. F. Baker. 1985*b*. Guanine nucleotides and Ca-dependent exocytosis. *FEBS Letters.* 189:345–349.

Knight, D. E., and N. T. Kesteven. 1983. Evoked transient intracellular free Ca^{2+} changes and secretion in isolated bovine adrenal medullary cells. *Proceedings of the Society of London, Series B.* 218:177–199.

Knight, D. E., and E. Koh. 1984. Ca^{2+} and cyclic nucleotide-dependence of amylase release from isolated rat pancreatic acinal cells rendered permeable by intense electric fields. *Cell Calcium.* 5:401–418.

Knight, D. E., V. Niggli, and M. C. Scrutton. 1984. Thrombin and activators of protein kinase C modulate secretory responses of permeabilised human platelets induced by Ca^{2+}. *European Journal of Biochemistry.* 143:437–446.

Knight, D. E., and M. C. Scrutton. 1980. Direct evidence for a role for Ca^{2+} in amine storage granule secretion by human platelets. *Thrombosis Research.* 20:437–446.

Knight, D. E., and M. C. Scrutton. 1984*a*. The relationship between intracellular second messengers and platelet secretion. *Biochemical Society Transactions.* 12:969–972.

Knight, D. E., and M. C. Scrutton. 1984*b*. Cyclic nucleotides control a system which regulates Ca^{2+}-sensitivity of platelet secretion. *Nature.* 309:66–68.

Knight, D. E., and M. C. Scrutton. 1986. Gaining access to the cytosol: the technique and some applications of electropermeabilization. *Biochemical Journal.* 234:497–506.

Knight, D. E., D. A. Tonge, and P. F. Baker. 1985. Inhibition of exocytosis in bovine adrenal medullary cells by botulinum toxin type D. *Nature.* 317:719–721.

Kraft, A. S., and W. B. Anderson. 1983. Phorbol esters increase the amount of Ca^{2+} phospholipid-dependent protein kinase associated with plasma membrane. *Nature.* 301:621–623.

Lingg, C., R. Fischer-Colbrie, W. Schmidt, and H. Winkler. 1983. Exposure of an antigen of chromaffin granules in cell surface during exocytosis. *Nature.* 301:610–611.

Llinas, R., T. L. M. McGuiness, C. S. Leonard, M. Sugimori, and P. Greengard. 1985. Intraterminal injection of synapsin I or calcium/calmodulin-dependent protein kinase II alters neurotransmitter release at the squid giant synapse. *Proceedings of the National Academy of Sciences.* 83:3035–3039.

Llinas, R., I. Z. Steinberg, and K. Walton. 1981. Presynaptic calcium currents in squid giant synapse. *Biophysical Journal.* 33:289–321.

Lucy, J. A., and Q. F. Ahkong. 1986. An osmotic model for the fusion of biological membranes. *FEBS Letters.* 199:1–11.

Lundberg, J. M., and T. Hökfelt. 1983. Co-existence of peptides and classical neurotransmitters. *Trends in Neurosciences.* 6:325–333.

Navonne, F., P. Greengard, and P. De Camillo. 1984. Synapsin I in nerve terminals: selective association with small synaptic vesicles. *Science.* 226:1209–1211.

Neher, E., and W. Almers. 1986. Fast calcium transients in rat peritoneal mast cells are not sufficient to trigger exocytosis. *EMBO Journal.* 5:51–53.

Neher, E., and A. Marty. 1982. Discrete changes of cell membrane capacitance observed under conditions of enhanced secretion in bovine adrenal medullary cells. *Proceedings of the National Academy of Sciences.* 79:6712–6716.

Nishizuka, Y. 1984. The role of protein kinase C in cell surface signal transduction and tumour promotion. *Nature.* 308:693–698.

Oetting, M., M. LeBoff, L. Swiston, J. Preston, and E. Brown. 1986. Guanine nucleotides are potent secretagogues in permeabilised parathyroid cells. *FEBS Letters.* 208:99–104.

Ornberg, R. L., and T. S. Reese. 1981. Beginning of exocytosis captured by rapid freezing of *Limulus* amoebocytes. *Journal of Cell Biology.* 90:40–54.

Palade, G. E. 1975. Intracellular aspects of the process of protein secretion. *Science.* 189:347–358.

Penner, R., E. Neher, and F. Dreyer. 1986. Intracellularly injected tetanus toxin and its fragment B inhibit exocytosis in bovine adrenal chromaffin cells. *Nature.* 324:76–78.

Peterfreund, R. A., and W. W. Vale. 1983. Phorbol diesters stimulate somatostatin secretion in cultured brain cells. *Endocrinology.* 20:200–208.

Pollard, H. B., C. J. Pazoles, and C. E. Creutz. 1977. A role for anion transport in the regulation and release from chromaffin granules and exocytosis from cells. *Journal of Supramolecular Structure.* 7:277–285.

Rink, T. J., and T. Hallam. 1984. What turns platelets on? *Trends in Biochemical Sciences.* 12:215–219.

Rink, T. J., A. Sanchez, and T. Hallam. 1983. Diacylglycerol and phorbol esters stimulate secretion without raising cytoplasmic free calcium in human platelets. *Nature.* 305:317–319.

Schmidt, W., A. Patzak, G. Lingg, H. Winkler, and H. Plattner. 1983. Membrane events in adrenal chromaffin cells during exocytosis: a freeze-etch analysis after rapid cryofixation. *European Journal of Cell Biology.* 32:31–37.

Schweitzer, E. S., and M. P. Blaustein. 1980. The use of antibody and complement to gain access to the interior of presynaptic nerve terminals. *Experimental Brain Research.* 38:443–453.

Shoback, D. M., J. Thatcher, R. Leombruno, and E. M. Brown. 1984. Relationship between parathyroid hormone secretion and cytosolic calcium concentration in dispersed parathyroid. *Proceedings of the National Academy of Sciences.* 81:3113–3117.

Simpson, L. L. 1985. Molecular pharmacology of botulinum toxin and tetanus toxin. *Annual Reviews of Pharmacology and Toxicology.* 25:155–188.

Steinhardt, R. A., and J. M. Alderton. 1982. Calmodulin confers calcium sensitivity on secretory exocytosis. *Nature.* 295:154–155.

Sudhof, T. C., J. H. Walker, and U. Fritsche. 1985. Characterization of calelectrin, a Ca^{2+}-binding protein isolated from the electric organ of *Torpedo marmorata. Journal of Neurochemistry.* 44:1302–1307.

Thelestam, M., and R. Möllby. 1979. Classification of microbial, plant and animal cytolysins based on their membrane-damaging effects on human fibroblasts. *Biochimica et Biophysica Acta.* 557:156–169.

Torri-Tarelli, F., F. Grohovaz, R. Fesce, and B. Ceccarelli. 1985. Temporal coincidence between synaptic vesicle fusion and quantal secretion of acetylcholine. *Journal of Cell Biology.* 101:1386–1399.

Trifaro, J. M., and R. L. Konigsberg. 1983. Microinjection of calmodulin antibodies into chromaffin cells provides direct evidence for a role of calmodulin in the secretory process. *Federation Proceedings.* 42:456.

Vacquier, V. D. 1975. The isolation of intact cortical granules from sea urchin eggs: calcium ions trigger granule discharge. *Developmental Biology.* 43:62–74.

Warren, G. B. 1985. Membrane traffic and organelle division. *Trends in Biochemical Sciences.* 10:439–443.

Whitaker, M. J., and P. F. Baker. 1983. Calcium-dependent exocytosis in an *in vitro* secretory granule plasma membrane preparation from sea urchin eggs and the effects of some inhibitors of cytoskeletal function. *Proceedings of the Royal Society of London, Series B.* 218:397–413.

White, J., M. Kielan, and A. Helenius. 1983. Membrane fusion proteins of enveloped animal viruses. *Quarterly Reviews of Biophysics.* 16:151–196.

White, J., K. Matlin, and A. Helenius. 1981. Cell fusion by Semliki forest, influenza and vesicular stomatitis viruses. *Journal of Cell Biology.* 89:674–679.

White, J. R., T. Ishizaka, K. Ishizaka, and R. I. Sha'afi. 1984. Direct demonstration of increased intracellular concentration of free calcium as measured by Quin-2 in stimulated rat peritoneal mast cells. *Proceedings of the National Academy of Sciences.* 81:3978–3982.

Wilson, S. P., and N. Kirshner. 1983. Calcium-evoked secretion from digitonin-permeabilized adrenal medullary chromaffin cells. *Journal of Biological Chemistry.* 258:4994–5000.

Zieseniss, E., and M. Plattner. 1985. Synchronous exocytosis in *Paramecium* cells involves very rapid reversible dephosphorylation of a 65-kD phosphoprotein in exocytosis-competent strains. *Journal of Cell Biology.* 101:2028–2035.

Regulation of Cytosolic Free Calcium

Chapter 2

The Plasma Membrane in the Control of the Signaling Function of Calcium

Ernesto Carafoli and Stefano Longoni

Laboratory of Biochemistry, Swiss Federal Institute of Technology (ETH), Zurich, Switzerland

Introduction

The very low Ca permeability of plasma membranes insulates the intracellular milieu from the high concentrations of Ca in the extracellular spaces. In most eukaryotic cells, the inwardly directed gradient of ionized Ca is of the order of 10^4. This is very convenient for the signaling function of Ca. The signaling function requires that the free Ca concentration in the ambient surrounding the cytosolic targets change rapidly, reversibly, and significantly: the large gradient of Ca ensures that even a minor (controlled) increase in the Ca permeability of the plasma membrane leads to significant Ca penetration and to significant changes in the free Ca concentration around the intracellular targets.

The plasma membrane, however, controls only a minor portion of the Ca used by cells to satisfy the total demands of their Ca-dependent functions (Carafoli et al., 1982). Quantitatively, the largest portion is mobilized from internal stores. The minor amounts of Ca imported from the extracellular spaces through the plasma membrane are of great importance (sometimes essential) to cells, since they trigger cascades of events that may even result in the liberation of a massive amount of Ca from intracellular stores (Fabiato and Fabiato, 1975).

In addition to being the barrier through which the "trigger" Ca penetrates into cells, the plasma membrane is by definition the final controller of the Ca balance between the intra- and extracellular spaces. Its Ca-importing and -exporting systems are poised in such a way as to produce a final set point, which in adult cells results in an ~10,000-fold difference in the ionized Ca concentration that normally exists between intra- and extracellular spaces. Three Ca-transporting systems are assumed to exist in most eukaryotic plasma membranes: a channel that mediates the import of Ca, a Ca-ATPase that mediates its export, and an Na/Ca exchanger that normally functions in the export direction, but which could also mediate Ca influx, particularly in some excitable tissues (e.g., heart). The relative importance of the three systems varies from tissue to tissue, and it may even be questioned whether a given system is molecularly the same in all cell types: a Ca channel with the properties conventionally defined in excitable tissues does not appear to exist in nonexcitable cells (e.g., the erythrocyte; Varecka and Carafoli, 1982). The liver Ca-ATPase appears to be different (Lotersztajn et al., 1981) from most other plasma membrane Ca-ATPases (Carafoli and Zurini, 1982).

The long-term algebraic sum of the Ca-transporting operation of the three systems under physiological conditions must by necessity be zero, since Ca overload and Ca deprivation do not normally occur in cells. The overall operation of the two Ca-exporting systems will thus offset the Ca-importing function of the Ca channel. The Ca-ATPase operates as a system designed to continuously eject Ca from cells. It interacts with Ca with adequate affinity even in the submicromolar concentrations present in most cytosols, but its total transport capacity is low. By contrast, the Na/Ca exchanger presumably only ejects Ca with reasonable efficiency when its concentration in the cytosol increases to levels adequate for its low Ca affinity.

In this succinct overview, some recent developments in the area of the Ca-ATPase and the Na/Ca exchanger will be described and discussed. The Ca channel will not be considered, since it is the topic of another presentation in this volume.

The Ca-transporting ATPase of Plasma Membranes

The discovery of the ATPase dates back to 1966 (Schatzmann, 1966). Subsequent work has characterized the enzyme as one of the members of the so-called E_1-E_2 transport ATPases, i.e., enzymes that form a phosphorylated intermediate in the reaction cycle and are inhibited by vanadate. The studies of Gopinath and Vincenzi (1977) and Jarrett and Penniston (1977) have established that the enzyme is stimulated by calmodulin, which increases the affinity of the ATPase for Ca down to a K_m below 0.5 μM. The observation that calmodulin stimulates the ATPase has opened the way to its isolation in a functionally competent state using calmodulin affinity chromatography columns (Niggli et al., 1979). The purified enzyme has been studied in detail, mainly in the laboratories of Carafoli and Penniston (see Carafoli and Zurini, 1982, for a review). Among the properties that have been established in the purified enzyme are (*a*) the transport of Ca in reconstituted

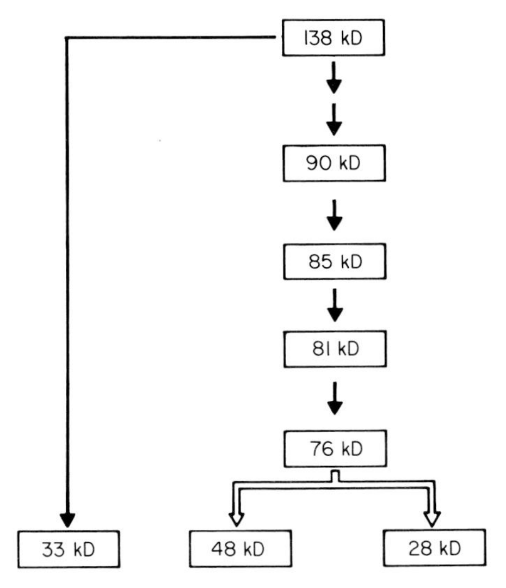

Figure 1. Proteolysis of the purified Ca-ATPase of the erythrocyte plasma membrane by trypsin.

systems and its Ca/ATP stoichiometry of 1:1, (*b*) the obligatory exchange of protons, with a probable 2:1 electroneutral stoichiometry during the transport of Ca, and (*c*) the alternative stimulation of the enzyme in the absence of calmodulin by acidic phospholipids and long-chain polyunsaturated fatty acids. The stimulation of the purified enzyme in the absence of calmodulin by controlled proteolysis, already observed in situ (Taverna and Hanahan, 1980; Enyedi et al., 1980), is of particular interest since it has permitted the mapping of functional domains in the ATPase molecule (Zurini et al., 1984; Benaim et al., 1984). Treatment with trypsin produces a number of transient and limit polypeptides (Fig. 1). Among the transient polypeptides, a group ranging in M_r from 90 to 76 kD still contains Ca-stimulated ATPase activity and pumps Ca into reconstituted liposomes (the lowest-M_r member of the group, the 76-kD polypeptide, has not yet been reconstituted). Analysis of the polypeptides of this group reveals that the sequence of production is 90, 85,

81, and 76 kD. The first two products still bind calmodulin; the last two do not. Protracted trypsin proteolysis splits the 76-kD polypeptide in a process that apparently yields two fragments of M_r 48 and 26 kD. The former is still able to form an acyl-phosphate when incubated with Ca and ATP, but no information is available on its ability to transport Ca (Zurini et al., 1984). It is of interest that treatment with trypsin almost immediately removes a polypeptide of M_r ~33 kD, which binds hydrophobic probes better than any other fragment of the molecule, and thus presumably contains its most hydrophobic segment(s) (Zurini et al., 1984). Since this fragment is evidently not involved in the transmembrane Ca-transport reaction (the 90–81-kD polypeptides, which do not contain it, still transport Ca across lipid bilayers), it is possible that its role is to anchor the ATPase molecule to the membrane without contributing to the "Ca channel" proper. The latter would then be formed by other, possibly less hydrophobic, transmembrane loops (Fig. 2). A similar situation has been postulated to prevail in other E_1-E_2

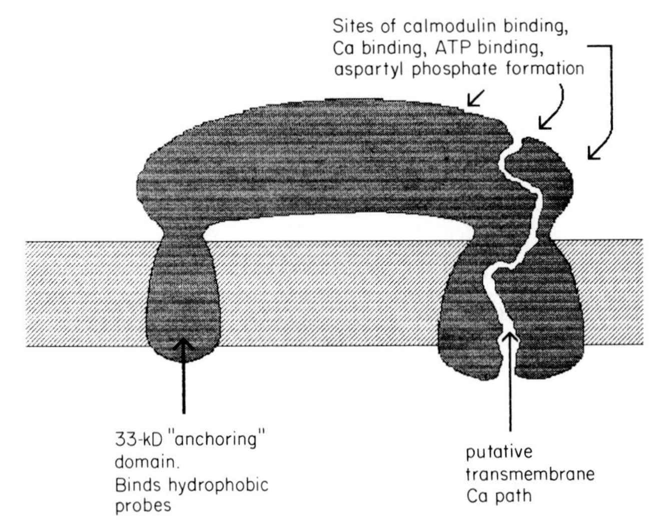

Figure 2. A hypothetical scheme of the architecture of the Ca-ATPase in the plasma membrane.

transport ATPases, e.g., the Ca-ATPase of sarcoplasmic reticulum (MacLennan et al., 1985) and the proton-pumping ATPase of the yeast plasma membrane (Serrano et al., 1986).

The effects of calmodulin on the Ca-dependent ATP hydrolysis and on the liposomal Ca pumping by the 90-, 85-, and 81-kD fragments are interesting. The 90-kD fragment exhibits properties that are practically indistinguishable from those of the intact ATPase; i.e., it binds calmodulin and responds to it with an increase in its Ca affinity. In the absence of calmodulin, the activity of the fragment appears repressed, as is the case for the intact enzyme. The 85-kD fragment binds calmodulin but fails to respond to it; i.e., its ATPase- and Ca-pumping activities are not stimulated or are stimulated only marginally: the fragment remains repressed in spite of the presence of the activator. The 81-kD fragment does not bind calmodulin and has fully expressed activity in the absence of the activator. It is of interest that the stimulation by acidic phospholipids is maintained in this fragment in spite of the absence of calmodulin activation. A scheme of the architecture of the calmodulin-interacting domain of the ATPase, based on the results described above, proposes that access to the active site of the molecule is limited by a 9-kD polypeptide that consists of the calmodulin-binding site proper and of a peripheral sequence that does not bind calmodulin but is essential for the expression of its stimulation (Benaim et al., 1984). Calmodulin activation is visualized as resulting from a conformational change of the 9-kD sequence that facilitates access to the active site.

In agreement with findings on other ATPases of the E_1-E_2 group, the Ca-ATPase of plasma membranes also appears to undergo conformational changes in the transition from the E_1 to the E_2 state. These changes can be revealed using a number of methods: intrinsic-fluorescence and circular-dichroism procedures have recently been used in our laboratory (Krebs et al., 1987), and have documented large modifications of the total α-helix content of the molecule, as well as more subtle changes, probably confined to the region containing its active site, during the E_1-E_2 transition. The latter changes can be conveniently explained with the occlusion of Ca in a "pocket" of the molecule, where one or more tryptophans would experience a different level of hydrophobicity. ATP would "liberate" the occluded Ca, returning the pocket to its original conformation (i.e., to the original hydrophobicity level).

The Na/Ca Exchanger of Plasma Membranes

Before 1979, Na/Ca exchange had been studied almost exclusively in "intact" heart (Reuter and Seitz, 1968) and nervous tissue preparations (Blaustein and Hodgkin, 1968). In 1979, Reeves and Sutko succeeded in demonstrating the reaction in a preparation of vesicles obtained from heart plasma membranes; their finding opened the way to a number of studies, mechanistic and/or structural, that were not possible in intact tissues (see Blaustein and Nelson, 1982, and Philipson and Nishimoto, 1982, for recent reviews). It has now been established that the system functions electrogenically, exchanging 3 Na for 1 Ca, and that it has a lower Ca affinity ($K_m > 2$–$10\ \mu M$), but a much higher maximal Ca transport velocity (>15 nmol/mg of membrane protein·s) than the Ca-ATPase. A number of treatments and compounds activate (EGTA, fatty acids, limited proteolysis) or inhibit (Mg^{2+}) the exchange process. The activation/deactivation of the reaction by a phosphorylation/dephosphorylation cycle is also of interest.

The exchanger has been solubilized from the membrane environment with the aid of detergents and reassembled in liposomes together with most of the solubilized proteins (Miyamoto and Racker, 1980). Although the procedure does not result in the purification of the exchanger, it provides a very important tool in purification attempts. Since the exchanger is enzymatically silent, and specific inhibitors that would enable it to be tagged have not yet been identified, purification approaches must of necessity rely on the reconstitution of the exchange activity for the success of the procedures to be assessed.

Purification studies on heart sarcolemma (Hale et al., 1984; Soldati et al., 1985) and brain synaptosomes (Barzilai et al., 1984) have employed a number of techniques, including the exposure of sarcolemma extracts to proteases (Hale et al., 1984) and the exposure of heterogeneous populations of solubilized and reconstituted synaptosomal plasma membrane vesicles to an Na gradient in the presence of Ca. Hale et al. (1984) have concluded that the heart plasma membrane exchanger is a glycoprotein of M_r 82 kD, whereas Barzilai et al. (1984) have identified it with a protein of M_r 70 kD. Work in our laboratory (Soldati et al., 1985) has correlated the exchange activity of heart sarcolemmal Triton X-100 extracts with the enrichment of specific proteins after rate-zonal centrifugations on sucrose gradients. It has been found that the reconstituted exchange activity correlates best with a protein of M_r ~33 kD: the latter protein has thus been tentatively identified with the exchanger. On the basis of this tentative identification, in a more recent series of experiments, polyclonal antibodies have been raised in rabbits using the 33-kD protein eluted from preparative SDS-polyacrylamide gels (Longoni, 1986; Longoni, S., and E. Carafoli, manuscript in preparation). The purified gammaglobulins inhibit, albeit only up to 50%, the Na/Ca exchange activity of heart sarcolemmal vesicles (Fig. 3). As expected, when applied to

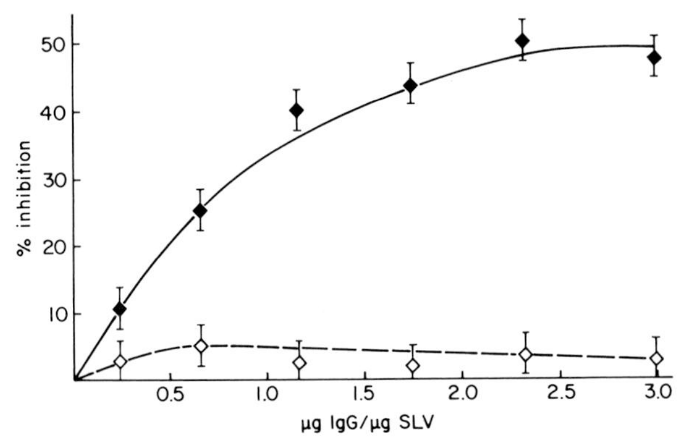

Figure 3. Inhibition of the Na/Ca exchanger of dog heart sarcolemma vesicles (SLV) by antibodies (IgG) against the proteins of the 30-kD region of heart sarcolemma. The preparation of the sarcolemmal vesicles, their solubilization, the preparation of polyclonal antibodies against the proteins of the 30-kD region of preparative SDS-polyacrylamide gels, and the measurement of the Na/Ca exchange process are described in Soldati et al. (1985), Longoni (1986), and Longoni and Carafoli (manuscript in preparation). (◆) Purified immune IgG; (◇) purified preimmune IgG.

nitrocellulose-transferred proteins of heart sarcolemma SDS-polyacrylamide gels, they react with a protein of M_r 33 kD (Fig. 4). Interestingly, however, they also interact with a protein of M_r ~70 kD and, under nonreducing conditions, also with a protein in the 140-kD range. Under reducing conditions, the reactivity with the 140-kD band disappears, and that with the component of M_r 70 kD weakens. These results indicate that the basic unit of the Na/Ca exchanger may indeed be a component of M_r ~33 kD; they also indicate a disulfide-bond–linked transition from monomer to dimer (70 kD) to tetramer (140 kD), as indicated in Fig. 5. We hope that work now in progress will clarify which of these forms is the active exchanger species. Other experiments currently underway are aimed at the production of more specific antibodies, since those used in the experiments described have been raised using an electroeluted SDS-polyacrylamide zone that contains, in addition to the main 33-kD band, a number of contaminants.

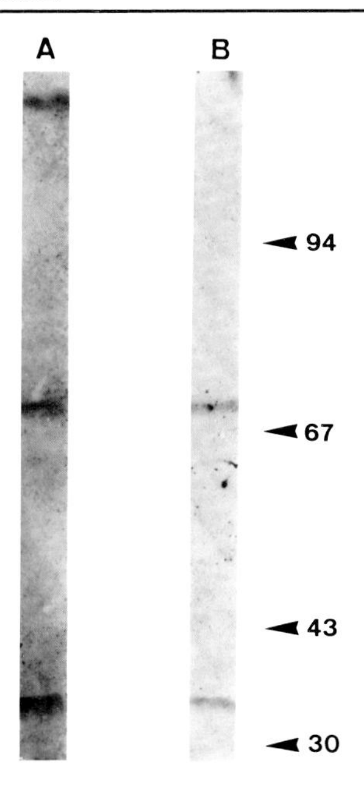

Figure 4. Reactivity of purified immunoglobulins obtained with the proteins of the 30-kD region of heart sarcolemma with nitrocellulose-transferred sarcolemmal vesicle extracts. Details are given in the legend to Fig. 3. Further details on the transfer of the proteins to nitrocellulose and on the incubation with the immunoglobulins can be found in Longoni (1986). (*A*) No dithiothreitol was present in the medium; (*B*) 3 mM dithiothreitol was present.

Summary

Eukaryotic plasma membranes contain three Ca-transporting systems: a Ca channel, an ATPase, and an Na/Ca exchanger. The ATPase is a high-affinity, low-capacity system, which continuously pumps Ca out of cells. The Na/Ca exchanger is a low-affinity, high-capacity system, which is particularly active in excitable cells. The exchanger probably functions in both the Ca efflux and influx directions.

The Ca-ATPase is a single polypeptide of M_r 138 kD, which is activated by calmodulin or, in its absence, by acidic phospholipids, polyunsaturated fatty acids, and limited proteolytic treatments. Trypsin produces a number of fragments, some of which (M_r 90, 85, and 81 kD) function as ATPases and transport Ca across reconstituted bilayer membranes. Trypsin proteolysis in the presence of different

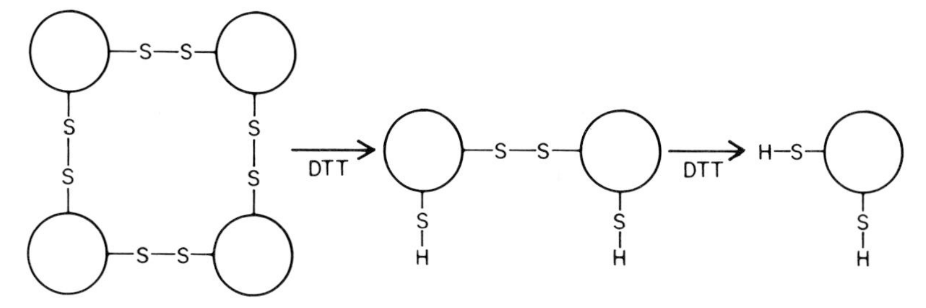

Figure 5. A hypothetical scheme of the subunit composition of the plasma membrane Na/Ca exchanger.

effectors has permitted us to locate the calmodulin-interacting domain of the enzyme in a 9-kD peripheral sequence that consists of a 4-kD calmodulin-binding subdomain and a subdomain of M_r 5 kD, which is essential for the expression of calmodulin stimulation.

The Na/Ca exchanger of plasma membranes has not yet been identified with certainty. On the basis of purification attempts using different approaches, probable M_r's of 82, 70, or 33 kD have been proposed. Antibodies raised against the 33-kD protein partially inhibit the exchange activity of heart sarcolemma vesicles. They interact with the 33-kD protein, but also, under nonreducing conditions, with proteins of M_r ~70 and ~140 kD. Under reducing conditions, the reactivity with the latter component disappears. It is suggested that the monomeric M_r of the exchanger is 33 kD, and that intermolecular disulfide bridges associate monomers into dimeric and tetrameric forms.

References

Barzilai, A., R. Spanier, and H. Rahamimoff. 1984. Isolation, purification, and reconstitution of the Na^+ gradient-dependent Ca^{2+} transporter (Na^+-Ca^{2+} exchanger) from brain synaptic plasma membranes. *Proceedings of the National Academy of Sciences.* 81:6521–6525.

Benaim, G., M. Zurini, and E. Carafoli. 1984. Different conformational states of the purified Ca^{2+}-ATPase of the erythrocyte plasma membrane revealed by controlled trypsin proteolysis. *Journal of Biological Chemistry.* 259:8471–8477.

Blaustein, M. P., and A. L. Hodgkin. 1968. The effect of cyanide on calcium efflux in squid axons. *Journal of Physiology.* 198:46P. (Abstr.)

Blaustein, M. P., and M. Nelson. 1982. Na^+-Ca^{2+} exchange: its role in the regulation of cell calcium. *In* Membrane Transport of Calcium. E. Carafoli, editor. Academic Press, Inc., New York. 217–236.

Carafoli, E., P. Caroni, M. Chiesi, and K. Famulski. 1982. Ca^{2+} as a metabolic regulator: mechanisms for the control of its intracellular activity. *In* Metabolic Compartmentation. H. Sies, editor. Academic Press, Inc., New York. 521–547.

Carafoli, E., and M. Zurini. 1982. The calcium ATPase of plasma membranes. Purification, reconstitution, and properties. *BBA Reviews in Bioenergetics.* 683:279–301.

Enyedi, A., B. Sarkadi, I. Szasz, G. Bot, and G. Gardos. 1980. Molecular properties of the red cell calcium pump. II. Effects of calmodulin, proteolytic digestion, and drugs on the calcium-induced membrane phosphorylation by ATP in inside-out red cell membrane vesicles. *Cell Calcium.* 1:299–310.

Fabiato, A., and F. Fabiato. 1975. Contractions induced by a calcium-triggered release of calcium from the sarcoplasmic reticulum of single skinned cardiac cells. *Journal of Physiology.* 249:469–495.

Gopinath, R. M., and F. F. Vincenzi. 1977. Phosphodiesterase protein activator mimics red blood cell cytoplasmic activator of (Ca^{2+} + Mg^{2+}) ATPase. *Biochemical and Biophysical Research Communications.* 77:1203–1209.

Hale, C. C., R. S. Slaughter, D. C. Ahrens, and J. P. Reeves. 1984. Identification and partial purification of the cardiac sodium-calcium exchange protein. *Proceedings of the National Academy of Sciences.* 81:6569–6573.

Jarrett, H. W., and J. T. Penniston. 1977. Partial purification of the Ca^{2+}-Mg^{2+} ATPase activator from human erythrocytes. Its similarity to the activator of the 3′:5′-cyclic nucleotide phosphodiesterase. *Biochemical and Biophysical Research Communications.* 77:1210–1216.

Krebs, J., M. Vasak, A. Scarpa, and E. Carafoli. 1987. Conformational differences between the E_1 and E_2-states of the Ca^{2+}-ATPase of the erythrocyte plasma membrane as revealed by circular dichroism and fluorescence spectroscopy. *Biochemistry.* In press.

Longoni, S. 1986. The Na^+/Ca^{2+} exchanger of heart sarcolemma. Identification and purification attempt. Ph.D. dissertation. ETH No. 8109, Zurich, Switzerland. 131 pp.

Lotersztajn, S., J. Hanoune, and F. Pecker. 1981. A high affinity, calcium-stimulated, Mg^{2+}-dependent ATPase in rat liver plasma membranes. Dependence on an endogenous protein activator different from calmodulin. *Journal of Biological Chemistry.* 256:11209–11215.

MacLennan, D. H., C. J. Brandl, B. Korczak, and N. M. Green. 1985. Amino acid sequence of a Ca^{2+} + Mg^{2+}-dependent ATPase from rabbit muscle sarcoplasmic reticulum, deduced from its complementary DNA sequence. *Nature.* 316:696–700.

Miyamoto, H., and E. Racker. 1980. Solubilization and partial reconstitution of the Ca^{2+}/Na^+ antiporter from the plasma membrane of bovine heart. *Journal of Biological Chemistry.* 255:2656–2658.

Niggli, V., J. T. Penniston, and E. Carafoli. 1979. Purification of the (Ca^{2+} + Mg^{2+}) ATPase from human erythrocyte membranes using a calmodulin affinity column. *Journal of Biological Chemistry.* 254:9955–9958.

Philipson, K. D., and A. Y. Nishimoto. 1982. Stimulation of Na^+-Ca^{2+} exchange in cardiac sarcolemmal vesicles by proteinase pretreatment. *American Journal of Physiology.* 243:C191–C195.

Reeves, J. P., and J. L. Sutko. 1979. Sodium-calcium exchange in cardiac membrane vesicles. *Proceedings of the National Academy of Sciences.* 76:590–594.

Reuter, H., and N. Seitz. 1968. The dependence of Ca^{2+} efflux from cardiac muscle on temperature and external ion composition. *Journal of Physiology.* 195:451–470.

Schatzmann, H. J. 1966. ATP-dependent Ca^{++} extrusion from human red cells. *Experientia.* 22:364–365.

Serrano, R., M. C. Kielland-Brandt, and G. R. Fink. 1986. Yeast plasma membrane ATPase is essential for growth and has homology with (Na^+ + K^+), K^+, and Ca^{2+} ATPases. *Nature.* 319:689–693.

Soldati, L., S. Longoni, and E. Carafoli. 1985. Solubilization and reconstitution of the Na^+/Ca^{2+} exchanger of cardiac sarcolemma. Properties of the reconstituted system and tentative identification of the protein(s) responsible for the exchange activity. *Journal of Biological Chemistry.* 260:13321–13327.

Taverna, R. D., and D. J. Hanahan. 1980. Modulation of human erythrocyte Ca^{2+}/Mg^{2+} ATPase activity by phospholipase A_2 and proteases. A comparison with calmodulin. *Biochemical and Biophysical Research Communications.* 94:652–659.

Varecka, L., and E. Carafoli. 1982. Vanadate-induced movements of Ca^{2+} and K^+ in human red blood cells. *Journal of Biological Chemistry.* 257:7414–7421.

Zurini, M., J. Krebs, J. T. Penniston, and E. Carafoli. 1984. Controlled proteolysis of the purified Ca^{2+}-ATPase of the erythrocyte membrane. A correlation between the structure and the function of the enzyme. *Journal of Biological Chemistry.* 259:618–627.

Chapter 3

The Role of the Plasma Membrane and Intracellular Organelles in Synaptosomal Calcium Regulation

David G. Nicholls, Talvinder S. Sihra, and Jose Sanchez-Prieto

Department of Biochemistry, University of Dundee, Dundee, Scotland

Introduction

Isolated nerve terminals (synaptosomes) produced by the gentle homogenization of brain tissue not only provide a model system for studying the uptake, metabolism, storage, and release of neurotransmitters, but also function as metabolically autonomous "mini-cells." In physiological media, their Na^+ pump maintains a plasma membrane potential of some −60 mV (Scott and Nicholls, 1980; Rugolo et al., 1986). The plasma membrane possesses carriers for the accumulation of neurotransmitters and precursors by Na^+ cotransport (for review, see Kanner, 1983), together with multiple pathways for the uptake and extrusion of Ca^{2+}, which will be discussed in this chapter. Studies with indicators of free cytoplasmic Ca^{2+} indicate that this parameter is maintained well below 1 μM in the polarized preparation (Ashley et al., 1984; Richards et al., 1984; Hansford and Castro, 1985; Nachshen, 1985).

The synaptosomal cytoplasm contains the complete glycolytic pathway and relies heavily upon glycolysis for energy production (Kauppinen and Nicholls, 1986*a*). The guinea pig synaptosomes discussed in this chapter are highly aerobic, and >90% of the total glycolytic activity in the guinea pig preparation takes place in synaptosomes containing functional mitochondria, as judged by the 10-fold stimulation of glycolysis upon activation of the Pasteur effect (Kauppinen and Nicholls, 1986*a*).

Guinea pig synaptosomes show a fivefold respiratory control (Scott and Nicholls, 1980; Kauppinen and Nicholls, 1986*a*) and this spare respiratory capacity can be utilized, for example, to drive the cycling of Na^+ across the plasma membrane in the presence of the Na^+ channel activator veratridine (Scott and Nicholls, 1980). The in situ mitochondria maintain a membrane potential of 150 mV (Scott and Nicholls, 1980) and can reversibly accumulate Ca^{2+} (Åkerman and Nicholls, 1981*a*).

Synaptosomes are thus far more "intact" than is commonly assumed and provide a suitable system for studying the integration of the plasma and mitochondrial Ca^{2+}-transport pathways in the regulation of cytosolic free Ca^{2+} and the role of this parameter in the initiation of transmitter release. In this chapter, we shall review the evidence for the different modes of plasma and mitochondrial Ca^{2+} transport, and their roles in the metabolic and secretory functions of the terminal.

Ca^{2+} Transport by Isolated Brain Mitochondria

The chemiosmotic mechanism by which mitochondria link respiration to ATP synthesis (for review, see Nicholls, 1982) also provides the driving force for the movement of Ca^{2+} across the mitochondrial membrane (Nicholls and Crompton, 1980; Nicholls and Åkerman, 1982). The pumping of H^+ out of the matrix by the respiratory chain leads to a membrane potential ($\Delta\psi_m$) across the mitochondrial inner membrane of 150–180 mV, negative in the matrix. The inner-membrane uniport carrier for Ca^{2+} allows the electrogenic entry of the divalent cation. Since the equilibrium distribution of a divalent cation across a membrane increases 10-fold for each 30 mV of membrane potential, the Ca^{2+}-uniporter in isolation would therefore lead to the irreversible sequestration of Ca^{2+} within the matrix. However, mitochondria also possess independent pathways for Ca^{2+} efflux from the matrix (Crompton et al., 1976), which in brain mitochondria is a Ca^{2+}/Na^+ exchange

(Crompton et al., 1978; Nicholls, 1978). The driving force for Ca^{2+} uptake via the "efflux" pathway is smaller than that for the uniporter, mainly because the exchanged Na^{+} compensates for the charge on the Ca^{2+}. As a result, a steady state cycling occurs that is driven by protons extruded by the respiratory chain but requires only a small fraction of the mitochondrial energy supply.

Although the ion movements during steady state cycling are symmetrical across the inner mitochondrial membrane, the kinetics of the individual pathways are distinctive. The activity of the uniporter increases as the cube of the free external Ca^{2+} (Zoccarato and Nicholls, 1982). Indeed, when Ca^{2+} is >5 μM, the entire respiratory capacity of the mitochondrion is utilized to accumulate Ca^{2+}. The only major exception to this appears to be heart, where the V_{max} of the Ca^{2+}-uniporter is limited.

The kinetics of the efflux pathways are difficult to determine in absolute terms, since the relationship between free and total Ca^{2+} in the matrix is not constant, but depends on the anion content of the matrix, particularly phosphate (Zoccarato and Nicholls, 1982). In heart mitochondria, the matrix free Ca^{2+} increased linearly with the Ca^{2+} load over the range 1–5 nmol Ca^{2+}/mg protein, but with <1 part per 1,000 as free cation (Hansford, 1985). At higher matrix Ca^{2+} loads (>10 nmol/mg), the activity of the efflux pathway stabilizes at a rate that shows a reciprocal relationship with the free phosphate in the matrix; this suggests that the formation of a Ca^{2+}-phosphate complex prevents any further rise in matrix free Ca^{2+} (Zoccarato and Nicholls, 1982).

The two pathways allow for the precise regulation of transmembrane Ca^{2+} distribution. At steady state, the external Ca^{2+} is maintained, which allows the uniporter to exactly balance the efflux pathway. Thus, at high matrix contents (>10 nmol/mg), this "set point" becomes independent of matrix Ca^{2+}, and isolated mitochondria act as perfect buffers of Ca^{2+}. For most mitochondria in the presence of physiological concentrations of Na and Mg, the set point lies in the region 0.5–3 μM, which is at the upper limit of the normal range of Ca^{2+} determined in intact cells. The very steep dependence of the uniporter upon Ca^{2+} means that any increase in free Ca^{2+} beyond the set point is rapidly sequestered by the mitochondrion.

When there is insufficient matrix Ca^{2+} to complex with phosphate (<10 nmol/mg), there is no set point, the activity of the efflux pathway increases with matrix Ca^{2+} content (Hansford, 1985), and both extramitochondrial and matrix free Ca^{2+} concentrations will increase when more Ca^{2+} is added to the system. Three matrix enzymes can be regulated by Ca^{2+} (Denton et al., 1980). Pyruvate dehydrogenase exists in an inactive phosphorylated form (PDH_b) and as a catalytically active dephosphoenzyme (PDH_a). The dephosphorylation is catalyzed by a phosphatase that is activated by Ca^{2+} in the range 0.1–10 μM. The effect of increasing free Ca^{2+} in the matrix is thus to increase the V_{max} of the PDH complex. The other two enzymes, NAD-linked isocitrate dehydrogenase (Denton et al., 1978) and 2-oxoglutarate dehydrogenase (McCormack and Denton, 1979), show an increase in their affinity for isocitrate and 2-oxoglutarate, respectively, over the same range of Ca^{2+}.

Mitochondrial Ca^{2+} transport pathways can thus either regulate matrix free Ca^{2+} (at low matrix Ca^{2+} loads) or set an upper limit to the cytosolic free Ca^{2+} (at higher matrix Ca^{2+}). While it is frequently maintained that only one of these functions can be of physiological relevance, this is unnecessarily restrictive, since the transport pathways would automatically switch from one function to the other, depending upon the circumstances within the cell.

Ca^{2+} Transport by Intrasynaptosomal Mitochondria

After preparation, which involves Ca^{2+} chelators and Ca^{2+}-free media, synaptosomes contain only 0.5 nmol Ca^{2+}/mg protein (Scott et al., 1980). In the presence of 1.3 mM Ca^{2+}, they reaccumulate 5–10 nmol Ca^{2+}/mg protein within 30 min. Rapid disruption and fractionation of synaptosomes by digitonin in the presence of EGTA and ruthenium red shows that ~50% of this Ca^{2+} is within the digitonin-resistant pellet (Scott et al., 1980; Åkerman and Nicholls, 1981*a*). The pellet Ca^{2+} was reduced by 85% in synaptosomes, where the mitochondrial membrane potential ($\Delta\psi_m$) and mitochondrial ATP synthesis were specifically collapsed by rotenone plus oligomycin, but not when oxidative phosphorylation was abolished, and $\Delta\psi_m$ was retained, by oligomycin alone. This indicated that the Ca^{2+} in the digitonin-resistant pellet was retained by the mitochondrial membrane potential, rather than by ATP (which would eliminate the possibility that the pellet Ca^{2+} was within digitonin-resistant endoplasmic reticulum).

When synaptosomes are depolarized by elevated K^+ or by veratridine, the total Ca^{2+} content increases rapidly by 30–50% owing to activation of the voltage-dependent Ca^{2+} channels (Blaustein, 1975). With high K^+, essentially all the increased Ca^{2+} is further translocated into the mitochondria, which provides a direct indication, in this in vitro system, of their Ca^{2+}-buffering capabilities (Åkerman and Nicholls, 1981*a*).

While the "buffering" mode for mitochondrial Ca^{2+} transport can thus be clearly demonstrated in the isolated synaptosome, the "dehydrogenase controlling" mode is more elusive. The depolarization-induced Ca^{2+} entry into synaptosomes does cause a slight increase in the proportion of PDH_a in the mitochondria (Schaffer and Olson, 1980; Hansford and Castro, 1985), although the synaptosomes have an exceedingly high initial content of PDH_a (90% in the former study and 68% in the latter). Pyruvate (either exogenous or generated by glycolysis) is an excellent substrate for intact synaptosomes, and the availability of the substrate has a marked effect upon the magnitude of $\Delta\psi_m$ in situ (Kauppinen and Nicholls, 1986*a*). The ease with which synaptosomes can be completely depleted of Ca^{2+} (Scott et al., 1980) provides a model for testing the role of Ca^{2+} as an activator of pyruvate oxidation in a "cellular" environment. This is important, since the ability of Ca^{2+} to activate the PDH complex in vitro requires the unphysiological conditions of an absence of pyruvate and very high ATP/ADP ratios. Rather disturbingly, no difference whatsoever can be detected in the ability of pyruvate to maintain synaptosomal $\Delta\psi_m$, ATP/ADP ratios, or respiration in the presence or absence of intrasynaptosomal Ca^{2+} (Kauppinen and Nicholls, 1986*b*). Thus, it does not appear that Ca^{2+} regulation of pyruvate oxidation occurs in the near-physiological environment of the intact synaptosome.

A major role for endoplasmic reticulum in synaptosomal Ca^{2+} regulation has been proposed, based on the ability of hypotonic lysates of synaptosomes to catalyze the ATP-dependent uptake of Ca^{2+} (Blaustein et al., 1978). However, since the same preparation techniques produce plasma membrane vesicles that can be identified by marker enzymes (Michaelis et al., 1983) and by the coexistence in the same membrane of the (Na^+ + K^+)-ATPase, a Ca^{2+}-transporting ATPase (Gill et al., 1981), and amino acid uptake pathways (Kanner, 1983), the observed ATP-dependent Ca^{2+} transport may be across inverted plasma membrane vesicles rather than endoplasmic reticulum.

Synaptosomal Plasma Membrane Ca^{2+} Transport

In recent years, it has become apparent that the plasma membranes of excitable cells possess two pathways, a Ca^{2+}-translocating ATPase and an Na^+/Ca^{2+} exchanger, both of which are capable in theory of extruding Ca^{2+} from the cell against the considerable Ca^{2+} electrochemical gradient that exists across the membrane under physiological conditions. Early studies with intact synaptosomes emphasized a role for Na^+/Ca^{2+} exchange in the extrusion of Ca^{2+} from the cytosol (Blaustein and Oborn, 1975). However, the Na^+ dependence for Ca^{2+} efflux could only be demonstrated in the unphysiological condition of external Ca^{2+} depletion (Blaustein and Ector, 1976). At the same time, a Ca^{2+}-ATPase activity was detected, which now appears to be largely localized on the plasma membrane (see above).

To assess the role of the two pathways in the extrusion of Ca^{2+}, it would be advantageous to inhibit each in turn and to see the effect on the ability of the terminal to regulate Ca^{2+}. Vanadate is a potent inhibitor of the plasma membrane Ca^{2+}-ATPase, but is ineffective in intact synaptosomes since it does not permeate the plasma membrane. An alternative strategy is to lower the ATP/ADP ratio in the cytoplasm to deprive the Ca^{2+}-ATPase of ATP. When this is done by the combination of iodoacetate and protonophore, the net accumulation of Ca^{2+} by synaptosomes is increased (Åkerman and Nicholls, 1981*c*), which suggests the inhibition of an ATP-dependent Ca^{2+}-extrusion mechanism. Controls showed that Na^+-pump inhibition did not underlie the increased Ca^{2+} uptake.

An alternative approach is to abolish the driving force for Ca^{2+} extrusion by the Na^+/Ca^{2+} exchanger, i.e., the Na^+ electrochemical gradient, by ouabain plus veratridine (Snelling and Nicholls, 1985). Under these conditions, there is an increased Ca^{2+} uptake, but this is almost entirely a consequence of the activation of voltage-dependent Ca^{2+} channels as a result of the plasma membrane depolarization. Ca^{2+} stabilizes at $\sim 1\ \mu M$ (Hansford and Castro, 1985) and Ca^{2+} can be pumped out across the plasma membrane as fast as it leaks into the terminal (Snelling and Nicholls, 1985). These experiments indicate that, at least in the nerve terminal, the Ca^{2+}-ATPase is the dominant influence at the plasma membrane. However, calmodulin antagonists, such as trifluoroperazine, cannot be used to investigate the regulation of the Ca^{2+}-ATPase in situ, since they are potent uncouplers of the mitochondria within the synaptosomes (Snelling and Nicholls, 1984).

A steady state cycling of Ca^{2+} between independent uptake and efflux pathways occurs across the plasma membrane. Even with polarized synaptosomes, there is a continuous leak of Ca^{2+} into the terminal, which is balanced by the efflux (Snelling and Nicholls, 1985). Since the Ca^{2+}-ATPase in the plasma membrane shares with the mitochondrial Ca^{2+}-uniporter the property of being highly dependent on Ca^{2+}, in the steady state the plasma membrane would seek to maintain a precisely regulated cytoplasmic free Ca^{2+} at which the activity of the Ca^{2+}-ATPase would match that of the influx pathway. In the nerve terminal, this balance occurs at

~0.2 μM (Hansford and Castro, 1985). A threefold increase in inward Ca^{2+} flux during chronic plasma membrane depolarization (Snelling and Nicholls, 1985) appears to raise this plasma membrane set point to 0.5 μM (Hansford and Castro, 1985).

Depolarization-induced Ca^{2+} Entry

It has long been established that plasma membrane depolarization induces a rapid uptake of Ca^{2+} into the synaptosome (Blaustein, 1975) through voltage-dependent Ca^{2+} channels that can be inhibited by high concentrations of verapamil. The detailed kinetics of Ca^{2+} entry show an exceedingly rapid phase, lasting only 1–2 s, followed by a continuous Ca^{2+} entry, which is elevated beyond the rate seen in polarized synaptosomes (Nachshen and Blaustein, 1980; Drapeau and Blaustein, 1983).

Cytosolic Free Ca^{2+}

Measurements of intrasynaptosomal cytoplasmic free Ca^{2+} using the fluorescent Ca^{2+}-indicator quin2 indicate a resting Ca^{2+} of ~0.2 μM (Richards et al., 1984; Ashley et al., 1984; Hansford and Castro, 1985; Nachshen, 1985). With high KCl, there is a rapid increase to 0.5 μM, while with veratridine, a value of 0.8 μM is slowly achieved over a 5-min period (Hansford and Castro, 1985). It should be emphasized, first, that the indicator technique measures the averaged free Ca^{2+} over the whole cytoplasm and so will not register higher concentrations localized, perhaps transiently, close to the plasma membrane, and second, that the very presence of the indicator, with its requirement to chelate Ca^{2+} in order to measure it, buffers and diminishes the extent of the increase in Ca^{2+}.

As discussed above, cytoplasmic Ca^{2+} will stabilize at a value at which uptake and efflux across the plasma membrane balance. Neither the mitochondrion nor any endoplasmic reticulum can play a role in defining the steady state Ca^{2+}. Instead, the mitochondrial matrix and endoplasmic reticular lumen must adjust their Ca^{2+} content until uptake and efflux from the cytoplasm balance.

The Coupling of Ca^{2+} to Secretion in the Synaptosome

A dependence on extracellular Ca^{2+} for the depolarization-induced release of a putative transmitter or hormone has long been accepted as an important criterion of exocytosis (Douglas, 1974). Studies with isolated synaptosomes have established that the rapid kinetics of release of labeled dopamine (Drapeau and Blaustein, 1983) correlate well with the entry of Ca^{2+} in the first few seconds after depolarization. However, studies of putative amino acid neurotransmitters such as GABA or glutamate have been complicated by the need to distinguish "exocytotic" release upon depolarization from the relatively slower depletion of the extensive cytoplasmic pools of these putative amino acid transmitters. Many agents that have been proposed to initiate Ca^{2+}-dependent exocytosis have been seen on closer analysis merely to induce the release of cytoplasmic pools of the amino acids after a decrease in the Na^{+} electrochemical gradient across the plasma membrane (Sihra et al., 1984). Recently, however, we (Nicholls and Sihra, 1986) have described a continuous fluorometric assay for released glutamate, which has enabled us to follow the authentic exocytotic release of this neurotransmitter and its coupling to Ca^{2+}.

Fig. 1 shows that in the presence, but not the absence, of external Ca^{2+}, depolarization with KCl induces a rapid release of ~2 nmol of glutamate per milligram of synaptosomal protein. The characteristics of this release clearly indicate an exocytotic mode, and are as follows (Nicholls and Sihra, 1986; Nicholls, D. G., T. S. Sihra, and J. Sanchez-Prieto, manuscript submitted for publication). (*a*) Ca^{2+}-dependent release is very rapid and is largely complete before significant glutamate is lost from the cytoplasm by reversal of the uptake pathway. (*b*) Glutamate release shows a sigmoidal dependence of $\Delta\psi_p$, with a threshold at −40 mV, which is consistent with the need for activation of voltage-dependent channels (Fig. 2). (*c*) Ba^{2+} and Sr^{2+} can substitute for external Ca^{2+}. (*d*) Ca^{2+}-dependent

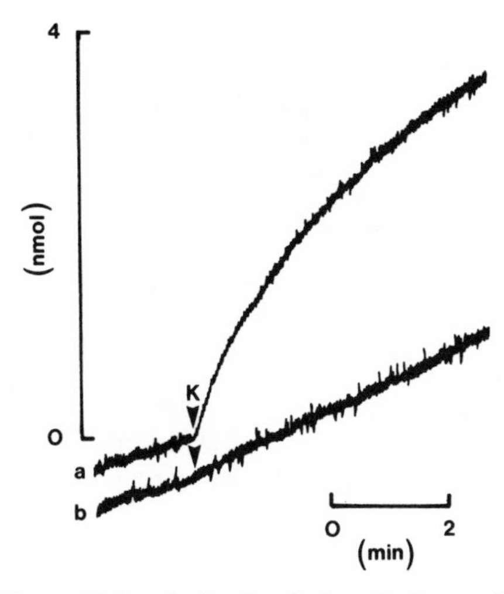

Figure 1. Dependence of the rapid depolarization-induced release of L-glutamate from guinea pig synaptosomes upon the presence of external Ca^{2+}. Synaptosomes were incubated in a stirred cuvette in the presence of glutamate dehydrogenase and $NADP^+$. Depolarization was induced by 30 mM KCl and the released glutamate was monitored by the increase in fluorescence of the reduced NADPH. (*a*) With 1.3 mM Ca^{2+}; (*b*) with 0.5 mM EGTA. For further details, see Nicholls and Sihra (1986).

glutamate release is inhibited by verapamil at high concentrations. (*e*) Partial depletion of the cytoplasmic pool of glutamate by a brief exchange with D-aspartate has no effect on Ca^{2+}-dependent release. (*f*) Preincubation of synaptosomes before depolarization decreases Ca^{2+}-independent cytoplasmic release but enhances Ca^{2+}-dependent release, which is consistent with a translocation from cytoplasm to vesicle. (*g*) Veratridine induces Ca^{2+}-dependent glutamate release. (*h*) Glutamate release is dependent on cytoplasmic ATP.

One surprising feature is that the Ca^{2+} stored in the intrasynaptosomal mitochondria is not able to induce the characteristic pattern of exocytotic glutamate release. Protonophores, such as FCCP, cause a rapid release of Ca^{2+} from the mitochondria and an increase in cytosolic free Ca^{2+} (Heinonen et al., 1984), which is ultimately pumped out of the synaptosome (Åkerman and Nicholls, 1981*a*). However, no glutamate release occurs (Fig. 3). This is not due to a protonophore-

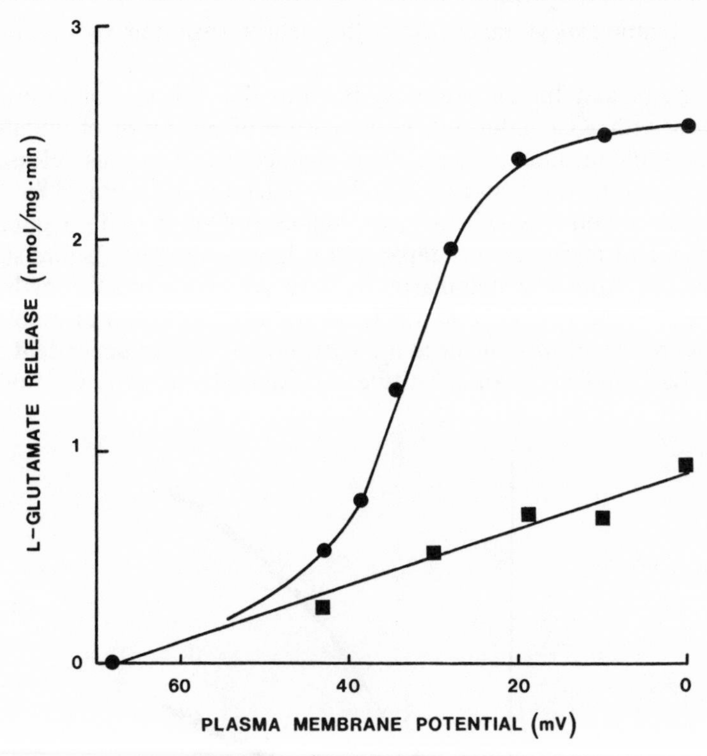

Figure 2. Voltage dependence for the release of glutamate in the presence or absence of exogenous Ca^{2+}. Synaptosomes were incubated as described in Fig. 1, with modifications (Nicholls and Sihra, 1986), in the presence of varying KCl. The plasma membrane potential was calculated from the observed ^{86}Rb gradient, and glutamate release was monitored as in Fig. 1. (●) With 1.3 mM Ca^{2+}; (■) with 0.5 mM EGTA.

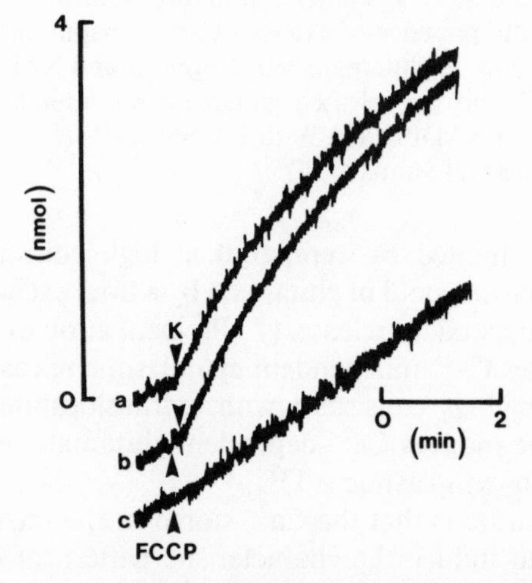

Figure 3. The failure of protonophores to induce glutamate exocytosis. Synaptosomes were incubated as described in Fig. 1. Where indicated, 1 μM FCCP (protonophore) or 30 mM KCl was added. Note that FCCP does not inhibit the subsequent KCl-induced efflux in trace *b*.

induced block of the exocytotic mechanism, since the subsequent addition of KCl leads to a normal release, but rather implies that actual transport of Ca^{2+} across the plasma membrane is essential.

The $Ca^{2+}/2H^+$ ionophore ionomycin causes the rapid release of glutamate in a Ca^{2+}-dependent manner. Ca^{2+} ionophores have a number of side effects, in that they depolarize the plasma membrane and deplete the cytoplasm of ATP (Åkerman and Nicholls, 1981*b*). However, a range of concentrations of ionomycin can be chosen such that plasma membrane depolarization is insufficient to activate voltage-dependent Ca^{2+} channels, and ATP depletion is insufficient to inhibit exocytosis. Under these conditions, the extent of exocytosis is closely comparable to that seen with KCl (Fig. 4).

Increasing ionomycin concentrations should raise the cytoplasmic free Ca^{2+} proportionately. We have previously described a technique whereby cytoplasmic free Ca^{2+} can be estimated by using Ca^{2+} ionophores (Åkerman and Nicholls,

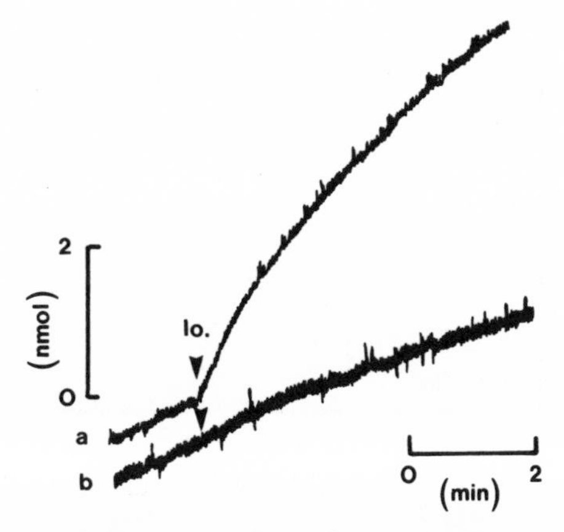

Figure 4. Ionomycin-induced glutamate release. Synaptosomes were incubated as described in Fig. 1. Where indicated, 50 μM ionomycin (Io.) was added; (*a*) with 1.3 mM Ca^{2+}; (*b*) with 0.5 mM EGTA.

1981*b*). Briefly, ionomycin will dissolve also in the mitochondrial inner membrane and greatly potentiate the Ca^{2+} efflux pathway. The rate of Ca^{2+} cycling, and hence energy dissipation, at the mitochondrial inner membrane becomes a function of the activity of the mitochondrial Ca^{2+}-uniporter, which in turn is a function of the free cytoplasmic Ca^{2+} that has been produced by the action of the ionophore at the plasma membrane. The respiration of the synaptosomes therefore increases when the ionophore is added, owing to this energy load. Calibration with isolated mitochondria allows a given extent of respiratory stimulation in the presence of ionomycin to be equated with a given free Ca^{2+} concentration. Fig. 5 shows the relationship between the cytoplasmic free Ca^{2+} and the glutamate release obtained by this method.

Ca^{2+} Fluxes during Synaptosomal Exocytosis

An inward flux of Ca^{2+} through voltage-dependent Ca^{2+} channels could readily raise Ca^{2+} close to the cytoplasmic face to 5–10 μM to activate Ca^{2+}-dependent exocytosis. Both the plasma membrane and the mitochondria would then immediately start to remove Ca^{2+} from the cytoplasm. The mitochondrion is likely to be the dominant process in this initial lowering of Ca^{2+}, since its entire respiratory capacity can be diverted to the accumulation of the cation, whereas the plasma membrane Ca^{2+}-ATPase has a much more restricted V_{max}, and the Na^+/Ca^{2+} exchange seems relatively inactive in the guinea pig preparation (Snelling and Nicholls, 1985). The internal mitochondria should be capable of accumulating Ca^{2+} at an initial rate of ~50 nmol/mg·min synaptosomal protein from a cytoplasm with >5 μM Ca^{2+} (Nicholls, 1978), whereas the plasma membrane Ca^{2+}-ATPase would be ~50 times slower (Snelling and Nicholls, 1985).

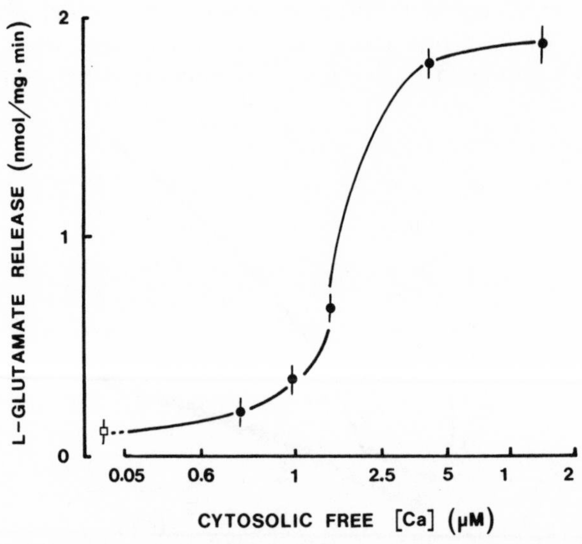

Figure 5. Cytoplasmic free Ca^{2+} required for the ionomycin-induced release of glutamate. Synaptosomes were exposed to a range of ionomycin concentrations in the presence of exogenous Ca^{2+} and the glutamate release was monitored. In a parallel experiment, the respiratory stimulation caused by the ionophore was followed in an oxygen electrode. This was in turn related to the free Ca^{2+} concentration required to induce the same stimulation in isolated brain mitochondria incubated under plausibly physiological conditions. For further details, see Åkerman and Nicholls (1981*b*).

The mitochondrion would continue to lower Ca^{2+} until its set point was attained at ~1 μM. Because of the extremely high capacity of the mitochondrion to accumulate Ca^{2+}, this cycle could be repeated many times, although eventually the cell would demand a sufficient recovery period for the plasma membrane to restore homeostasis. The plasma membrane Ca^{2+}-ATPase would continue to lower Ca^{2+} to 0.1–0.3 μM. Under these conditions, the mitochondrial Ca^{2+}-uniporter would not be able to balance the mitochondrial efflux pathway, and a net efflux of Ca^{2+} from the matrix would occur. The plasma membrane Ca^{2+}-ATPase would thus eventually drain all the excess Ca^{2+} from the mitochondrial matrix and initial conditions would be regained. In this model, the mitochondrion could be envisaged as the central sink for the Ca^{2+} entering across the plasma membrane to initiate exocytosis.

Acknowledgments

The expert technical assistance of Mr. Craig Adam is gratefully acknowledged.

Work from our laboratory is supported by the British Medical Research Council. J.S.-P. was supported by a short-term EMBO Fellowship.

References

Åkerman, K. E. O., and D. G. Nicholls. 1981*a*. Intra-synaptosomal compartmentation of Ca during depolarization-induced Ca uptake across the plasma membrane. *Biochimica et Biophysica Acta*. 645:41–48.

Åkerman, K. E. O., and D. G. Nicholls. 1981*b*. Ca transport by intact synaptosomes: influence of ionophore A23187 on plasma membrane potential, plasma membrane Ca transport, mitochondrial membrane potential, cytosolic free Ca concentration and noradrenaline release. *European Journal of Biochemistry*. 155:67–73.

Åkerman, K. E. O., and D. G. Nicholls. 1981*c*. ATP depletion increases Ca^{2+} uptake by synaptosomes. *FEBS Letters*. 135:212–214.

Ashley, R. H., M. J. Brammer, and R. Marchbanks. 1984. Measurement of intrasynaptosomal free calcium by using the fluorescent indicator quin2. *Biochemical Journal*. 219:149–158.

Blaustein, M. P. 1975. Effects of potassium, veratridine and scorpion venom on calcium accumulation and transmitter release by nerve terminals in vitro. *Journal of Physiology*. 247:617–655.

Blaustein, M. P., and A. C. Ector. 1976. Carrier-mediated Na^{+}-dependent and Ca^{2+}-dependent Ca^{2+} efflux from pinched-off presynaptic nerve terminals (synaptosomes) in vitro. *Biochimica et Biophysica Acta*. 419:295–308.

Blaustein, M. P., and C. J. Oborn. 1975. The influence of Na^{+} on Ca^{2+} fluxes in pinched-off nerve terminals in vitro. *Journal of Physiology*. 247:657–686.

Blaustein, M. P., R. W. Ratzlaff, and N. K. Kendrick. 1978. The regulation of intra-cellular Ca^{2+} in presynaptic nerve endings. *Annals of the New York Academy of Sciences*. 301:195–211.

Crompton, M., M. Capano, and E. Carafoli. 1976. The sodium-induced efflux of calcium from heart mitochondria. A possible mechanism for the regulation of mitochondrial calcium. *European Journal of Biochemistry*. 69:453–462.

Crompton, M., R. Moser, H. Lüdi, and E. Carafoli. 1978. The interrelations between the transport of sodium and calcium in mitochondria from various mammalian tissues. *European Journal of Biochemistry*. 82:25–31.

Denton, R. W., J. G. McCormack, and N. J. Edgell. 1980. Role of calcium ions in the regulation of intramitochondrial metabolism. Effects of Na, Mg and ruthenium red on the Ca^{2+}-stimulated oxidation of oxoglutarate and on pyruvate dehydrogenase activity in intact rat heart mitochondria. *Biochemical Journal*. 190:107–117.

Denton, R. M., D. A. Richards, and J. G. Chin. 1978. Calcium ions and the regulation of NAD-linked isocitrate dehydrogenase from the mitochondria of rat heart and other tissues. *Biochemical Journal*. 176:899–906.

Douglas, W. W. 1974. Involvement of calcium in exocytosis and the exocytosis-vesiculation sequence. *In* Calcium and Cell Regulation. R. M. S. Smellie, editor. The Biochemical Society, London. 1–28.

Drapeau, P., and M. P. Blaustein. 1983. Initial release of ^{3}H-dopamine from rat striatal synaptosomes: correlation with calcium entry. *Journal of Neuroscience.* 3:703–713.

Gill, D. L., E. F. Grollman, and L. D. Kohn. 1981. Calcium transport mechanisms in membrane vesicles from guinea pig brain synaptosomes. *Journal of Biological Chemistry.* 256:184–192.

Hansford, R. G. 1985. Relation between mitochondrial calcium transport and control of energy metabolism. *Reviews of Physiology, Biochemistry and Pharmacology.* 102:1–72.

Hansford, R. G., and F. Castro. 1985. Role of pyruvate dehydrogenase interconversion in brain mitochondria and synaptosomes. *Biochemical Journal.* 227:129–136.

Heinonen, E., K. E. O. Åkerman, and K. Kaila. 1984. Depolarization of the mitochondrial membrane potential increases free cytosolic Ca^{2+} in synaptosomes. *Neuroscience Letters.* 49:33–37.

Kanner, B. 1983. Bioenergetics of neurotransmitter transport. *Biochimica et Biophysica Acta.* 726:293–316.

Kauppinen, R. A., and D. G. Nicholls. 1986*a*. Synaptosomal bioenergetics: the role of glycolysis, pyruvate oxidation and responses to hypoglycaemia. *European Journal of Biochemistry.* 158:159–165.

Kauppinen, R. A., and D. G. Nicholls. 1986*b*. Pyruvate utilization by synaptosomes is independent of calcium. *FEBS Letters.* 199:222–226.

McCormack, J. G., and R. M. Denton. 1979. The effects of calcium ions and adenine nucleotides on the activity of pig heart oxoglutarate dehydrogenase complex. *Biochemical Journal.* 180:533–544.

Michaelis, E. K., M. L. Michaelis, H. H. Chang, and T. E. Kitos. 1983. High-affinity Ca^{2+}-stimulated Mg-dependent ATPase in rat brain synaptosomes, synaptic membranes and microsomes. *Journal of Biological Chemistry.* 258:6106–6114.

Nachshen, D. A. 1985. Regulation of cytosolic calcium concentration in presynaptic nerve endings isolated from rat brain. *Journal of Physiology.* 363:87–101.

Nachshen, D. A., and M. P. Blaustein. 1980. Some properties of potassium-stimulated calcium influx in presynaptic nerve endings. *Journal of General Physiology.* 76:709–728.

Nicholls, D. G. 1978. Ca transport and proton electrochemical potential in mitochondria from cerebral cortex and rat heart. *Biochemical Journal.* 170:511–522.

Nicholls, D. G. 1982. Bioenergetics: an Introduction to the Chemiosmotic Theory. Academic Press, Inc., New York. 196 pp.

Nicholls, D. G., and K. E. O. Åkerman. 1982. Mitochondrial calcium transport. *Biochimica et Biophysica Acta.* 683:57–88.

Nicholls, D. G., and M. Crompton. 1980. Mitochondrial calcium transport. *FEBS Letters.* 111:261–268.

Nicholls, D. G., and T. S. Sihra. 1986. Synaptosomes possess an exocytotic pool of glutamate. *Nature.* 321:772–773.

Richards, C. D., J. C. Metcalfe, G. A. Smith, and T. R. Hesketh. 1984. Changes in free calcium levels and pH in synaptosomes during transmitter release. *Biochimica et Biophysica Acta.* 803:215–220.

Rugolo, M., J. O. Dolly, and D. G. Nicholls. 1986. The mechanism of action of β-bungarotoxin at the presynaptic plasma membrane. *Biochemical Journal.* 223:519–523.

Schaffer, W. T., and M. S. Olson. 1980. The regulation of pyruvate oxidation during membrane depolarization of rat brain synaptosomes. *Biochemical Journal.* 192:741–751.

Scott, I. D., K. E. O. Åkerman, and D. G. Nicholls. 1980. Ca^{2+} transport by intact synaptosomes, intrasynaptosomal compartmentation and the role of the mitochondrial membrane potential. *Biochemical Journal.* 192:873–880.

Scott, I. D., and D. G. Nicholls. 1980. Energy transduction in intact synaptosomes: influence of plasma-membrane depolarization on the respiration and membrane potential of internal mitochondria determined in situ. *Biochemical Journal.* 186:21–33.

Sihra, T. S., I. G. Scott, and D. G. Nicholls. 1984. Ionophore A23187, verapamil, protonophores and veratridine influence the release of GABA from synaptosomes by modulation of the plasma membrane potential rather than the cytosolic calcium. *Journal of Neurochemistry.* 43:1624–1630.

Snelling, R., and D. G. Nicholls. 1984. The calmodulin antagonists trifluoperazine and R24571 depolarize the mitochondria within guinea pig cerebral cortical synaptosomes. *Journal of Neurochemistry.* 42:1552–1557.

Snelling, R., and D. G. Nicholls. 1985. Calcium efflux and cycling across the synaptosomal plasma membrane. *Biochemical Journal.* 226:225–231.

Zoccarato, F., and D. G. Nicholls. 1982. The role of phosphate in the regulation of the independent calcium-efflux pathway of liver mitochondria. *European Journal of Biochemistry.* 127:333–338.

Chapter 4

Calcium-permeable Channels in Vascular Smooth Muscle: Voltage-activated, Receptor-operated, and Leak Channels

Christopher D. Benham and Richard W. Tsien

Department of Physiology, Yale University School of Medicine, New Haven, Connecticut

Introduction

Ca entry across the cell membrane is important as a mechanism of signal transduction in many cell systems. Driven by a steep electrochemical gradient, Ca entry may produce a rapid rise in intracellular Ca to trigger the release of hormones and transmitters, regulate metabolic processes, activate Ca-sensitive membrane ion channels, or initiate contraction. In light of the varied effects of Ca entry in a wide range of cell types, it would be reasonable to suppose that Ca entry can be mediated by a variety of different pathways.

There is little dispute about the existence of voltage-dependent Ca channels in a wide variety of excitable cells. Thanks to the development of new recording methods and new chemical probes, much is known about the properties of voltage-dependent Ca channels (Hagiwara and Byerly, 1983; Tsien, 1983; Reuter, 1983; Trautwein and Pelzer, 1985; Tsien et al., 1987*a*, *b*). However, voltage-dependent Ca channels are only one of a number of classes of Ca entry pathways that have been postulated. In smooth muscle, for example, studies of ion flux, membrane potential, and muscle tension led both Bolton (1979) and Van Breemen et al.

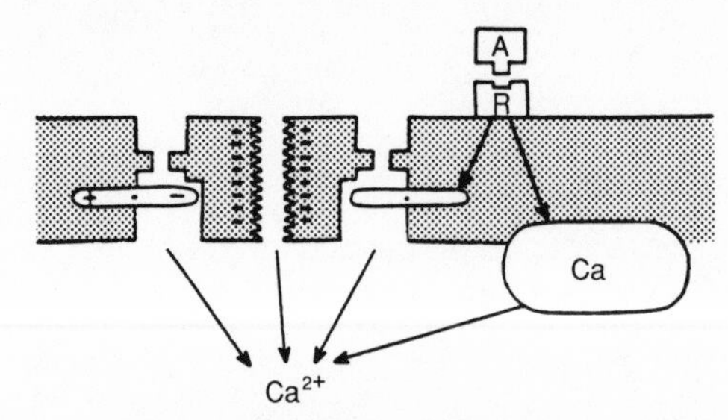

Figure 1. Schematic representation of proposed pathways for Ca delivery to the myoplasm of smooth muscle cells. Pathways for Ca influx across the surface membrane include (left to right): voltage-gated Ca channels, a Ca "leak" pathway, and receptor-operated calcium channels. Receptor occupation by agonists can also induce release of internal Ca stores (adapted from Cauvin et al., 1983).

(1979) to propose that Ca entry is mediated by voltage-activated Ca channels and two other kinds of Ca entry pathways. As illustrated schematically in Fig. 1, it was hypothesized that "receptor-operated channels" open in response to neurotransmitters and other stimulant substances, and that "leak channels" allow a steady resting influx of Ca. In this scheme of Van Breemen et al., receptor-operated channels and leak channels are distinct from voltage-activated channels or intracellular Ca stores that release Ca to mediate contraction in the absence of extracellular Ca. The concept of receptor-operated channels had already been formulated by Somlyo and Somlyo (1968, 1971), as part of the explanation for "pharmacomechanical coupling" in smooth muscle. Receptor-operated Ca channels have also been invoked to explain a variety of findings in liver (Exton, 1982; Reinhart et al., 1984), exocrine gland cells (Putney, 1978), neutrophils (Korchak et al., 1984), and other cell types.

In contrast to the present agreement about voltage-activated Ca channels, the existence of other kinds of Ca entry pathways remains much more controversial. In the absence of direct experimental recordings under voltage clamp, the concept of receptor-operated channels has not been universally accepted. Other authors have argued that receptor-operated Ca channels are not necessary to explain the original experimental findings that led to their postulation (e.g., Cohen et al., 1986). For example, observations of neurotransmitter effects on K-depolarized smooth muscle (Evans et al., 1958) can be attributed to Ca release from internal stores or to modulation of conventional voltage-dependent Ca channels that are incompletely inactivated. Findings of Ca entry or depolarizing responses even at relatively negative potentials and even in the presence of dihydropyridines can be explained by the participation of dihydropyridine-resistant, low–voltage-activated Ca channels (Cohen et al., 1986) or Ca-permeable cation channels gated by intracellular Ca (von Tscharner et al., 1986). Observations of Ca entry or contractile activation in the absence of any detectable depolarization or even hyperpolarization (Droogmans et al., 1977) could be explained by the participation of Ca-activated K channels (e.g., Benham et al., 1986). Arguments along these lines are difficult to settle without an electrophysiological demonstration of receptor-operated channels that minimizes the confounding effects of other permeability mechanisms.

In this chapter, we present direct evidence for the existence of all three categories of Ca-permeable channels, based on patch-clamp experiments in single smooth muscle cells of the rabbit ear artery. This preparation serves as an example of how resting Ca entry or Ca entry in response to depolarization and/or neurotransmitters controls cytoplasmic free Ca and thereby regulates cellular function (contraction and control of blood pressure). We report single channel recordings of (*a*) voltage-sensitive Ca channels activated by membrane depolarization, (*b*) receptor-operated Ca-permeable channels activated by ATP, and (*c*) "leak" channels that are active even without depolarization.

Methods

Adult rabbits (1.5–2.5 kg) were killed by cervical dislocation or by intravenous pentobarbitone injection. Single smooth muscle cells from the central ear artery were dispersed as previously described (Benham and Bolton, 1986). Briefly, the ear artery was dissected free of connective tissue, cut into strips, and incubated for 90 min in low-Ca saline containing collagenase (Cooper Biomedical, Inc., Malvern, PA), elastase (Sigma Chemical Co., St. Louis, MO), and bovine serum albumin. Cells were separated by trituration and stored at 4°C until use.

Membrane current recordings were made at 21°C using standard patch-clamp techniques (Hamill et al., 1981). Current records were filtered at 1 kHz and sampled at 5 kHz for analysis by a PDP-11/23 computer. Linear leak and capacitative currents have been subtracted from the whole-cell current record of potential-operated channels.

The recording pipettes (1–5 MΩ) contained an intracellular solution of the following composition (mM): 130 CsCl, 5 HEPES, 10 EGTA, 3 $MgCl_2$, and 0.5 ATP, buffered to pH 7.2 with TEA-OH. The physiological salt solution used for the cell dissociation and for recording the Ca currents in Fig. 1 contained (mM): 130 NaCl, 5 KCl, 10 glucose, 10 HEPES, 1.2 $MgCl_2$, and 1.5 $CaCl_2$, buffered to pH 7.3 with NaOH. Ca was reduced to 10 μM (added Ca) for the enzyme-containing dissociation media. The isotonic Ba solution contained (mM): 110 $BaCl_2$, 10 glucose, and 10 HEPES, buffered to pH 7.3 with TEA-OH. The junction potential between the pipette and 110 mM Ba solutions was found to be −7 mV (pipette negative), but has not been corrected for. Bay K 8644 and Bay R 5417 were gifts of Dr. A. Scriabine, Miles Laboratories, New Haven, CT.

Results

Two Components of Voltage-gated Ca Current in Whole-Cell Recordings

To record inward Ca channel currents, we voltage-clamped arterial smooth muscle cells with Cs-containing patch pipettes to reduce outward K currents, the dominant membrane conductances in these preparations. When cells were bathed in a physiological salt solution containing 1.5 mM Ca, small but clear inward currents could be detected (Fig. 2*A*). With depolarizing voltage pulses from a holding

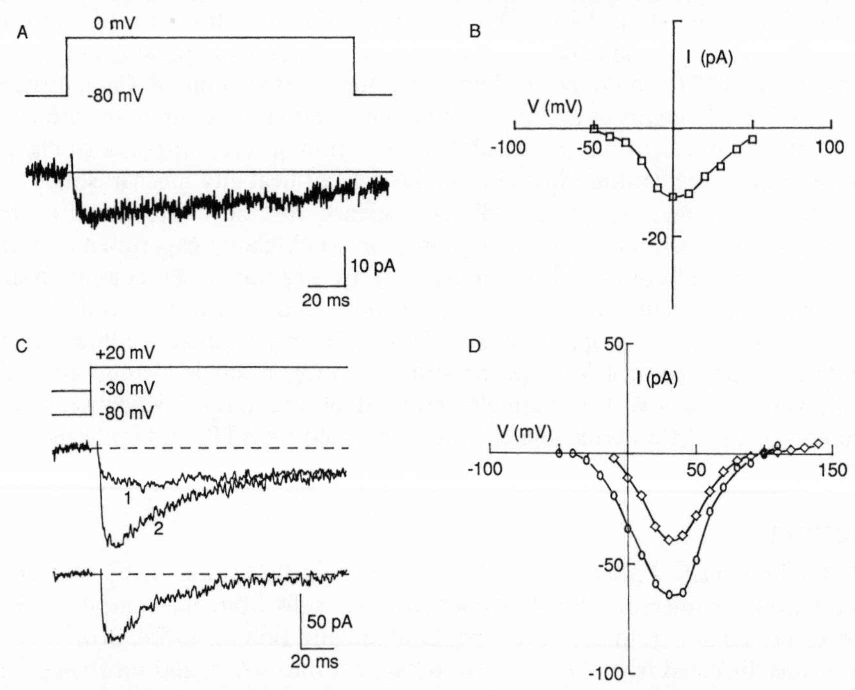

Figure 2. Whole-cell Ca and Ba currents in ear artery cells. (*A*) Inward current record from a cell in physiological saline containing 1.5 mM Ca. (*B*) Peak current-voltage relationship for Ca currents in the same cell: HP = −80 mV. (*C*) Currents evoked by test pulses to +20 mV from HP = −30 mV (record 1) and HP = −80 mV (record 2) in another cell with 110 mM external Ba. The lower record shows the difference current signal obtained by subtracting record 1 from record 2. (*D*) Peak current-voltage relationships for the same cell as in *C*, evoked from HP = −30 mV (diamonds) and HP = −90 mV (circles). (From Benham et al., 1987*b*.)

potential (HP) of −80 mV, inward current was evoked at test potentials (TP) positive to −50 mV, and reached a maximal amplitude of 10–15 pA at ~0 mV (*B*). The extrapolated reversal potential was positive to +50 mV, as expected for a Ca-selective channel. Most of our recordings were carried out with 110 mM Ba in the bathing solution (Fig. 2*C*). This greatly increased the Ca channel current, reduced the possibility of contaminating outward currents, and improved the stability of the recording. Cells remained relaxed and often gave peak inward currents of 100 pA or more for as long as 30 min. In these smooth muscle cells, unlike many other cell types, the addition of ATP (0.5 mM) to the pipette solution was sufficient to prevent rundown.

With 110 mM Ba as the charge carrier, it was possible to distinguish two components of inward Ca channel current that showed similarities to the currents designated as L and T in cardiac cells and neurons (Fig. 2*C*). Thus, our studies parallel those of Bean et al. (1986) and Loirand et al. (1986). A sustained current (component L) was seen in isolation in response to voltage steps from HP = −30 mV to TP = +20 mV. An additional, rapidly inactivating current (component T) was recruited by using a more negative holding potential (HP = −80 mV); both currents were strongly divalent cation selective, reversing polarity near +100 mV (Fig. 2*D*). The amplitudes of the inactivating and sustained current components were roughly equal in the whole-cell recordings illustrated here. However, wide variations in the relative contributions were seen and some cells showed one type of current almost exclusively. L current was absent in cells that appeared to be damaged during the dispersal procedure; in healthy cells, the L currents were often five times larger than the largest T current ever seen in isolation.

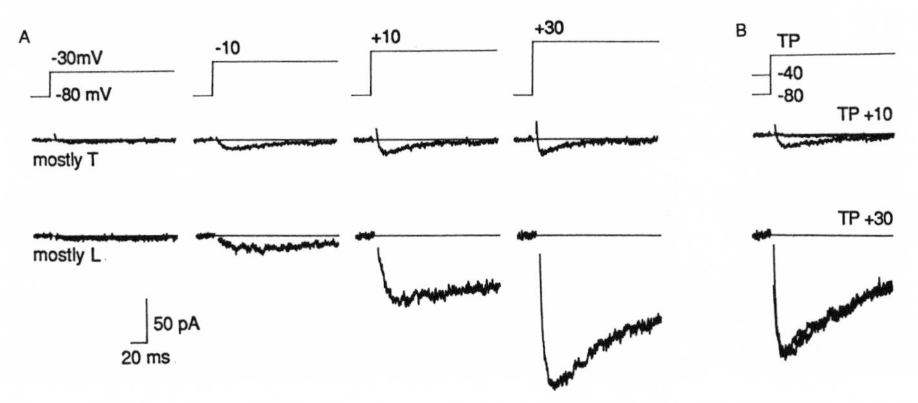

Figure 3. Whole-cell currents in T- and L-rich cells with 110 mM external Ba. (*A*) Current records from a T-rich cell (upper traces) and an L-rich cell (lower traces), evoked by increasingly strong test depolarizations. HP = −80 mV throughout. (*B*) Effect of changing the holding potential from −40 to −80 mV in the same two cells as in *A*. (From Benham et al., 1987*b*.)

Kinetic Properties of T and L Currents

The properties of the individual current components were best characterized in cells that showed one predominant current component. Fig. 3 displays a family of records from a cell showing T current only (upper row), along with records from another cell showing mainly L current (lower row). T current was partially activated at TP = −30 mV and reached a peak at +10 mV. Even at −10 mV, inactivation was sufficiently fast so that the current declined completely by the end of the 150-ms test pulse. If the holding potential was changed from −80 to −40 mV, this transient current was entirely inactivated (Fig. 3*B*). L currents were activated over a range of test depolarizations somewhat more positive than the activation range for T currents, and reached a peak at +30 mV. This difference can be seen in the peak current-voltage relationships for these two cells (Fig. 4, *A* and *B*). At a given test potential, L current showed less inactivation than T current. During the 150-ms test depolarization, some L current inactivation was seen beyond TP = +10 mV; at +30 mV, the L current inactivated by ~50% over the course of the test pulse. In contrast to T current, peak L current was reduced very little by changing the holding potential from −80 to −40 mV (Fig. 3*B*, lower traces). The difference in the voltage dependence of inactivation of T and L currents is illustrated in Fig. 4*C*. Inactivation of T current is 50% complete at HP = −55 mV, while L current is half-inactivated at HP = −23 mV.

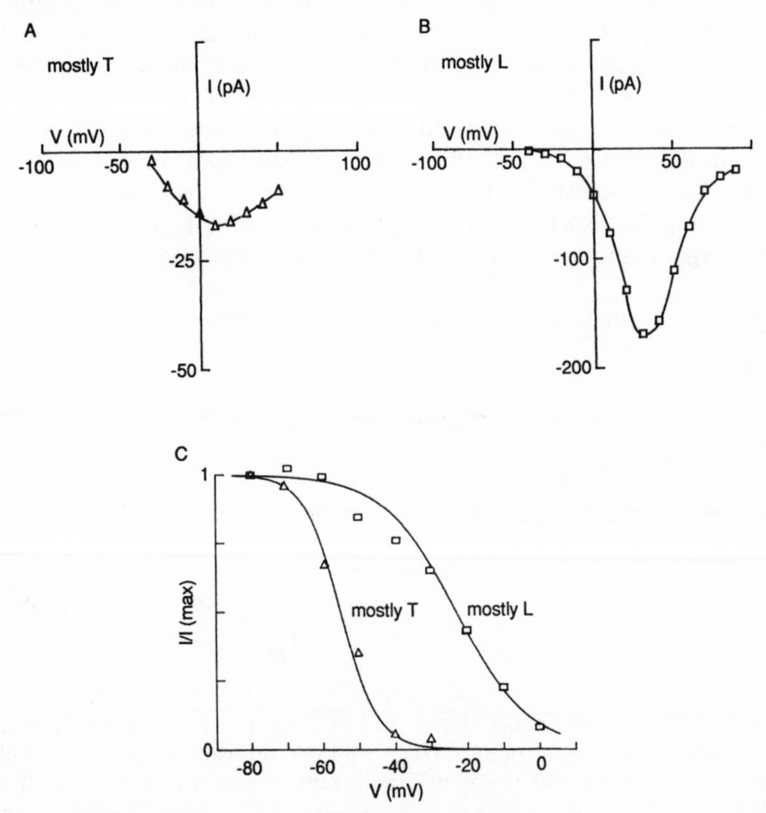

Figure 4. (*A* and *B*) Peak current-voltage plots (HP = −80 mV) for the same two cells shown in Fig. 3. (*C*) Steady state inactivation plotted for a T-rich cell (triangles) and an L-rich cell (squares). The data were fitted by eye by curves of the form $I/I_{max} = 1/\{1 - \exp[(V - V_H)/k]\}$, where V_H is the potential at which half of the current was inactivated. For the T-rich cell, I_{max} = 25 pA, V_H = −55 mV, and k = 5 mV. For the L-rich cell, I_{max} = 42 pA, V_H = −23 mV, and k = 10 mV. (From Benham et al., 1987*b*.)

Pharmacological Differences between T and L Currents

We found that T and L currents in arterial smooth muscle cells could be distinguished by their very different sensitivity to dihydropyridines, like T and L currents in other systems (e.g., Bean, 1985; Nowycky et al., 1985; Nilius et al., 1985). We examined the effects of both antagonist and agonist dihydropyridine compounds on L currents. The application of 5 μM (−)PN202-791, an antagonist compound, almost completely blocked L current, even when the current was evoked from a relatively negative holding potential (Benham et al., 1987*b*). With lower concentrations of antagonist, it was apparent that the degree of block depended strongly on the holding potential, as previously reported in smooth muscle cells (Bean et al., 1986) and cardiac cells (Sanguinetti and Kass, 1984; Bean, 1984). Exposure to the dihydropyridine agonists BAY K 8644 or BAY R 5417, the (+)isomer of Bay K 8644, caused a large increase in L current amplitude, a shift in the peak current to more negative potentials, and a more rapid inactivation (Benham et al., 1987*b*; see also Bean et al., 1986). In contrast to L currents, T currents were very little affected by dihydropyridine agonists or antagonists. The T current that remained in the presence of the dihydropyridine Ca antagonist had a very similar time course to T current in the absence of drug, and the voltage dependence of inactivation was essentially the same (Benham et al., 1987*b*; see also Bean et al., 1986).

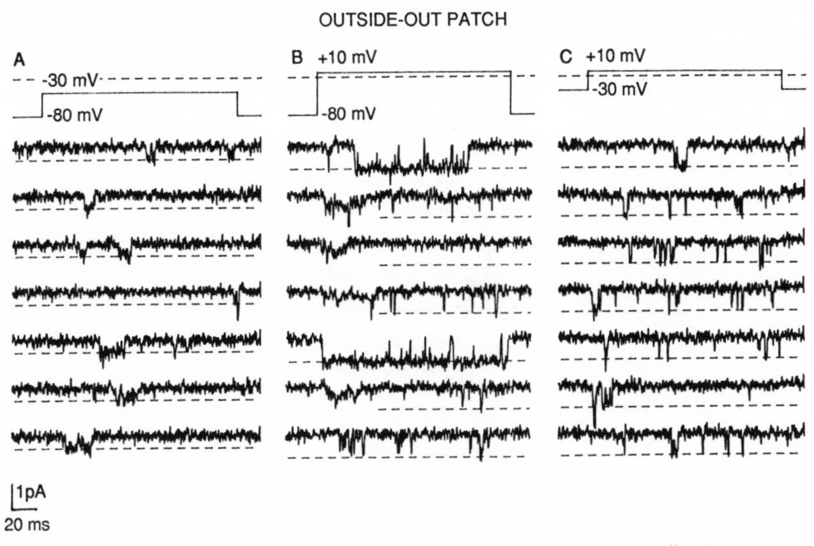

Figure 5. Single channel currents recorded in isolated patches with 110 mM Ba as the charge carrier. Representative current records were obtained from one patch with three stimulus protocols. In *C*, only L channel activity was evoked from a holding potential of −30 mV. *B* shows both types of currents with HP = −80 mV, and *A* shows only T channel openings at TP = −30 mV. (From Benham et al., 1987*b*.)

Unitary Currents Carried by T- and L-Type Channels

The whole-cell current recordings provided good evidence that ear artery cells contain two types of voltage-gated Ca channel. To confirm this at the level of individual Ca channels, we examined the properties of unitary currents in isolated outside-out membrane patches with 110 mM Ba in the bathing medium. Fig. 5 illustrates two types of Ca channel with voltage- and time-dependent properties corresponding to components T and L in whole-cell recordings. Channel activity corresponding to L current was elicited by pulses from a relatively depolarized holding potential (−30 mV) to a test potential of +10 mV (Fig. 5*C*). The unitary currents had an amplitude of ~1 pA and were evenly distributed throughout the test pulse. Test pulses from HP = −80 mV to TP = −30 mV (Fig. 5*A*) evoked another kind of channel activity that we attribute to T-type channels. In this case, the unitary currents were only ~0.5 pA in amplitude, despite the increased driving force for Ba entry at −30 mV, which indicates that the unitary conductance is much smaller than in the case of the L-type channel activity in the right panel.

Fig. 5*B* shows that pulses from HP = −80 to TP = +10 mV evoke activity of both types of channel in combination, which is consistent with the behavior of T and L currents in whole-cell recordings. At this more positive test potential, the openings of the small-conductance channels are bunched near the beginning of the sweep, as expected for a rapidly inactivating T current, while openings of the large-conductance channel are seen throughout the pulse, as in panel *C*. Two of the sweeps in *B* provide an illustration of unusually long openings of the large-conductance channel, a pattern of gating (mode 2) that occurred in <1% of all sweeps. Although not illustrated here, mode 2 activity of the L-type channels was greatly promoted by the presence of dihydropyridine agonists. These observations

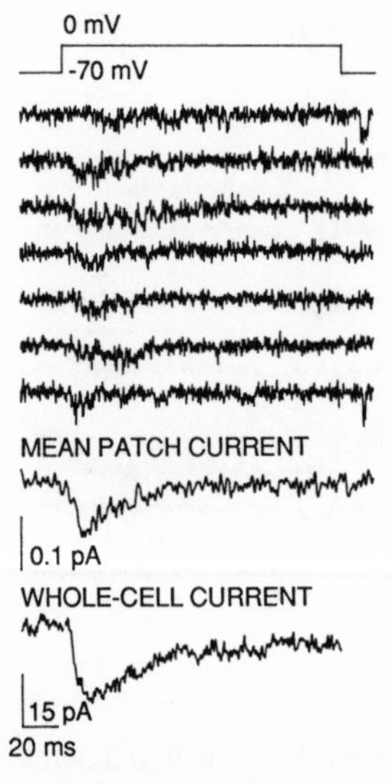

Figure 6. Whole-cell and unitary T-type Ca channel currents from the same cell. The bathing solution contained 5 μM (−)PN202-791 to minimize L-type Ca channel activity. The whole-cell current (bottom trace) was recorded before the creation of an outside-out patch and subsequent recording of unitary currents (noisy traces). The mean patch current is the average of 143 individual records. Note the similar time courses of the mean patch current and the whole-cell current. (From Benham et al., 1987*b*.)

are very much in line with findings for L-type channels in other preparations, including A10 cells, a cell line derived from rat aortic smooth muscle (Fox et al., 1986).

The large-conductance channels were also sensitive to dihydropyridine Ca antagonists; in the presence of (−)PN202-791, only small-amplitude unitary currents were observed. We took advantage of the dihydropyridine insensitivity of T-type channels to look for stronger evidence of a correspondence between unitary openings with a small conductance and inactivating T currents in whole-cell recordings. A whole-cell current was evoked in a cell in the presence of 5 μM (−)PN202-791 and then an outside-out patch was pulled from the same cell (Fig. 6). The same stimulus protocol was then used to evoke patch currents. The average patch current showed decay with a similar time course to the whole-cell current obtained from the same cell. This provides strong evidence that the small-conductance channels underlie the T current in whole-cell recordings.

Single channel current-voltage plots were constructed from measurements of unitary amplitudes at different potentials (Fig. 9). The data indicate single channel conductances of 7 pS for the small channel and 26 pS for the large channel. These values are very similar to those obtained for T and L channels in dorsal root ganglion neurons (Nowycky et al., 1985) and cardiac cells (Nilius et al., 1985). There is no indication of the intermediate-conductance N-type Ca channels found

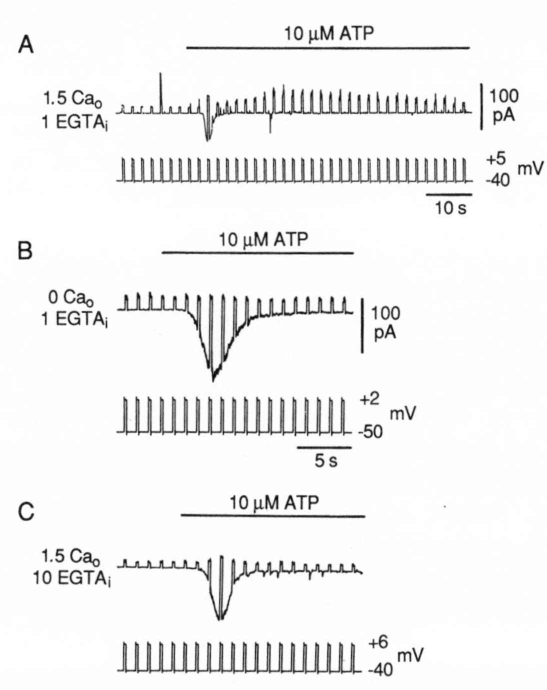

Figure 7. Responses of ear artery cells under whole-cell voltage clamp to bath application of 10 μM ATP: activation of Ca-activated K current as a bioassay of changes in intracellular free Ca. The external solution was normal saline (*A* and *C*) or Ca-free saline (*B*). The intracellular solution included 130 mM KCl and EGTA at the concentrations indicated. Depolarizing voltage pulses 500 ms in duration were applied every 1–2 s. (*A*) ATP evokes an inward current at −40 mV and then an outward current at +5 mV in the presence of extracellular Ca. (*B*) In the absence of extracellular Ca (1 mM EGTA), the secondary rise of outward current is abolished. (*C*) The outward current is also abolished by increasing the internal Ca buffering. (From Benham et al., 1987*a*.)

in various neuronal preparations. As Nilius et al. (1985) observed in heart cells, we found that L-type channel activity declined within minutes after patch excision, whereas T-type channels remained active as long as the patch remained stable.

Receptor-operated Channels Activated by ATP

To look for direct evidence for receptor-operated Ca channels, we chose to study the effects of ATP. This neurotransmitter is released by sympathetic neurons at the junction with vascular smooth muscle and acts on purinergic receptors on post-junctional cells. It has been suggested that ATP is responsible for the generation of excitatory junction potentials (see Burnstock and Kennedy, 1986, for review).

Fig. 7 shows membrane current responses of single ear artery cells to ATP under varying conditions of extracellular and intracellular Ca. Depolarizing pulses are used to measure membrane conductance and to promote the opening of Ca-activated K channels (Benham et al., 1986), which serve as a bioassay of changes in cytoplasmic free Ca. Panel *A* shows results obtained with a near-physiological level of external Ca (1.5 mM) and mild Ca buffering (1 mM EGTA) of the pipette (internal) solution. The cell was held at −40 mV and depolarized to +5 mV every 2 s. Under these conditions, bath application of 10 μM ATP caused (*a*) a phasic increase in membrane conductance associated with a transient inward current at −40 mV and an increase in outward current at +5 mV. This response quickly decayed, and was followed by (*b*) a delayed increase in outward current at +5 mV, with relatively little current at the holding potential. In the absence of extracellular Ca (Fig. 7*B*), the initial inward current was little affected, but the increase in outward current was abolished, which indicates that the outward current was dependent on extracellular Ca. The inclusion of 10 mM EGTA in the pipette to increase intracellular Ca buffering also abolished the outward current (Fig. 7*C*).

The abolition of the delayed increase in outward currents by removal of external Ca or buffering of internal Ca suggests that Ca entry directly or indirectly associated with the ATP-activated inward current is responsible for promoting Ca-activated K current. The ATP-activated inward current has a reversal potential close to 0 mV. Additional ion substitution experiments (Benham et al., 1987*a*) demonstrate that the reversal potential of the ATP-activated current is little affected by exchanging extracellular Na with K or Cs, but is shifted to more negative values if the monovalent cations are replaced by impermeant substitutes (e.g., Tris). Replacing Cl with gluconate had no effect on the current. These results suggest that the inward current is generated by a nonselective cation channel and raise the possibility that the channel may be significantly permeable to Ca.

Our work at Yale has been directed at answering a number of questions that arise from the experiments of Benham et al. (1987*a*). Can Ca flux through the ATP-activated channel be measured directly? Does ATP act through a second messenger such as intracellular Ca itself? What is the single channel conductance of receptor-operated channels and how does it compare with voltage-gated Ca channels in the same cells?

Direct Demonstration of Ca Permeation

To investigate the divalent cation permeability of the ATP-activated pathway, we recorded currents from outside-out patches with isotonic Ca in the bathing solution and strong Ca buffering (10 mM EGTA plus 1 mM BAPTA) of the internal solution. Fig. 8*A* illustrates a typical experiment. The membrane potential was held at very negative voltages where voltage-gated channels are not activated, and ATP was applied by applying pressure to a puffer pipette. Under these conditions, as in experiments with Tyrode's solution, ATP evoked a prominent transient inward current. The decline is due to desensitization, inasmuch as a second application of 10 μM ATP 1 min later had no effect. Since Ca is the only significant cationic charge carrier in the external solution, such recordings indicate that the ATP-activated channels must be permeable to Ca. In this respect, the channels in smooth muscle may differ from ATP-activated currents in cardiac atrial cells (Friel and Bean, 1986) and sensory neurons (Krishtal et al., 1983), which seem to lack significant Ca permeability (but see Sharma and Sheu, 1986).

Fig. 8 *B* shows results from a very similar experiment, in which current records were taken on a more rapid time base to allow the resolution of unitary openings. Traces lasting 400 ms are separated by intervals of 800 ms. ATP was applied from the puffer pipette between the first and second traces and evoked a large inward current, which declined rapidly to a level where the underlying unitary current activity could be resolved. The unitary current amplitude was ~0.5 pA at −100 mV with Ca as the charge carrier.

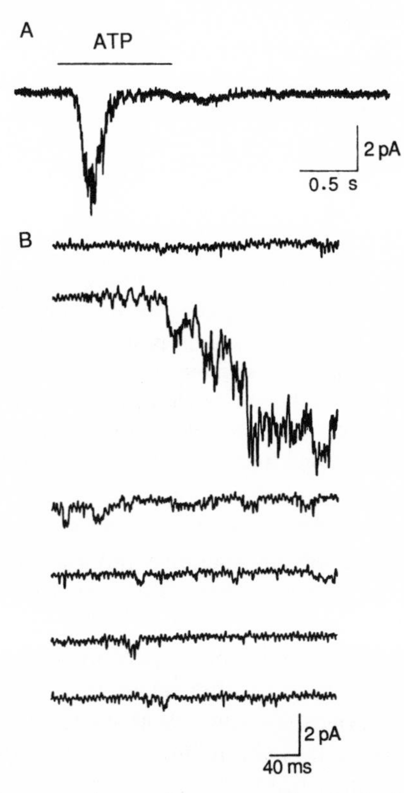

Figure 8. ATP-activated current in outside-out patches bathed in 110 mM Ca while the internal solution contained 130 mM CsCl, 10 mM EGTA, 1 mM BAPTA, and 10 mM HEPES-TEA. (*A*) ATP was applied by pressure application from a patch pipette containing 10 μM ATP positioned ~100 μm from the cell. HP = −120 mV. (*B*) ATP was applied in the same way to another cell held at −100 mV, between the first and second traces. Each 400-ms trace is separated by an interval of 800 ms. The ATP current reached a peak of 9 pA. Note the unitary currents that are resolved once the ATP response has largely desensitized.

To facilitate comparison of the ATP-activated channels with the two types of voltage-gated Ca channels described earlier in this chapter, we recorded currents from outside-out patches with 110 mM Ba as the external charge carrier. Fig. 9 *A* shows unitary Ba currents activated by ATP in an outside-out patch at −90 mV. ATP was applied between the second and third traces shown and evoked unitary currents 0.65 pA in amplitude. The unitary current-voltage relationships for ATP-activated channels and T- and L-type channels are compared in Fig. 9 *B*. It is clear that with 110 mM Ba as the charge carrier, the ATP-activated channels can be

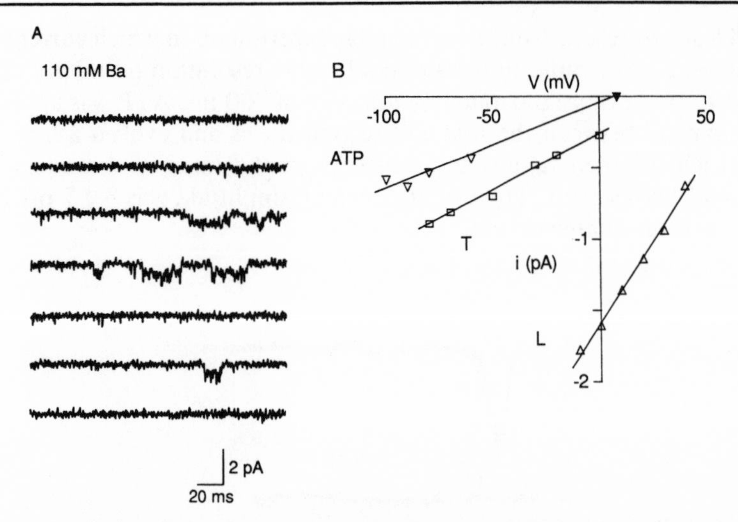

Figure 9. (*A*) ATP-activated unitary currents with 110 mM Ba as the charge carrier at HP = −90 mV. ATP was applied by passive diffusion from the puffer pipette containing 100 μM ATP. (*B*) Plot of single channel current against potential for an ATP channel (downward triangles), voltage-gated T-type channels (squares), and L-type channels (upward triangles). The data were obtained from three different patches with 110 mM Ba as the charge carrier. The data points were averages of 3–10 measurements of unitary amplitude. The filled triangle indicates the reversal potential for the ATP current as determined from whole-cell recordings.

clearly distinguished by their lower conductance and less positive reversal potential. Together with other lines of evidence, these results suggest that receptor-operated current and voltage-gated Ca currents arise from different channels.

Fig. 10 compares ATP-activated unitary currents carried by Ca, Ba, and Na. All three recordings were carried out with a steady holding potential of −140 mV, far negative to the range of potentials where voltage-gated Ca channels are known to open. The unitary Na current is about three- to fourfold larger than either the Ca or Ba currents at this potential. In this respect, the receptor-operated channels resemble voltage-gated Ca channels (see Tsien et al., 1987*b*, for references).

Although Na ions give larger unitary currents than Ca or Ba ions, the ATP-

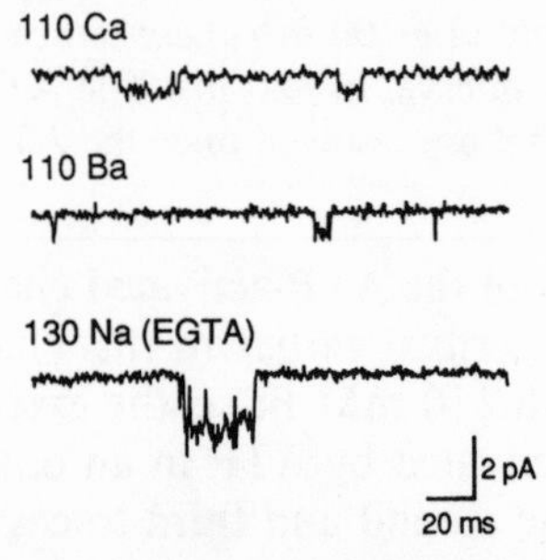

Figure 10. Single channel currents activated by ATP in three patches bathed in 110 mM Ca, 110 mM Ba, and 130 mM Na (1 mM EGTA). The internal solution was the same as in Fig. 7. All records were obtained at HP = −140 mV.

activated channels show a preference for divalent cations over monovalent cations, as judged by reversal potential measurements. In experiments to be presented elsewhere, we found that the ATP-activated current reversed at +12 mV with 110 mM Ca outside (and 130 Cs inside), which indicates a mild selectivity in favor of Ca. To evaluate the relative permeabilities for the main physiological cations at near-physiological levels of external Ca, we determined the reversal potential of the ATP-activated current with 2 mM Ca plus 135 mM glucosamine outside and 130 mM Na inside. The average value, −39 mV, translates to a P_{Ca}/P_{Na} of 3.3 when evaluated with Fatt and Ginsborg's (1958) modification of the Goldman-Hodgkin-Katz equation.

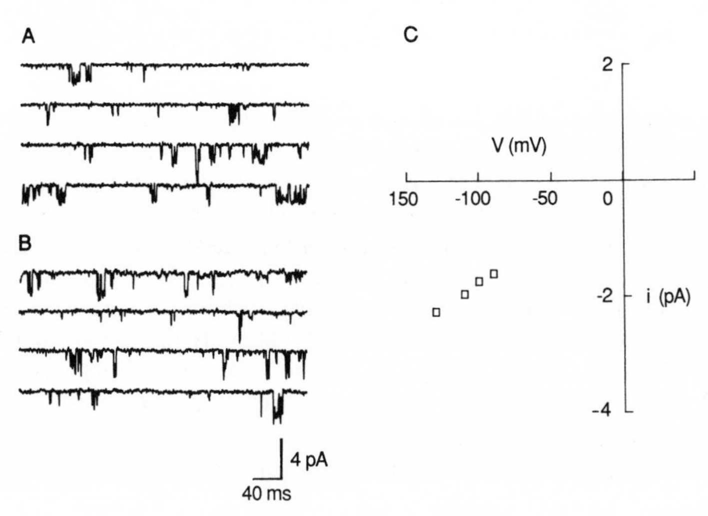

Figure 11. Leak channels opening spontaneously at negative potentials with 110 mM Ba as the charge carrier. Outside-out patch with Cs-containing pipette. (*A* and *B*) Unitary currents at steady holding potentials of −100 mV (*A*) and −130 mV (*B*). (*C*) Single channel current-voltage relationship. The straight line indicates a slope conductance of 16 pS.

"Leak" Channels Permeable to Divalent Cations

Radioactive isotope studies of Ca influx have shown that there is a significant influx of Ca in quiescent vascular smooth muscle in the absence of direct stimulation of voltage-gated or receptor-operated pathways (Van Breemen et al., 1979). Although the resting Ca leak pathway has been depicted as a channel in Fig. 1, no direct evidence has been presented to indicate whether it is a membrane channel or a transport mechanism. While recording activity in membrane patches bathed in 110 mM Ba, we have occasionally observed inward unitary currents that appear spontaneously at negative membrane potentials. Records from one such experiment are shown in Fig. 11. The unitary amplitude was ~1.6 pA at −90 mV and 2.25 pA at −130 mV; the associated current-voltage relationship showed a unitary conductance of 16 pS. On the other hand, the degree of channel opening was not strongly affected by the holding potential. Further experiments are needed to explore the ionic selectivity of this channel; from the results with isotonic Ba as the charge carrier, it is already clear that the channel is different from the ATP-activated channels, which have a conductance of 5–6 pS under these conditions. The "leak" channels were seen in <10% of the patches; if the Ca flux at physiological levels is like that of voltage-gated channels, even a sparse distribution of leak channels might be sufficient to account for the resting Ca influx in quiescent cells.

Discussion

This is the first paper to provide direct evidence in an excitable cell for the coexistence of three kinds of channels with selectivity for divalent ions: voltage-gated, ligand-gated, and leak channels. Our experiments strongly support the hypothesis that these pathways have distinctive properties at the single channel level.

Voltage-gated Ca Channels

We used a combination of whole-cell and single channel recordings to demonstrate the existence of two types of voltage-activated Ca channels in vascular smooth muscle cells. Our results in ear artery are qualitatively similar to whole-cell recordings of Ca currents in mesenteric artery (Bean et al., 1986), azygous vein (Sturek and Hermsmeyer, 1986), and portal vein (Loirand et al., 1986), and whole-cell and single channel recordings in coronary arteries (Isenberg and Klockner, 1985). The two types of Ca channel in these vascular smooth muscle preparations show a strong resemblance to T- and L-type channels in sensory neurons (see Nowycky et al., 1985, for references) and heart cells (Bean, 1985; Nilius et al., 1985). As in other preparations, L-type Ca channels in ear artery cells are characterized by (*a*) a single channel conductance of ~25 pS in 110 mM Ba, (*b*) an average current that activates with relatively strong depolarizations and inactivates relatively slowly, (*c*) a clear responsiveness to dihydropyridine agonists and antagonists, and (*d*) a tendency to show progressive rundown with intracellular dialysis or patch excision. T-type Ca channels in the artery cells are like those in other systems with respect to (*a*) a single channel conductance of 8 pS in 110 mM Ba, (*b*) an average current that activates with relatively negative test depolarizations and inactivates relatively rapidly, (*c*) no pronounced responsiveness to dihydropyridine agonists and antagonists, and (*d*) a relatively greater ability to survive the dispersal procedure, intracellular dialysis, and patch excision.

Like similar results from the single channel recordings of Isenberg and Klockner (1985) in bovine vascular cells, our findings are significantly different from those of Worley et al. (1986) and Caffrey et al. (1986). Worley et al. recorded two types of Ca channel activity in membrane patches excised from rabbit mesenteric artery cells, which showed unitary conductances of 8 and 15 pS with 80 mM external Ba. Both channel types were almost completely inhibited by 50 nM nisoldipine. Caffrey et al. (1986) also found a Bay K 8644–sensitive channel with a conductance of ~15 pS (110 mM external Ba) in aortic smooth muscle cells. These results are quite different from ours. In our hands, the 8-pS channel was very insensitive to dihydropyridines; the dihydropyridine-sensitive L channel had a considerably higher unitary conductance of 26 pS. The 15-pS channel observed by Worley et al. and Caffrey et al. was also different from the N-type channel in DRG neurons, which is resistant to dihydropyridines (Nowycky et al., 1985).

Ligand-gated Ca-permeable Channels

One of our main conclusions is that ATP-activated channels are different from voltage-gated Ca channels in several respects.

Voltage dependence. Unlike any known voltage-gated channel in arterial smooth muscle cells or other excitable cells, the ATP-activated channels can be opened by ligands at membrane potentials as negative as −140 mV. The probability of opening shows no striking voltage dependence, although whole-cell recordings do reveal some relaxations after large voltage steps.

Unitary currents. With 110 mM Ba as the external charge carrier, the unitary current-voltage relationship of the ATP-activated channels clearly sets them apart from voltage-gated channels (T- and L-type) or "leak" Ca channels in the same cells. The ATP-activated channels are found in virtually every outside-out patch, sometimes in large numbers, and may be the most common type of Ca-permeable channel in these preparations.

Pharmacological sensitivity. Unlike voltage-gated Ca channels, ATP-activated channels gave sizable currents even in the presence of 5 μM nifedipine or 0.5 mM Cd. We find that ATP-activated channels rapidly desensitize in response to a maintained exposure to ATP, while voltage-gated channels remain active. Operationally, either kind of channel can be studied in isolation of the other.

Our results suggest that the ATP-activated channels and voltage-gated channels may have at least one fundamental similarity: both kinds of channels transfer Na ions more rapidly than Ca ions, even though reversal potential measurements indicate that P_{Ca} is greater than P_{Na}. These findings can be understood in terms of a pore with high-affinity divalent cation–binding sites, in which "sticky" ions translocate less rapidly but appear more permeable in reversal potential measurements (see Tsien et al., 1987*b*, for review).

Distinctions between Direct and Indirect Ligand Activation

Von Tscharner et al. (1986) looked for ligand-gated Ca channels in neutrophils as an explanation of Ca entry activated by fMet-Leu-Phe. Instead of a Ca-selective channel directly activated by ligands, they found evidence for an indirect mechanism involving ligand-activated release of Ca from internal stores, and activation by intracellular Ca of a Ca-permeable, nonselective cation channel. The mechanism of ATP activation in ear artery cells is quite different in that (*a*) bath application of ATP fails to activate channels isolated in cell-attached patches, (*b*) ATP activation occurs despite strong buffering of internal Ca by EGTA plus BAPTA in whole cells or outside-out patches, and (*c*) ATP activation is not prevented by prior application of caffeine to unload intracellular Ca stores.

While an indirect mechanism of channel activation can be ruled out for ATP-activated channels, it could be quite important in explaining Ca influx activated by norepinephrine. Although hypotheses about receptor-operated channels originated in studies of the action of norepinephrine in vascular smooth muscle, we avoided norepinephrine in our initial studies in view of the known complexity of its actions (Bolton and Large, 1986). In ear artery cells, evidence already exists that norepinephrine releases intracellular Ca stores and causes opening of Ca-activated K channels (Benham and Bolton, 1986). This raises the possibility that norepinephrine-induced inward cation currents are also mediated by a rise in intracellular

Ca, very much like the inward cation currents activated by fMet-Leu-Phe in neutrophils (von Tscharner et al., 1986). Norepinephrine also produces additional modulatory effects on voltage-gated Ca channels, increasing the amplitude of L currents (Aaronson et al., 1986). Actions of norepinephrine on internal stores and voltage-gated Ca channels raise complications that need to be sorted out in future efforts to identify a receptor-operated channel activated by norepinephrine. One promising approach is to apply caffeine to deplete internal stores and to introduce strong Ca buffers to suppress internal Ca transients. It will be interesting to see whether norepinephrine still activates inward currents under these conditions, and if such currents are carried by the same channels that are opened by ATP.

Comparisons between ATP-activated Channels and Other Ligand-gated Channels

There are a number of similarities between ATP-activated channels in arterial smooth muscle and channels activated by acetylcholine at the motor endplate (Lewis and Stevens, 1979; Adams et al., 1980), by *N*-methyl-D-aspartate (NMDA) in spinal or hippocampal neurons (Nowak et al., 1984; MacDermott et al., 1986), and by internal cyclic GMP in vertebrate photoreceptors (Haynes et al., 1986; Zimmerman and Baylor, 1986). Each of these ligand-activated channels shows significant Ca permeability and larger unitary conductances with Na than with Ca.

The ATP-activated channels in smooth muscle are most like NMDA receptor channels in their substantial selectivity for Ca over Na, with values of P_{Ca}/P_{Na} of ~4 (this paper; MacDermott et al., 1986), more than an order of magnitude greater than acetylcholine receptor channels (Adams et al., 1980). On the other hand, the ATP-activated channels are relatively insensitive to Mg, unlike NMDA-activated channels (Nowak et al., 1984; Mayer et al., 1984). Under physiological conditions, depolarization causes relief from this Mg block, giving the NMDA-activated channels a striking voltage dependence not found for the ATP-activated channels.

Divalent Cation–selective Leak Channels

Leak channels that are active in the absence of depolarization or ligand have received the least attention of all the various kinds of Ca-permeable channels. Our recordings of divalent cation–permeable channels open at resting potentials provide the first direct evidence for the leak pathway hypothesized by Cauvin et al. (1983) in smooth muscle. Recent experiments demonstrate the existence of Ca-permeable leak channels in other membrane systems. Kuno et al. (1986) have recently reported a voltage-independent channel in helper T lymphocytes that has a 7-pS slope conductance and a strongly positive reversal potential with 110 mM Ba as the external charge carrier. Cardiac sarcolemmal membranes incorporated into planar bilayers express a Ca-permeable channel open at resting membrane potentials (Rosenberg et al., 1987). The channels have an apparent open time of hundreds of milliseconds near −80 mV and a slope conductance of ~10 pS with ~100 mM Ba or Ca as the divalent charge carrier.

It will be interesting to look for Ca-selective leak channels in other excitable cells, given that the level of intracellular free Ca in quiescent cells is generally much higher than expected from the operation of Na/Ca exchangers or Ca pumps in the absence of a countervailing leak. If the transfer rate through individual channels is high, as appears to be the case for leak channels found so far, even a small number of leak channels might have a profound effect on the resting level of intracellular free Ca.

Acknowledgments

This work was supported by U.S. Public Health Service grant HL13306, and by grants from Marion Laboratories and Smith Kline & French (United Kingdom).

References

Aaronson, P. I., C. D. Benham, T. B. Bolton, P. Hess, R. J. Lang, and R. W. Tsien. 1986. Two types of single channel and whole cell calcium or barium currents in single smooth muscle cells of rabbit ear artery and the effects of noradrenaline. *Journal of Physiology.* 377:36P. (Abstr.)

Adams, D. J., D. M. Dwyer, and B. Hille. 1980. The permeability of endplate channels to monovalent and divalent metal cations. *Journal of General Physiology.* 75:493–510.

Bean, B. P. 1984. Nitrendipine block of cardiac calcium channels: high-affinity binding to the inactivated state. *Proceedings of the National Academy of Sciences.* 81:6388–6392.

Bean, B. P. 1985. Two kinds of calcium channels in canine atrial cells. Differences in kinetics, selectivity and pharmacology. *Journal of General Physiology.* 86:1–30.

Bean, B. P., M. Sturek, A. Puga, and K. Hermsmeyer. 1986. Calcium channels in muscle cells isolated from rat mesenteric arteries: modulation by dihydropyridine drugs. *Circulation Research.* 59:229–235.

Benham, C. D., and T. B. Bolton. 1986. Spontaneous transient outward currents in single visceral and vascular smooth muscle cells of the rabbit. *Journal of Physiology.* 381:385–406.

Benham, C. D., T. B. Bolton, N. G. Bryne, and W. A. Large. 1987*a*. Action of extracellular adenosine triphoshate in single smooth muscle cells dispersed from the rabbit ear artery. *Journal of Physiology.* In press.

Benham, C. D., P. Hess, and R. W. Tsien. 1987*b*. Two types of calcium channels in single smooth muscle cells from rabbit ear artery studied with whole-cell and single channel recordings. *Circulation Research.* In press.

Benham, C. D., T. B. Bolton, R. J. Lang, and T. Takewaki. 1986. Calcium-activated potassium channels in single smooth muscle cells of rabbit jejunum and guinea-pig mesenteric artery. *Journal of Physiology.* 371:45–67.

Bolton, T. B. 1979. Mechanisms of action of transmitters and other substances on smooth muscle. *Physiological Reviews.* 59:607–718.

Bolton, T. B., and W. A. Large. 1986. Are junction potentials essential? Dual mechanism of smooth muscle cell activation by transmitter released from autonomic nerves. *Quarterly Journal of Experimental Physiology.* 71:1–28.

Burnstock, G., and C. Kennedy. 1986. A dual function for adenosine 5′-triphosphate in the regulation of vascular tone. *Circulation Research.* 58:319–330.

Caffrey, J. M., I. R. Josephson, and A. M. Brown. 1986. Ca channels of amphibian stomach and mammalian aorta smooth muscle cells. *Biophysical Journal.* 49:1237–1242.

Cauvin, C., R. Loutzenhizer, and C. Van Breemen. 1983. Mechanisms of calcium antagonist induced vasodilation. *Annual Review of Toxicology.* 23:373–396.

Cohen, C. J., R. A. Janis, D. G. Taylor, and A. Scriabine. 1986. Where do Ca^{2+} antagonists act? *In* Calcium Antagonists in Cardiovascular Disease. L. H. Opie, editor. Raven Press, New York. 151–163.

Droogmans, G., L. Raeymaekers, and R. Casteels. 1977. Elecro- and pharmacomechanical coupling in the smooth muscle cells of the rabbit ear artery. *Journal of General Physiology.* 70:129–148.

Evans, D. H. L., H. O. Schild, and S. Thesleff. 1958. Effects of drugs on depolarized plain muscle. *Journal of Physiology.* 143:474–485.

Exton, J. H. 1982. Molecular mechanisms involved in α-adrenergic responses. *Trends in Pharmacological Sciences.* 3:111–115.

Fatt, P., and B. L. Ginsborg. 1958. The ionic requirements for the production of action potentials in crustacean muscle fibres. *Journal of Physiology.* 142:516–543.

Fox, A. P., P. Hess, J. B. Lansman, B. Nilius, M. C. Nowycky, and R. W. Tsien. 1986. Shifts between modes of calcium channel gating as a basis for pharmacological modulation of calcium influx in cardiac, neuronal and smooth-muscle-derived cells. *In* New Insights into Cell and Membrane Transport Processes. G. Poste and S. T. Crooke, editors. Plenum Publishing Corp., New York. 99–124.

Friel, D., and B. P. Bean. 1986. External ATP-activated currents in cardiac atrial cells. *Biophysical Journal.* 49:402*a*. (Abstr.)

Hagiwara, S., and L. Byerly. 1983. The calcium channel. *Trends in Neurosciences.* 6:189–193.

Hamill, O. P., A. Marty, E. Neher, B. Sakmann, and F. J. Sigworth. 1981. Improved patch-clamp techniques for high resolution current recording from cells and cell-free membrane patches. *Pflügers Archiv.* 391:85–100.

Haynes, L. W., A. R. Kay, and K.-W. Yau. 1986. Single cyclic GMP-activated channel activity in excised patches of rod outer segment membrane. *Nature.* 321:66–70.

Isenberg, G., and U. Klockner. 1985. Elementary currents through single Ca channels in smooth muscle cells isolated from bovine coronary arteries. *Pflügers Archiv.* 403:R23. (Abstr.)

Korchak, H. M., K. Vienne, L. E. Rutherford, and G. Wiessmann. 1984. Neutrophil stimulation: receptor, membrane and metabolic events. *Federation Proceedings.* 43:2749–2754.

Krishtal, O. A., S. M. Marchenko, and V. I. Pidoplichko. 1983. Receptor for ATP in the membrane of mammalian sensory neurones. *Neuroscience Letters.* 35:41–45.

Kuno, M., J. Goronzy, C. M. Weyand, and P. Gardner. 1986. Single-channel and whole-cell recordings of mitogen-regulated inward currents in human cloned helper T lymphocytes. *Nature.* 323:269–273.

Lewis, C. A., and C. F. Stevens. 1979. Mechanism of ion permeation through channels in a postsynaptic membrane. *In* Membrane Transport Processes. C. F. Stevens and R. W. Tsien, editors. Raven Press, New York. 3:133–151.

Loirand, G., P. Pacaud, C. Mironneau, and J. Mironneau. 1986. Evidence for two distinct calcium channels in rat vascular smooth muscle cells in short-term primary culture. *Pflügers Archiv.* 407:566–568.

MacDermott, A. B., M. L. Mayer, G. L. Westbrook, S. J. Smith, and J. L. Barker. 1986. NMDA-receptor activation elevates cytoplasmic calcium in cultured spinal cord neurones. *Nature.* 321:519–522.

Mayer, M. L., G. L. Westbrook, and P. B. Guthrie. 1984. Voltage-dependent block by Mg^{2+} of NMDA responses in spinal cord neurones. *Nature.* 309:261–263.

Nilius, B., P. Hess, J. B. Lansman, and R. W. Tsien. 1985. A novel type of cardiac calcium channel in ventricular cells. *Nature.* 316:443–446.

Nowak, L., P. Bregestovski, P. Ascher, A. Herbet, and A. Prochiantz. 1984. Magnesium gates glutamate-activated channels in mouse central neurones. *Nature.* 307:462–465.

Nowycky, M. C., A. P. Fox, and R. W. Tsien. 1985. Three types of neuronal calcium channel with different calcium agonist sensitivity. *Nature.* 316:440–443.

Putney, J. W., Jr. 1978. Stimulus-permeability coupling: role of calcium in the receptor regulation of membrane permeability. *Pharmacological Reviews.* 30:209–245.

Reinhart, P. H., W. M. Taylor, and F. L. Bygrave. 1984. The contribution of both extracellular and intracellular calcium to the action of α-adrenergic agonists in perfused rat liver. *Biochemical Journal.* 220:35–42.

Reuter, H. 1983. Calcium channel modulation by neurotransmitters, enzymes and drugs. *Nature.* 301:569–574.

Rosenberg, R. L., and R. W. Tsien. 1987. Calcium permeable channels from cardiac sarcolemma open at resting membrane potentials. *Biophysical Journal.* 51:29*a.* (Abstr.)

Sanguinetti, M. C., and R. S. Kass. 1984. Voltage-dependent block of calcium channel current in the calf cardiac Purkinje fiber by dihydropyridine calcium channel antagonists. *Circulation Research.* 55:336–348.

Sharma, V. K., and S.-S. Sheu. 1986. Micromolar extracellular ATP increases intracellular calcium concentration in isolated rat ventricular myocytes. *Biophysical Journal.* 49:351*a.* (Abstr.)

Somlyo, A. P., and A. V. Somlyo. 1968. Electromechanical and pharmacomechanical coupling in vascular smooth muscle. *Journal of Pharmacology and Experimental Therapeutics.* 159:129–145.

Somlyo, A. V., and A. P. Somlyo. 1971. Strontium accumulation by sarcoplasmic reticulum and mitochondria in vascular smooth muscle. *Science.* 174:955–958.

Sturek, M., and K. Hermsmeyer. 1986. Calcium and sodium channels in spontaneously contracting vascular muscle cells. *Science.* 233:475–478.

Trautwein, W., and D. Pelzer. 1985. Gating of single calcium channels in the membrane of enzymatically isolated ventricular myocytes from adult mammalian hearts. *In* Cardiac Electrophysiology and Arrhythmias. D. Zipes and J. Jalife, editors. Grune & Stratton, Orlando, FL. 31–42.

Tsien, R. W. 1983. Calcium channels in excitable cell membranes. *Annual Review of Physiology.* 45:341–358.

Tsien, R. W., A. P. Fox, P. Hess, E. W. McCleskey, B. Nilius, M. C. Nowycky, and R. L. Rosenberg. 1987*a.* Multiple types of calcium channel in excitable cells. *In* Proteins of Excitable Membranes. B. Hille and D. M. Fambrough, editors. John Wiley & Sons, New York. 167–187.

Tsien, R. W., P. Hess, E. W. McCleskey, and R. L. Rosenberg. 1987*b.* Calcium channels: mechanisms of selectivity, permeation and block. *Annual Review of Biophysics and Biophysical Chemistry.* 16:265–290.

Van Breemen, C., P. Aaronson, and R. Loutzenhizer. 1979. Sodium-calcium interactions in mammalian smooth muscle. *Pharmacological Reviews.* 30:167–208.

von Tscharner, V., B. Prod'hom, M. Baggiolini, and H. Reuter. 1986. Ion channels in human neutrophils activated by a rise in free cytosolic calcium concentration. *Nature.* 324:369–372.

Worley, J. F., III, J. W. Deitmer, and M. T. Nelson. 1986. Single nisoldipine-sensitive calcium channels in smooth muscle cells isolated from rabbit mesenteric artery. *Proceedings of the National Academy of Sciences.* 83:5746–5750.

Zimmerman, A. L., and D. A. Baylor. 1986. Cyclic GMP-sensitive conductance of retinal rods consists of aqueous pores. *Nature.* 321:70–72.

Chapter 5

Alterations in Intracellular Calcium Produced by Changes in Intracellular Sodium and Intracellular pH

L. J. Mullins and J. Requena

Department of Biophysics, University of Maryland School of Medicine, Baltimore, Maryland, and Instituto Internacional de Estudios Avanzados, Centro de Biociencias, Caracas, Venezuela

Responses of Ca Indicators Injected into an Axon

It has been known since the work of Hodgkin and Keynes (1957) that stimulation of an axon in seawater leads to an increased entry of Ca as measured with isotopes and that this increase is larger at higher concentrations of Ca in the seawater.

Using aequorin as an indicator of intracellular Ca, Baker et al. (1971) were able to show that stimulation increased the light emission from this indicator, as did the application of Na-free external solutions. Depolarization of squid axons had a complex response that could be separated into tetrodotoxin (TTX)-sensitive and TTX-insensitive parts.

This study largely confirmed earlier isotopic studies (Baker et al., 1969*b*) that showed that there was a Ca entry that was enhanced by low Na outside the axon and high Na inside the axon, as well as a reverse process where Na efflux was shown to depend in part on external Ca, with a mean of 4 Na emerging per Ca taken up (Baker et al., 1969*a*).

Such studies supported the view that there was a mechanism that exchanged Na for Ca across the axonal membrane with the requirement of an obligatory coupling of Na to Ca. There also appeared to be other pathways that allowed Ca movement with depolarization, such as the Na channel and a TTX-insensitive pathway that was thought to be a Ca channel.

Modification of the Response of a Ca Indicator

Because isotopic studies had shown that Ca efflux increased with hyperpolarization (Mullins and Brinley, 1975), it seemed possible that the Na/Ca exchange process was electrogenic and exchanged more Na charges than Ca charges. Such an arrangement would make Ca movement membrane potential dependent, not by opening membrane channels, but by running the exchanger in a direction that depended on the Na electrochemical gradient.

We decided to see to what extent we could modify the entry of Ca in response to stimulation and to steady depolarization in axons where (*a*) the internal Na was normal, and (*b*) internal Na had been reduced by prolonged stimulation of an axon in an Li seawater.

The results of this study were that a substantial reduction in $[Na]_i$ (*a*) had no effect on Ca entry in response to repetitive stimulation, and (*b*) virtually abolished the response to steady depolarization (Mullins and Requena, 1981). A further finding was that if the [Ca] in seawater was changed from 50 to 1 mM, there was no measurable response of injected aequorin to stimulation, while the response to steady depolarization was easily seen.

When a squid axon injected with arsenazo III is depolarized with 450 mM K solution, there is a change in [Ca], as shown in Fig. 1. The resting level of Ca in axons is of the order of 50 nM and this is increased severalfold by the depolarization. Of course, the Ca radial gradient resulting from Ca influx can be expected to be highly nonlinear, but the indicated [Ca] remains steady for many minutes and falls to resting levels as quickly as the mixing time of the experimental chamber allows the recovery of the resting membrane potential.

If an axon injected with arsenazo III is exposed to Li seawater and stimulated for 10^5 impulses, returned to seawater, and given a depolarizing test as in Fig. 1, there is no response of the Ca indicator to depolarization. Parallel studies with Na-

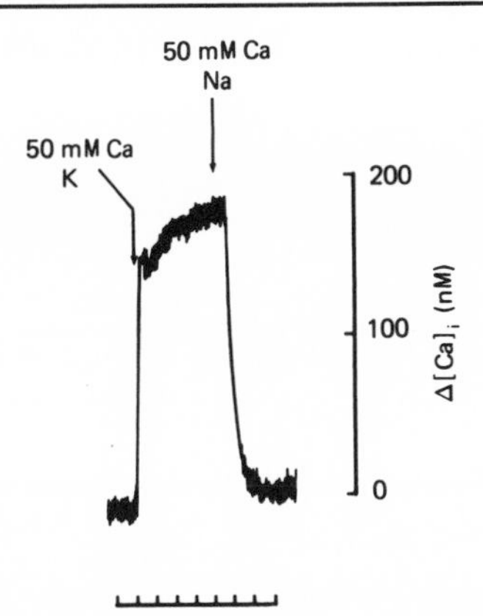

Figure 1. An arsenazo III–injected squid axon was depolarized in 450 mM KCl, 50 mM Ca solution, and the change in $[Ca]_i$ was measured. At the time indicated, 50 mM Ca, Na seawater was reapplied (Mullins et al., 1983).

sensitive electrodes in axons show that this stimulation in Li decreased $[Na]_i$ to about half its normal value. The effect is easily reversible and a further stimulation in Na seawater to return $[Na]_i$ to normal also results in an increase in Ca with depolarization.

The response of axons to stimulation at 60–100 impulses/s is different from that observed with steady depolarization on two counts: first, if $[Ca]_o$ is made 1 mM, there is no response to stimulation (in terms of an increase in Ca) using either arsenazo III or aequorin. Second, the response rises more slowly with stimulation at 60/s than it does upon depolarizing the axon with KCl. This is to be expected, since with stimulation at the frequency mentioned, the axon is depolarized <10% of the time compared with 100% with KCl.

An experiment illustrating this finding is shown in Fig. 2. Here, an axon stimulated in 50 mM Ca seawater gave an easily measured aequorin response. Next, the axon was depolarized with just enough $[K]_o$ to give about the same amplitude response. The K solution was removed, $[Ca]_o$ was made 1 mM, and the treatments described above were repeated. There was clearly no response to stimulation, as indeed one would not be expected, since responses to stimulation

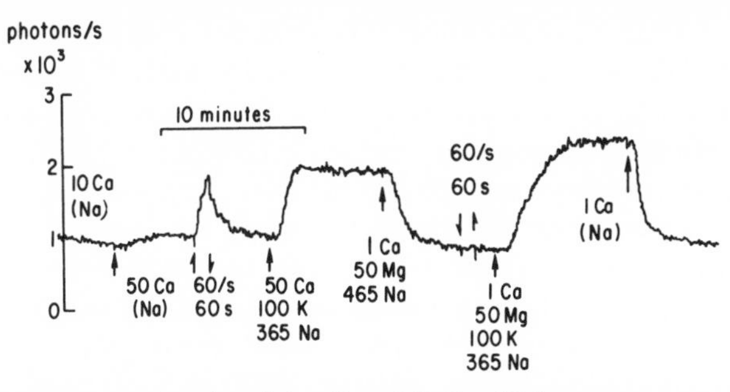

Figure 2. This axon was stimulated in 50 mM Ca (Na) seawater to produce the test stimulation response and then depolarized in 100 mM K (Na) seawater with 50 mM Ca. After this, the seawater was changed to 1 mM Ca (Na) (50 Mg) and a test stimulation then produced no response, whereas the subsequent 100 mM K depolarization produced a response similar to that in 50 mM Ca (Mullins and Requena, 1981).

are linear with $[Ca]_o$ and the expected response (1/50 of the earlier one) would be below the noise level. The response to depolarization, however, is the same (except for a slowed rise time), which suggests that the Ca entries for the two treatments are via different mechanisms.

Explaining the Effects of $[Na]_i$

Why should changing the internal Na concentration of an axon affect its response to depolarization? One can consider the following sorts of explanations: lowering $[Na]_i$ (*a*) increases the buffering of Ca by the axoplasm, (*b*) changes the sensitivity of the intracellular Ca indicator, and (*c*) decreases the amount of Ca that enters with depolarization.

An altered Ca buffering would seem to be most unlikely since the response of the axon to stimulation (a Ca entry mainly through Na channels) is unaffected by lowering $[Na]_i$ and one would have to invent complicated subsidiary explanations about having the Na channel carry more Ca as $[Na]_i$ is decreased. It is true that mitochondria have an Na/Ca exchange, but in axons with a normal $[Ca]_i$, mitochondrial buffering of Ca is not observable during depolarization.

The Ca indicator aequorin is not sensitive to changes in $[Na]_i$ in the range of 10–20 mM as judged by in vitro experiments, and we have also used the Ca indicator arsenazo III and shown the same sort of responses to altered $[Na]_i$ as observed with aequorin. Thus, it would appear that there is no basis for the notion that the indicator sensitivity is modified.

We are left, therefore, with the conclusion that a decrease in $[Na]_i$ decreases that rate at which Ca enters the axon with steady depolarization. This finding can be explained if we assume that the rate-limiting step for Ca entry is the formation of a compound at the inside surface of the membrane that involves multiple binding of Na^+, as in the following reaction:

$$\mathrm{Na} + X = \mathrm{Na}X,\ \mathrm{Na}X + \mathrm{Na} = \mathrm{Na}_2X,\ \ldots\ \mathrm{Na}_nX.$$

The Quantitative Dependence of Ca Entry on $[Na]_i$

By using arsenazo III as an indicator of $[Ca]_i$, and employing an Na-sensitive electrode to measure $[Na]_i$, it was possible to measure the change in $[Ca]_i$ with depolarization as $[Na]_i$ was increased. These measurements showed that at an $[Na]_i$ of 18 mM, there was no measurable response, while at an $[Na]_i$ of 37 mM, the response was saturated (Mullins et al., 1983). In between these values, the curve relating Ca entry to $[Na]_i$ rose steeply, with a Hill coefficient of ~7, which suggests a high degree of cooperativity to the reaction.

The results of this study are shown in Fig. 3, where the rate at which Ca-arsenazo absorbance rises with time is plotted against $[Na]_i$. Depolarization was to a constant level of −5 mV from a resting potential of −55 mV. Note that half-saturation of this curve is at 28 mM. Studies of the mean $[Na]_i$ of squid axons, as judged by Na electrode measurements, showed a value of 19 ± 1.6 mM; thus, freshly isolated axons have very little activation of their Na/Ca exchange system, but small increases in $[Na]_i$ (such as those brought on by stimulation) can be effective in greatly increasing activation.

Measurements of the dependence of Ca influx on $[Na]_i$ have been made

(DiPolo, 1979) using isotopic measurements and internal dialysis. DiPolo's results show a half-saturation of the reaction of ~55 mM, or twice the values obtained either with Na-free solutions or with K depolarization. In addition, the curves so obtained were initially rather linear, with [Na]$_i$ at high values for [Ca]$_i$, and clearly did not exhibit the high level of cooperativity shown in Fig. 3.

Steady depolarization does things other than activating Na/Ca exchange in a direction that carries Ca inward. It also opens ion channels and decreases the electrochemical driving force on cations moving inward. Therefore, it seemed useful to see whether the response of the entry of Ca as dependent on [Na]$_i$ could be obtained if, instead of depolarizing the membrane, we applied Na-free solutions. These are known to enhance Ca entry and this entry is the expected behavior of the Na/Ca exchange system. This is so because the Na electrochemical gradient is proportional to ($E_m - E_{Na}$) and a change in either ought to have the same effect in driving the exchange reaction.

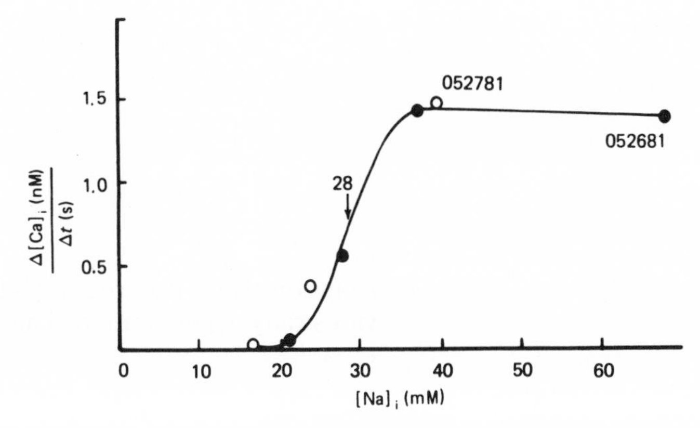

Figure 3. The initial rate of increase in [Ca]$_i$ upon the application of 3 mM Ca, 450 mM K is plotted (ordinate) as a function of [Na]$_i$ for two axons injected with arsenazo III (Mullins et al., 1983).

An experiment to test this is shown in Fig. 4. The axon was injected with aequorin and stimulated in Li to lower [Na]$_i$ to ~12 mM. Light was measured as the external solution was changed from seawater to Na-free seawater. After the response was obtained, the axon was depolarized with 450 mM K, 10 mM Ca seawater. After the results were recorded, the axon was stimulated to raise [Na]$_i$ and a second test was made as outlined above. The left-hand panel of Fig. 4 shows these responses to both Na-free solution and depolarization for values of Na$_i$ from 12 to 80 mM. The right-hand panel shows a plot of the light emission induced by Na-free solution and by depolarization and shows that they both fall on the same curve. It would appear, therefore, that both Na-free solutions and depolarization act to promote a Ca entry that has an identical dependence on [Na]$_i$.

Some Physiological Applications

Measurements similar to that cited above (Mullins et al., 1983) have been made by Lee and Dagostino (1982) and by Eisner et al. (1984) on cardiac cells. In both cases, Na-specific electrodes were used to measure [Na]$_i$, and muscle tension was used as an index of Ca release and/or Ca entry. In response to an action potential or a brief voltage-clamp pulse, a substantial amount of the Ca necessary for contraction comes from a store in the sarcoplasmic reticulum. The content of this store is in large measure set by surface membrane Na/Ca exchange, and this in turn is governed by $[Na]_i$ as indicated above. Thus, if $[Na]_i$ is high, then $[Ca]_i$ is high and the sarcoplasmic reticulum is able to store more Ca since it takes Ca from

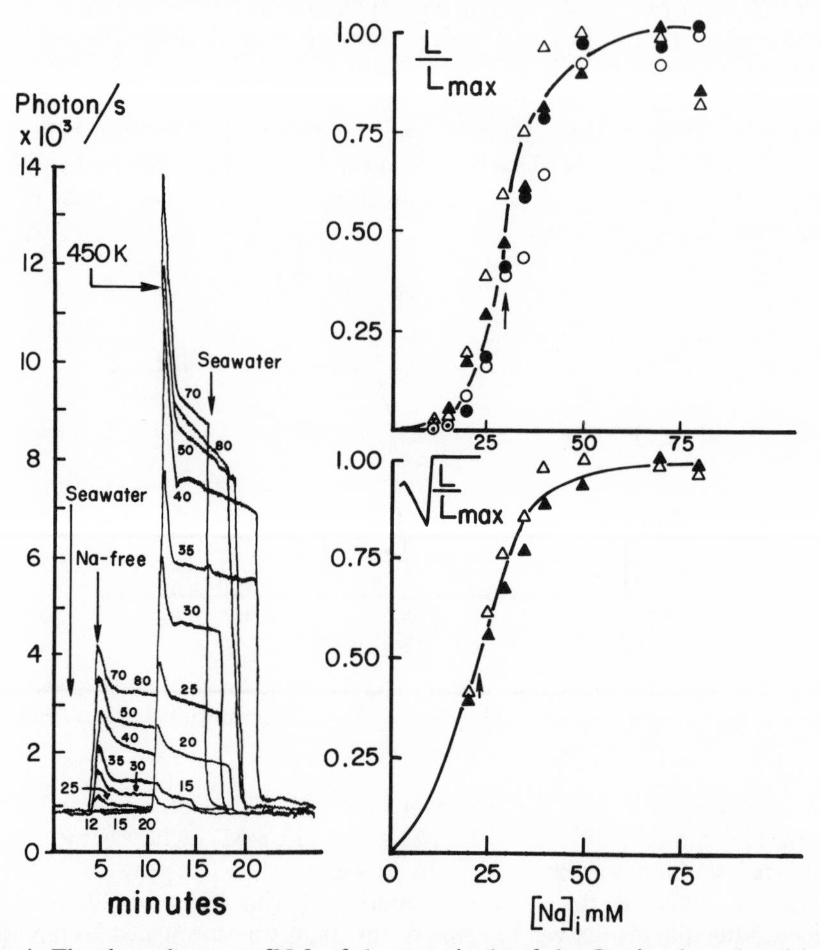

Figure 4. The dependence on $[Na]_i$ of the magnitude of the Ca signal produced by the removal of $[Na]_o$, followed immediately by a K depolarization. The left-hand panel shows the time course of records of aequorin signal superimposed for various $[Na]_i$ levels, all in the same axon and for 10 mM Ca. Records were synchronized at the onset of depolarization. The right-hand panel shows, as a function of $[Na]_i$, either the normalized (as a fraction of the maximum) magnitude of the increment in light (top) or the normalized square root (as a fraction of the maximum) (bottom) of the increment registered for the light signal. In the latter case, points were omitted for the low light emission range since this must be expected to be linear with $[Ca]_i$. The experimental points were: the peak of the rapid phasic response (▲) and the tonic or plateau (△) reached during a K depolarization and the peak (●) and the plateau (○) observed during the Tris (Na-free) episode. The axon was injected with aequorin only. Axon diameter, 0.600 mm. Membrane potential, −54 mV. 040683A. (Requena et al., 1986.)

the myoplasm during diastole. The expectation is, therefore, that cardiac tension ought to show a high degree of dependence on [Na]$_i$ and, in Fig. 5, data have been plotted on log-log coordinates and show that tension varies as more than the sixth power of [Na]$_i$. Note that a change in [Na]$_i$ of as little as 1 mM can more than double tension.

What Happens to Ca That Is Buffered?

Previous studies of the relationship between free and bound Ca in the axoplasm (Brinley et al., 1977) have shown that only 1 Ca in 1,000 that enter an axon goes

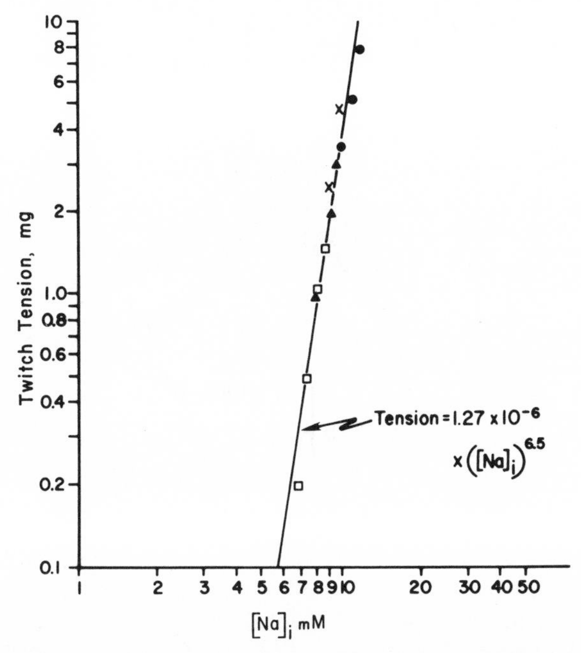

Figure 5. Relationship between twitch tension (T) and intracellular Na$^+$ activity (a^i_{Na}). T and a^i_{Na} were plotted on logarithmic coordinates. Data were taken from Lee and Dagostino (1982): O, from Fig. 5; ▲, from Fig. 6; ×, from Fig. 7; □, from Fig. 8. The straight line was determined by least-squares linear regression ($r = 0.97$).

to increase the free [Ca]. What happens to the other Ca ions? There do not appear to be soluble compounds in the axoplasm that simply complex with Ca at physiological levels of this ion, nor do the mitochondria appear to play any major storage role at physiological levels of [Ca]. The structure that accumulates Ca with high affinity appears to be the smooth endoplasmic reticulum and this structure appears to exchange Ca, at least in part, for protons.

Ca Entry Produces Acid

In a study of the change in pH$_i$ upon depolarization, we found (Mullins et al., 1983) that an easily detectable decrease in pH occurs. This finding suggests that the uptake of Ca by intracellular structures may, at least in part, be by an exchange of Ca for protons. At a pH$_i$ of 7.3, both Ca and H are present in about equal concentrations. As noted above, Ca buffering is such that only 1 Ca in 10^3 is free; for protons, only 1 in 10^4 is free, so that a small increase in Ca in the axoplasm such as that produced by a bioelectric event is ultimately transformed into a *very* small increase in $[H]_i$.

Making the Axoplasm Alkaline Reduces $[Ca]_i$

If the buffering of Ca occurs by an exchange of Ca for H^+, then it ought to be possible to increase the buffering of Ca by decreasing [H]. An experiment to examine this point is shown in Fig. 6, where an axon was injected with aequorin

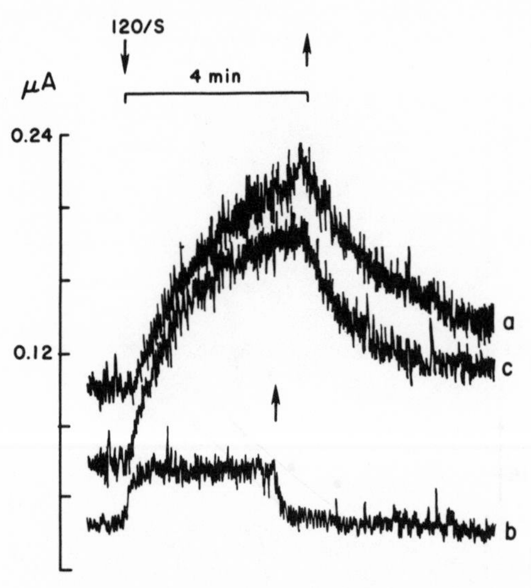

Figure 6. An axon in 50 mM Ca seawater was stimulated at 120/s for 240 s as shown in trace *a*. Seawater was then changed to 30 mM NH_4, 50 mM Ca for 10 min and stimulation was repeated at 120/s for 200 s (trace *b*). Finally, the axon was returned to 50 mM Ca, NH_4-free seawater for 6 min and stimulated again (trace *c*) (Mullins and Requena, 1979).

and stimulated at 120/s for 4 min. The result (trace *a*) is a slow rise in light emission and a fall upon terminating stimulation. Next, 10 mM NH_4Cl was added to the seawater and the stimulation was repeated after a time sufficient for pH_i to have stabilized at a level ~0.3 units higher than normal. The resting glow of the axon declined about fourfold and the response to stimulation rose much more rapidly and was fivefold smaller than that in the absence of NH_4 (trace *b*). Since the small pH change does not affect Ca entry with stimulation as measured with isotopes, nor does it affect the sensitivity of aequorin as a Ca indicator, it seems necessary to suppose that it is the buffering of Ca that has been enhanced by the change in pH_i. This enhanced buffering lowers both the resting level of $[Ca]_i$ and the increase in $[Ca]_i$ resulting from stimulation. The effect is entirely reversible since removal of NH_4 from seawater allows a restoration of the original pH_i and the response of the axon to stimulation. Trace *c* was taken before the internal pH had fully recovered and it shows that the resting glow was intermediate between the initial level and that when pH_i was ~7.6.

Since studies of the relationship between Ca entry and $[Na]_i$ showed little sensitivity of Ca entry with stimulation and a large dependence on $[Na]_i$ with steady depolarization, it is reasonable to conclude that most Ca entry with stimulation is via Na channels, and virtually all Ca entry with steady depolarization is via Na/Ca exchange.

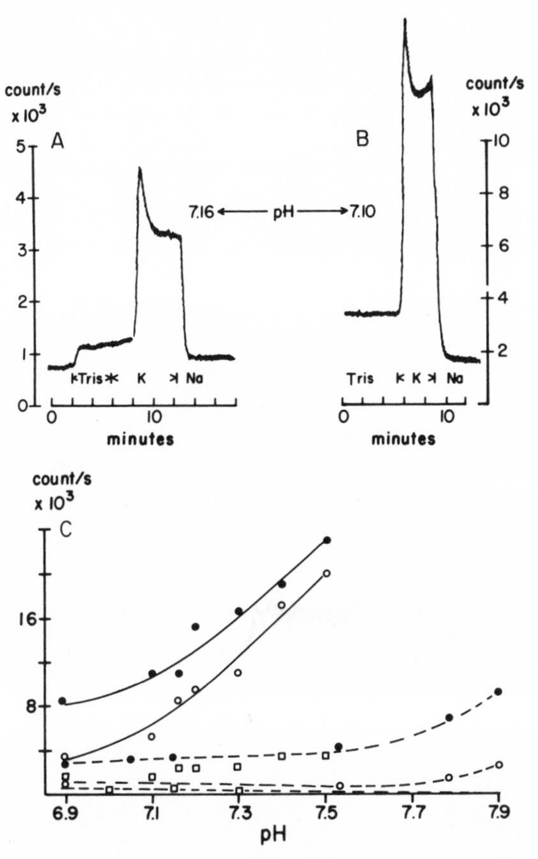

Figure 7. The effect of pH_i on the magnitude of the Ca signal of the removal of $[Na]_o$ followed by K depolarization. The axon was injected with aequorin only, impaled with a pH-sensitive microelectrode, and exposed throughout to 10 mM Ca_o. In *A* and *B*, two selected traces are shown that correspond to Ca signals caused by the substitution of Tris for $[Na]_o$ followed by a K depolarization before (trace at pH 7.16) and after (trace at pH 7.10) stimulation, a procedure used to elevate $[Na]_i$ above the physiological level. Note the different ordinates for the two traces. In *C*, the magnitude of the aequorin signal is plotted as a function of the pH_i and for the two levels of $[Na]_i$, elevated (continuous lines) and physiological (broken lines). The experimental points are: the peak of the rapid phasic response (●), the plateau (○) during a K depolarization, and the tonic level (□) during the Tris episode. Initial pH_i, 7.39; membrane voltage, −59 mV; diameter, 0.650 mm. 050683B. (Requena et al., 1986.)

Internal pH Affects Na/Ca Exchange

Mullins et al. (1983) studied the effect of pH_i on Ca entry with depolarization and concluded that it is strongly inhibited by a decrease in pH_i. Measurements with isotopes (DiPolo and Beaugé, 1982) also showed that efflux that is dependent on

$[Na]_o$ is strongly inhibited by acidification of the axoplasm. Studies with cardiac sarcoplasmic membrane vesicles (Bers and Ellis, 1982) confirm that Na/Ca exchange in this system is enhanced by alkaline pH and inhibited by acid. A similar sensitivity of intact cardiac muscle cells to internal pH changes (insofar as their Na/Ca exchange system is concerned) has also been observed (Vaughan-Jones et al., 1983).

It is important to distinguish between the various methods of measuring Na/Ca exchange as mentioned above and to evaluate the errors that may arise in connection with the measurements made. In the case of measurements of $[Ca]_i$ using intracellular indicators, the change in buffering of Ca by changes in pH_i will confuse these measurements. Thus, making the axoplasm alkaline will enhance buffering and make the actual change in $[Ca]_i$ an underestimate of the real change in Ca flux through the exchanger. With isotopes used in axons, since all measurements are made with internal dialysis, there will be a stronger competition between the EGTA buffer used in the dialysis medium to control $[Ca]_i$ and the intrinsic buffers of the axoplasm when pH_i is made alkaline. This effect might be expected to decrease Ca influx because of an enhanced ability of the axoplasmic buffers to capture incoming isotopic Ca, so that, again, isotopic methods underestimate the sensitivity of Na/Ca to pH changes.

A recent detailed study of the effects of pH_i on Ca entry with depolarization (Requena et al., 1986) is summarized in Fig. 7. A freshly isolated squid axon was subjected to depolarization and to Na-free seawater and the change in aequorin glow was measured. The results are shown in panel *A*. The axon was then subjected to stimulation to raise $[Na]_i$. A second test with Na-free solution and K depolarization gave the result shown in panel *B*. Note that the plateau in this panel is four times that of the initial one. pH_i was then manipulated while a glass electrode inside the axon gave the exact value, and these tests were performed over a range of pH_i from 6.9 to 7.9; the results are plotted in panel *C*. What this diagram suggests is that for axons with an enhanced $[Na]_i$, Ca entry after K depolarization approximately doubles in magnitude, going from pH_i 6.9 to 7.2, while there is scarcely any change in the response of the Na-free effect over the same range of pH values. If one recalls that a substantial increase in Ca buffering must also be expected to take place, it is clear that the enhancement of buffering and of the Na-free effect are of about the same magnitude and opposite in direction, so that the result is only the appearance of no effect. Similarly, the results at a lower $[Na]_i$ are that buffering changes (which decrease the signal) and increases in Ca entry (which increase the signal) approximately cancel. A similar finding was made with respect to the resting glow in the axon in the presence of normal seawater, as the tabulation below shows.

pH_i	Axon $[Na]_i$ Low	Axon $[Na]_i$ High
	Resting glow (*counts/s*)	
6.90	1,200	
7.05	900	1,700
7.15	700	
7.50	600	1,700
7.92	600	1,700

These results suggest that when $[Na]_i$ is low, the effect of pH_i on Ca buffering is likely to predominate, while when $[Na]_i$ is high, the effects of pH_i on Na/Ca exchange and on buffering tend to cancel.

References

Baker, P. F., M. P. Blaustein, A. L. Hodgkin, and R. A. Steinhardt. 1969*a*. The influence of calcium on sodium efflux in squid axons. *Journal of Physiology.* 200:431–458.

Baker, P. F., M. P. Blaustein, R. D. Keynes, J. Manil, T. I. Shaw, and R. A. Steinhardt. 1969*b*. The ouabain-sensitive fluxes of sodium and potassium in squid giant axons. *Journal of Physiology.* 200:459–496.

Baker, P. F., A. L. Hodgkin, and E. B. Ridgway. 1971. Depolarization and calcium entry in squid axons. *Journal of Physiology.* 218:709–855.

Bers, D. M., and D. Ellis. 1982. Intracellular calcium and sodium activity in sheep heart Purkinje fibres. Effects of changes of external sodium and intracellular pH. *Pflügers Archiv.* 393:171–178.

Brinley, F. J., Jr., T. Tiffert, A. Scarpa, and L. J. Mullins. 1977. Intracellular calcium buffering capacity in isolated squid axons. *Journal of General Physiology.* 70:355–384.

DiPolo, R. 1979. Calcium influx in internally dialyzed squid giant axons. *Journal of General Physiology.* 73:91–113.

DiPolo, R., and L. Beaugé. 1982. The effect of pH on Ca^{2+} extrusion mechanisms in dialyzed squid axons. *Biochimica et Biophysica Acta.* 688:237–245.

Eisner, D. A., W. J. Lederer, and R. D. Vaughan-Jones. 1984. The quantitative relationship between twitch tension and intracellular sodium activity in sheep cardiac Purkinje fibres. *Journal of Physiology.* 355:251–266.

Hodgkin, A. L., and R. D. Keynes. 1957. Movements of labelled calcium in squid giant axons. *Journal of Physiology.* 138:253–281.

Lee, C. O., and M. Dagostino. 1982. The effect of strophanthidin on Na_i and twitch tension. *Biophysical Journal.* 40:185–198.

Mullins, L. J., and F. J. Brinley, Jr. 1975. Sensitivity of calcium efflux from squid axons to changes in membrane potential. *Journal of General Physiology.* 65:135–152.

Mullins, L. J., and J. Requena. 1979. Calcium measurement in the periphery of an axon. *Journal of General Physiology.* 74:393–413.

Mullins, L. J., and J. Requena. 1981. The "late" channel calcium in squid axons. *Journal of General Physiology.* 78:683–700.

Mullins, L. J., T. Tiffert, G. Vassort, and J. Whittembury. 1983. Effects of internal sodium and hydrogen ions and of external calcium ions and membrane potential on calcium entry in squid axons. *Journal of Physiology.* 338:295–319.

Requena, J., L. J. Mullins, J. Whittembury, and F. J. Brinley, Jr. 1986. Dependence of ionized and total Ca in squid axons on Na_o-free or high-K_o conditions. *Journal of General Physiology.* 87:143–159.

Vaughan-Jones, R. D., W. J. Lederer, and D. A. Eisner. 1983. Ca^{2+} ions can affect intracellular pH in mammalian cardiac muscle. *Nature.* 301:522–524.

Chapter 6

Calcium and Magnesium Movements in Cells and the Role of Inositol Trisphosphate in Muscle

A. P. Somlyo, A. V. Somlyo, M. Bond, R. Broderick, Y. E. Goldman, H. Shuman, J. W. Walker, and D. R. Trentham

Pennsylvania Muscle Institute, University of Pennsylvania School of Medicine, Philadelphia, Pennsylvania

Introduction

The subcellular translocations of Ca^{2+} and the cellular messengers modulating these translocations play major roles in nearly every aspect of cell function and are, accordingly, central themes of modern cell physiology. Thus, the recognition of the important roles of Ca^{2+} in muscle and, more recently, in other cells, led to major questions bearing, respectively, on the identity of the intracellular organelles that are the sources and sinks of cytoplasmic free Ca^{2+} and on the mechanisms that couple surface membrane signals to intracellular Ca^{2+} release. Technical developments have been essential for directly addressing these questions, and the results described in the present communication, concerning the intracellular movements of Ca^{2+} and Mg^{2+} and the action of inositol 1,4,5-trisphosphate ($InsP_3$), are based predominantly on the development of two methods. The first of these, electron-probe x-ray microanalysis (EPMA), in conjunction with rapid freezing, permits the unambiguous and quantitative localization, at electron-microscopic resolution, of Ca^{2+}, Mg^{2+}, and other elements within cells (A. P. Somlyo, 1985*a*, 1986; A. P. Somlyo and Shuman, 1982). Rapid freezing and cryoultramicrotomy, in addition to preserving the in vivo distribution of diffusible elements, also provides the time resolution, on a millisecond time scale, for following subcellular ion movements triggered by physiological or pathological stimuli.

The time resolution of the action of physiological messengers, independent of diffusional delays, is also the contribution of the second method: laser-flash photolysis of inert, photolabile precursors of physiological messengers (Kaplan et al., 1978; Lester and Nerbonne, 1982; Hibberd et al., 1984). We have used this method to determine whether the Ca^{2+}-releasing action of $InsP_3$ in smooth (A. V. Somlyo et al., 1985*a*; Suematsu et al., 1984) and in striated (Volpe et al., 1985, 1986; Vergara et al., 1985) muscle is sufficiently fast to trigger normal contractions by delivering the message, across the gap between the surface membrane and the sarcoplasmic reticulum, to release Ca^{2+} (A. P. Somlyo, 1985*b*).

Sensitivity and Spatial Resolution: Quantitation of Low Ca^{2+} in Retinal Rods and Localization of Ca^{2+} and Cl^- in Atrial Specific Granules

EPMA is based on the fact that the ionization of atoms by fast electrons generates characteristic x-rays having energies identifying the excited atom. The interaction of incident fast electrons with atomic nuclei generates a background of continuum x-rays. Elemental quantitation of ultrathin sections with EPMA is generally based on the linear relationship between elemental concentrations and the ratio of the number of characteristic/continuum x-rays (Hall, 1971; Shuman et al., 1976; Kitazawa et al., 1983). Errors caused by overlap between the potassium K_β and the adjacent calcium K_α peak can be eliminated, for all practical purposes, by scrupulous calibration of the x-ray spectrometer and specific computer fitting methods (Kitazawa et al., 1982) that also permit removal of the potassium peaks to better display the calcium K_α signal. We illustrate this process with two examples, one of which contains very low concentrations of Ca^{2+} and requires high detection sensitivity, the other containing high [Ca^{2+}] and [Cl^-] in a small organelle, hence requiring good spatial resolution and the retention of the subcellular localization of a diffusible ion during rapid freezing, cryoultramicrotomy, and EPMA.

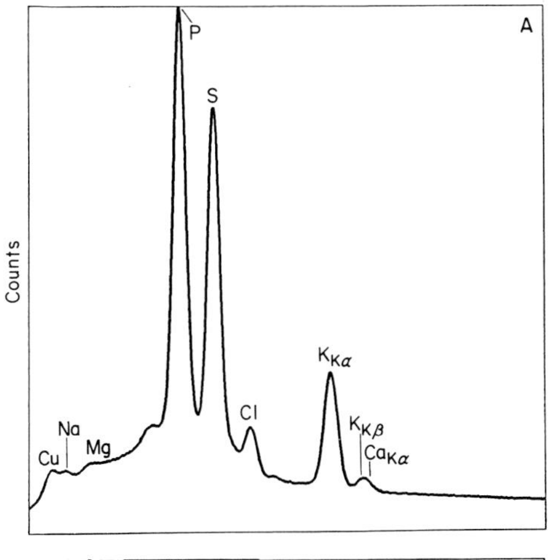

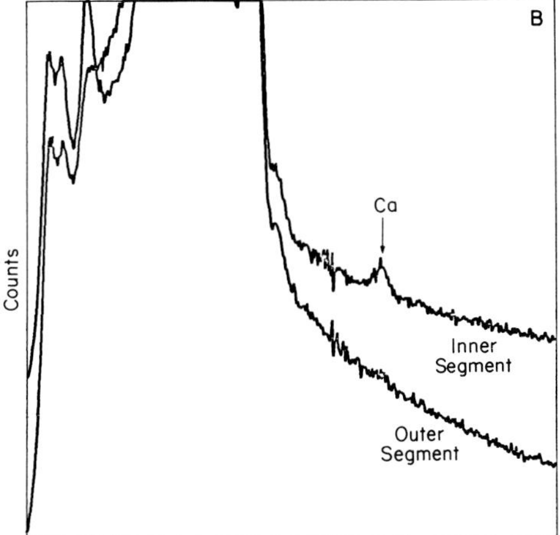

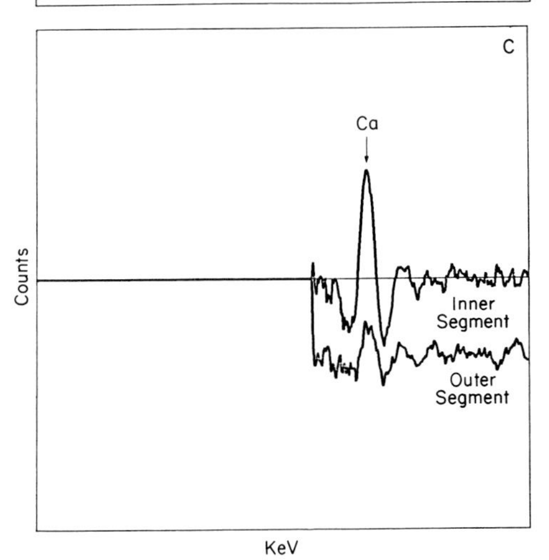

Figure 1. Averaged x-ray spectra collected from cryosections of 47 frog retinal rod outer segments; full scale, 5,093 counts. (*A*) "Raw" spectrum. (*B*) Same spectrum after subtraction of the K^+ peaks, with the computer fitting routine, to illustrate the Ca^{2+} peak. The spectrum is displayed at higher gain (full scale, 512 counts) than in *A*. The upper trace in this panel is the similarly processed, averaged spectrum of 26 rod inner segments. The two spectra are displaced relative to each other for display. The Ca concentrations illustrated in these averaged spectra were (mmol/kg dry wt ± SEM) 0.6 ± 0.13 (outer segment) and 4 ± 0.26 (inner segment). (*C*) The same spectrum as shown in *B* after digital filtering to remove background. The Ca^{2+} peaks are above the noise level in both spectra. (From A. P. Somlyo and Walz, 1985.)

Vertebrate Retinal Rod

The elemental composition of dark-adapted and illuminated frog and toad retinal rods was examined by EPMA in cryosections of rapidly frozen retinae (A. P. Somlyo and Walz, 1985). The averaged x-ray spectrum of the outer segment (Fig. 1) illustrates the very low, physiological Ca^{2+} content (0.3 mmol/kg dry wt) of the retinal disks that occupy most of the outer segment. The Ca^{2+} content did not change significantly upon illumination, although the decreased Na^{+} and Mg^{2+} and the increased K^{+} clearly showed the physiological light response (A. P. Somlyo and Walz, 1985). The low Ca^{2+} content of the retinal rod outer segment and the absence of a detectable, light-induced change in total Ca^{2+} are consistent with recent studies showing, with other methods, that illumination of rods does *not* increase free cytoplasmic $[Ca^{2+}]$ (Gold, 1986), as would be expected if retinal disks contained a high concentration of Ca^{2+} that was released by light.

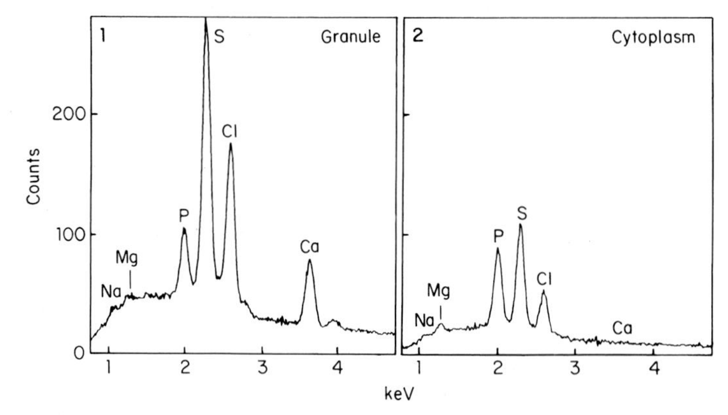

Figure 2. Averaged x-ray spectra obtained from cryosections of rat right atria. The K peaks have been subtracted by computer. Panel 1: granule; panel 2: paired cytoplasmic spectra obtained with the identical probe parameters as used for the collection of panel 1. Note the high Ca^{2+}, Cl^{-}, and S content of the granules.

The lower limits of EPMA quantitation of Ca^{2+} have also been explored in skinned skeletal muscle fibers exposed to EGTA-buffered Ca^{2+} (Kitazawa et al., 1982). Given the geometric efficiency of available x-ray detectors, the lower limit of practical detection sensitivity is ~300 μmol Ca^{2+}/kg dry wt with EPMA. Better sensitivity may be expected from electron energy loss spectroscopy (Shuman and Somlyo, 1987) and/or improved x-ray detectors.

Atrial Specific Granules

Atrial specific granules (ASGs) are ~0.2–0.4 μm myocardial organelles (Kisch, 1956; Jamieson and Palade, 1969; Sommer and Johnson, 1980), which store the prohormones of atrial natriuretic factors (ANFs), peptides secreted by the heart that have vasodilator and natriuretic activity (DeBold, 1986). Averaged x-ray spectra of, respectively, ASGs and the adjacent cytoplasm are shown in Fig. 2. As indicated by the characteristic x-ray continuum ratios, Ca^{2+}, Cl^{-}, and S are very significantly higher, and P is lower, in the granules than in the cytoplasm. The Ca^{2+} content of the granules is 71 ± 3 mmol/kg dry wt ± SEM (N = 90). We suspect that this represents Ca^{2+} binding to neighboring acidic groups in the

N-terminal portion of the prohormone sequence, similar to the acidic groupings in the Ca^{2+}-binding protein, calsequestrin (Fliegel et al., 1987). The calculated ratio (6.5) of $[Cl^-]_{granule}/[Cl^-]_{cytoplasm}$ (assuming that Cl^- is in solution both inside the granules and in the cytoplasm) is equivalent to an inside positive Cl equilibrium potential of +49 mV.

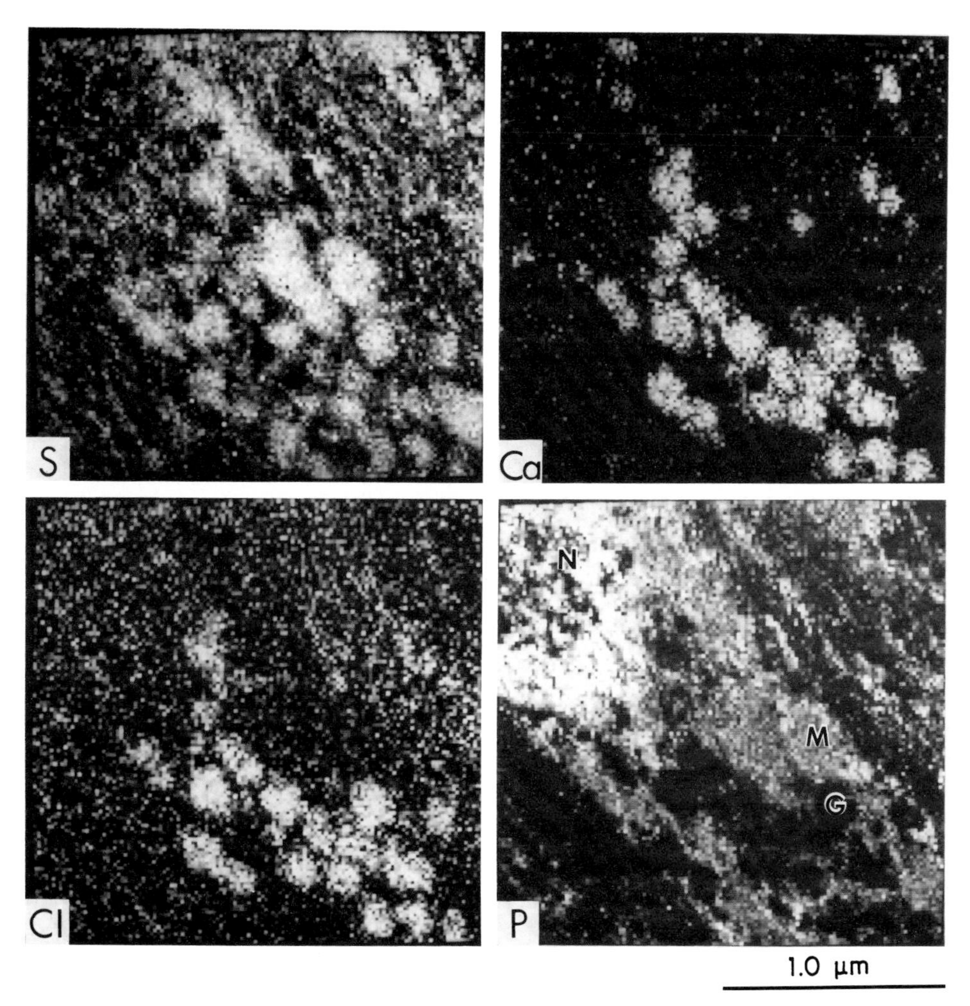

Figure 3. X-ray maps of a cryosection of rat right atrium showing the distribution of elements in a 2-μm^2 region of the nuclear pole, containing mitochondria (M), atrial granules (G), and condensed chromatin on the inside of the nuclear (N) envelope. The gray level intensities in the maps are adjusted to indicate the relative elemental distributions, with black being the lowest and white the highest number of atoms of each element.

Local concentrations can be visualized at electron-microscopic resolution and x-ray maps obtained by collecting x-ray spectra from each pixel of an area scanned by an electron beam and modulating the display intensity as a function of either the total number of atoms of a specific element or its concentration within the microvolume represented by each pixel. An x-ray map of a cryosection of rat atrium obtained in this manner is shown in Fig. 3 and illustrates the high concentration of Ca, S, and Cl and the low concentration of P in ASGs. The map was obtained with a field emission gun, with pixels of 15 nm diam, collected over a total period of 36 h. The demonstration of the high Ca^{2+} and Cl^- content of ASGs does not nearly tax the, at least, 8.7-nm spatial resolution of EPMA (A. P. Somlyo and Shuman, 1982) or the 3.4-nm resolution attained with energy-filtered electron microscopy (Shuman et al., 1986).

Ca^{2+} Localization in Muscle and Its Mobilization by $InsP_3$

Striated Muscle

Ca^{2+} can be readily localized with EPMA in the terminal cisternae (TC) of the sarcoplasmic reticulum (SR), where, in resting striated muscle, most of the cellular Ca^{2+} is stored (A. V. Somlyo et al., 1981). During a 1.2-s tetanus, ~60% of the Ca^{2+} content of the TC is released, accompanied by the uptake of Mg^{2+} and K^+. Mg^{2+} uptake into the SR also accompanies caffeine-induced (Yoshioka and Somlyo, 1984) and quinine-induced (Yoshioka and Somlyo, 1987) Ca^{2+} release. These findings, and the delayed post-tetanic return of Mg^{2+} from the TC into the cytoplasm (A. V. Somlyo et al., 1985*b*), suggest that the tetanic influx of Mg^{2+} into the SR is through the passive, relatively nonselective Ca^{2+} channels of the SR membrane. Since the Mg^{2+} and K^+ accumulated during tetanic Ca^{2+} release is insufficient to account for the electrical charge released as Ca^{2+}, it is likely that proton influx, undetectable by EPMA, provides the remaining charge balance within the SR (A. V. Somlyo et al., 1981, 1985*b*).

The very large amount of Ca^{2+} (~1 mmol/liter cell H_2O) released during a tetanus is distributed within the cytoplasm, and is equivalent to the concentration of Ca^{2+}-binding sites on troponin and on the soluble Ca^{2+}-binding protein parvalbumin (A. V. Somlyo et al., 1981). Indeed, EPMA revealed two kinetic components of the post-tetanic return of Ca^{2+} to the TC: the first component (~25% of the total Ca^{2+} released) accounts for the Ca^{2+} pumped back from the Ca^{2+}-specific sites on troponin to the SR during relaxation, while the second, slow component returns to the TC after relaxation, and its time course is consistent with the off rate of Ca^{2+} from parvalbumin (A. V. Somlyo et al., 1985*b*).

Smooth Muscle

The SR, in spite of its sparsity in some smooth muscles (Devine et al., 1972), plays a major role in the physiological regulation of cytoplasmic Ca^{2+} in smooth, as in striated, muscle. The ATP-supported uptake of Ca^{2+} (Endo et al., 1982) has been directly localized to the SR in smooth muscle with EPMA (A. P. Somlyo et al., 1979, 1982), and has also been identified as the Ca^{2+} pump of relatively pure, fragmented SR preparations (Wuytack et al., 1984). Ca^{2+} is stored both in the junctional SR that forms surface couplings (A. P. Somlyo and Somlyo, 1971) with the plasma membrane (Bond et al., 1984*a*) and in the central SR (Kowarski et al., 1985; A. P. Somlyo et al., 1979). Norepinephrine can release Ca^{2+} from the junctional (Bond et al., 1984*a*) and the central SR (Kowarski et al., 1985), in depolarized or in normally polarized smooth muscles, both in the presence and absence of extracellular Ca^{2+}. Approximately 50% of the Ca^{2+} stored in the central SR of the rabbit main pulmonary artery is released by maximal stimulation with norepinephrine (Kowarski et al., 1985). The lumen of the central SR communicates with that of the peripheral SR (Devine et al., 1972). Therefore, it is not known whether the central SR itself is the actual site of Ca^{2+} release. There is no evidence of significant physiological mitochondrial Ca^{2+} uptake in smooth muscle even after maximal contractions maintained for 30 min (A. P. Somlyo et al., 1979; Bond et al., 1984*b*; Kowarski et al., 1985).

The Time Course of Smooth and Striated Muscle Contraction Induced by Ca^{2+} Released from the SR by $InsP_3$ Flash-Photolyzed from Caged $InsP_3$

The release of intracellular Ca^{2+} by $InsP_3$ has been measured both with Ca^{2+}-selective electrodes in permeabilized rabbit main pulmonary artery (A. P. Somlyo et al., 1985*a*) and with $^{45}Ca^{2+}$ in isolated smooth muscle cells (Suematsu et al., 1984), and is sufficient to trigger contractions (A. V. Somlyo et al., 1985*a*). These studies, and others showing increased phosphatidyl inositol turnover during cholinergic stimulation of smooth muscle (Baron et al., 1984), suggested that $InsP_3$ is the physiological messenger of "pharmacomechanical Ca^{2+} release," the membrane potential–independent mechanism through which excitatory drugs release Ca^{2+} from the SR (A. V. Somlyo and Somlyo, 1968; for review, see A. P. Somlyo, 1985*c*). A major test of a messenger—whether its action is sufficiently rapid to account for its putative physiological role—remained to be demonstrated. Therefore, we determined the time course of contractions evoked by photolysis of caged $InsP_3$ (Goldman et al., 1986; Walker et al., 1987) in permeabilized rabbit main pulmonary artery smooth muscle, in which we had already established the SR as the major intracellular source of activator Ca^{2+} (Kowarski et al., 1985). The latency of tension development (~390 ms) following photolysis of caged $InsP_3$, as well as the rate of ensuing contraction, were comparable to, respectively, the latencies in intact smooth muscle and the rate of contraction following photolysis of caged ATP in the presence of Ca^{2+} (Fig. 4). When $InsP_3$ was introduced by photolysis rather than diffusion from the bath, the sensitivity to $InsP_3$ was increased by two orders of magnitude (i.e., 0.5 μM $InsP_3$ released upon photolysis induced a contraction of similar magnitude as that produced by 30 μM $InsP_3$ introduced by diffusion), in keeping with the high $InsP_3$ phosphatase activity measured in the main pulmonary artery (15–20 μM/s) (Walker et al., 1987). These findings support the notion that $InsP_3$ is the physiological messenger mediating transmitter-induced (pharmacomechanical) Ca^{2+} release from the SR of smooth muscle. The mechanism of electromechanical Ca^{2+} release in smooth, as in striated (see below) muscle, remains to be determined (A. P. Somlyo, 1985*b*).

High concentrations of $InsP_3$ can also stimulate contractions in skinned skeletal muscle fibers, and it has been suggested that $InsP_3$ may also be the transmitter mediating electromechanical coupling in skeletal muscle (Volpe et al., 1985, 1986; Vergara et al., 1985). The very slow responses to high concentrations of exogenous $InsP_3$ and the unphysiologically low free Mg^{2+} required for optimal activity have been tentatively ascribed to diffusional delays and to the action of $InsP_3$-ase (Vergara et al., 1985; Volpe et al., 1985). Our results do not support this interpretation.

The contractile response of a mechanically skinned frog skeletal muscle fiber to photolysis of caged $InsP_3$, in the presence of low free Mg^{2+} (40 μM), is shown in

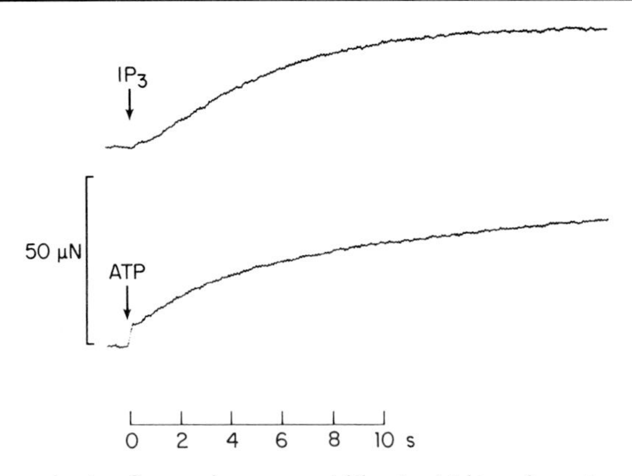

Figure 4. Tension transients of saponin-permeabilized rabbit main pulmonary artery smooth muscle, after liberation of $InsP_3$ (*A*) or ATP (*B*) by laser photolysis (arrows) from, respectively, caged $InsP_3$ (10 μM) and caged ATP (10 mM). Note the similar time courses of the contractions induced by photolysis of the two caged precursors. (From Goldman et al., 1986.)

Fig. 5. The peak contraction was not reached for ~10 s. The rate of contraction is about three orders of magnitude slower than the normal twitch of frog muscle and slower even than the contraction evoked by externally added caffeine (not shown). The concentration of photolyzed caged $InsP_3$ required to cause contraction was about two orders of magnitude higher in skeletal than in smooth muscle. Furthermore, the $InsP_3$-ase activity of the frog skeletal muscles was also significantly lower than that of smooth muscle. We cannot exclude the possibility that, when modulated by other nucleotides (e.g., GTP), $InsP_3$ could have a more pronounced and rapid effect in frog skeletal muscle, but the results obtained under the current conditions do not support the notion that $InsP_3$ is the physiological messenger of excitation-contraction coupling in striated muscle.

Mitochondrial Ca^{2+} in Liver and Other Cells

The in vivo concentration of mitochondrial Ca^{2+} has been the subject of some controversy, owing to the high probability of translocation of Ca^{2+} into mitochondria during cell fractionation (for review, see A. P. Somlyo, 1984). Unfortunately,

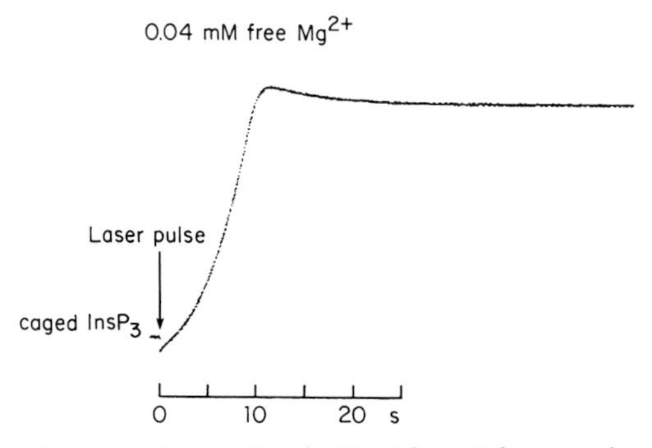

Figure 5. Tension development by mechanically skinned frog semitendinosus muscle fiber upon liberation of $InsP_3$ by laser-flash photolysis of caged $InsP_3$. Note the very slow time course of the contraction developed, even in the presence of the low free [Mg^{2+}].

perfusion of tissues with Ca-free solutions to avoid overestimates caused by Ca^{2+} translocation introduces the possibility of depleting normal, endogenous mitochondrial Ca (Bond et al., 1984*b*).

The concentration of mitochondrial Ca bears on the important physiological question of whether mitochondria can regulate cytoplasmic free Ca^{2+} or, alternatively, are metabolically regulated by small fluctuations in matrix free [Ca^{2+}] acting on Ca^{2+}-sensitive mitochondrial dehydrogenases (for review, see Denton and McCormack, 1980; Hansford, 1985). A high mitochondrial Ca^{2+} content (>10 mmol/kg mitochondrial dry wt) would be consistent with activation of the Ca^{2+} mitochondrial efflux pathway and mitochondrial regulation of cytoplasmic free [Ca^{2+}], but this would saturate the mitochondrial enzymes. On the other hand, low concentrations of total Ca (matrix free [Ca^{2+}] of 0–3 μM) would permit the modulation of mitochondrial enzymes, but not the regulation of cytoplasmic [Ca^{2+}] by mitochondria. It seemed particularly important to resolve the question of mitochondrial Ca content in the liver, an organ in which metabolic regulation plays a major role and is influenced by hormones that mobilize intracellular Ca^{2+} (Berridge and Irvine, 1984).

EPMA of rat livers rapidly frozen (Fig. 6) in anesthetized animals in situ indicated that the endogenous mitochondrial Ca^{2+} content is low (A. P. Somlyo et al., 1985*a*). Assuming that the ratio of intramitochondrial free to total Ca^{2+} is $\sim 10^{-3}$ (Coll et al., 1982; Hansford and Castro, 1982), the EPMA measurements indicate that the mitochondrial matrix free [Ca^{2+}] is ~0.2–0.4 μM. Indeed, the range of normal endogenous mitochondrial Ca^{2+} in a variety of cells (Table I) is below 3 mmol/kg dry wt. In none of these cells (liver, brain, retinal rods, muscle, etc.) do mitochondria contain a large proportion of intracellular Ca^{2+} or play a significant role in the *physiological* regulation of cytoplasmic Ca^{2+}. However, under pathological conditions, when cytoplasmic free Ca^{2+} rises to abnormally high levels, mitochondria can accumulate a large amount of Ca^{2+}, and could become pathological foci of calcification (e.g., A. P. Somlyo et al., 1978). Recent EPMA studies also indicate that pathological Ca^{2+} accumulation by mitochondria (up to ~300 mmol/kg dry wt) can be reversible (Broderick et al., 1986).

It remains to be determined whether mitochondrial Ca^{2+} can be increased sufficiently by physiological stimuli to activate mitochondrial enzymes. In a series of experiments in which the intact (nonperfused) liver of anesthetized rats was frozen, in control animals or after the injection of vasopressin alone (Bond et al., 1986) or with glucagon (infused over 10–12 min in anterior mesenteric or jugular veins), the average mitochondrial Ca^{2+} ranged between 0.7 and 2.2 mmol/kg dry wt (SEM = 0.2 for each group). However, taking into account animal-to-animal variations by analysis of variance, the differences between these values were not statistically significant. Specifically, mitochondria of animals injected over 10–12 min (total volume of 0.25 ml into jugular vein) with vasopressin (7×10^{-6} M) and glucagon (5×10^{-7} M) contained (mmol/kg ± SEM) 2.2 ± 0.2 Ca^{2+}, compared with the 1.7 ± 0.2 in diluent-injected animals. These values are markedly different from the 50 mmol Ca^{2+}/kg dry wt measured by one group (Altin and Bygrave, 1986) in mitochondria isolated from perfused livers of animals injected with vasopressin and glucagon. Similar measurements made by a second group (Assimacopoulos-Jeannet et al., 1986) showed up to ~5 mmol Ca^{2+}/kg dry wt. It is possible that perfusion with Krebs-Ringer solution sensitizes the liver to Ca^{2+}

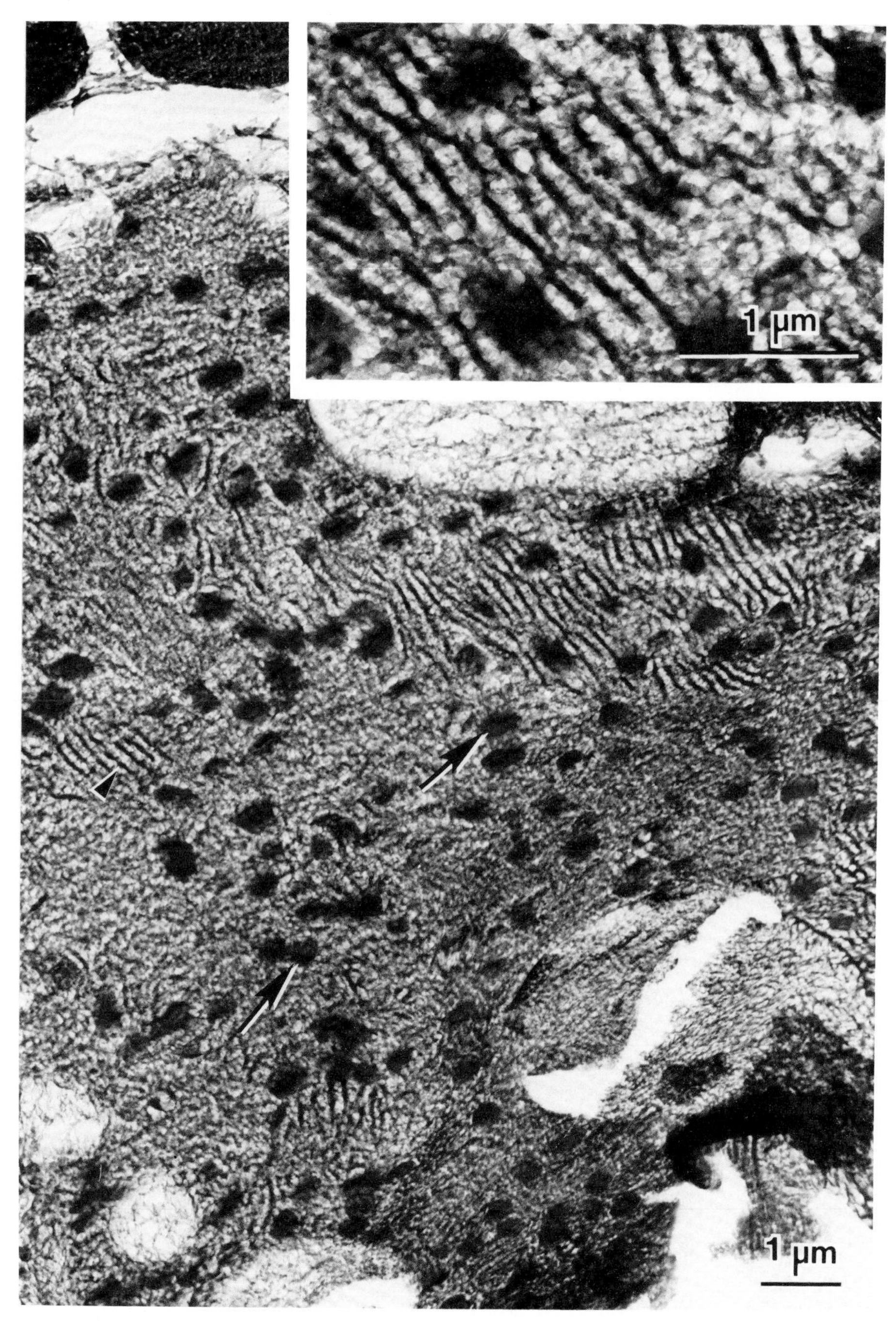

Figure 6. Electron micrograph of cryosection of rat liver frozen in situ. Mitochondria (arrows) and stacks of rough endoplasmic reticulum (ER) (arrowhead) are clearly identifiable. The inset shows a region of rough ER at high magnification. Note the close association of the rough ER with the mitochondria, which could interfere with separation of these organelles into isolated cell fractions.

loading, including mitochondrial Ca^{2+} loading, and some of the Ca^{2+} content ascribed by these authors to mitochondria may represent Ca^{2+} translocated during isolation or contained in another compartment (e.g., perimitochondrial endoplasmic reticulum) sedimenting with mitochondria. On the other hand, it would also be premature to exclude the possibility, on the basis of our measurements of total mitochondrial Ca^{2+}, that variations in mitochondrial matrix Ca^{2+} can contribute to metabolic regulation. Estimates of the $[Ca^{2+}]_{free}/[Ca]_{total}$ ratio based on in vitro experiments may not reflect the correct in vivo values, in which case the small, and statistically not significant, fluctuations in mitochondrial Ca^{2+} could give rise to significant changes in matrix free Ca^{2+}. Nevertheless, the possibility that changes in mitochondrial constituents other than Ca^{2+}, such as Mg^{2+} (see below) or ADP (Chance and Williams, 1956), play a major role in metabolic regulation of mitochondria deserves continued consideration.

TABLE I
Mitochondrial Ca Content In Situ

Animal	Tissue	Condition	
		*mmol/kg dry mitochondrion ± SEM**	
Rabbit	Portal vein	Relaxed	1.6±0.2
	" "	Contracted 30 min	2.3±0.4
	" "	0 Ca	0.1±0.2‡
	" "	Na-loaded	2.0±1.0
	Main pulmonary artery	Relaxed	2.6±0.31
	" " "	Contracted	2.0±0.32
	Heart (papillary)		0.5±0.2
Guinea pig	Portal vein	Relaxed	0.7±0.3
	" "	Contracted	1.1±0.3
Rat	Liver	In animal	0.8±0.1
	Brain cortex	In animal	1.5±0.26
Rana pipiens	Retinal rod	Dark-adapted	0.0±0.2
	" "	Illuminated	0.4±0.2

* Mmol/kg dry wt ≃ nmol/mg mitochondrial protein.
‡ Significantly different ($P < 0.01$) from relaxed rabbit portal vein.
Values shown are from electron probe studies referenced in the bibliography.

Mitochondrial Mg^{2+} Movements In Situ

EPMA of rapidly frozen tissues has revealed subcellular translocations of Mg^{2+} in several cell systems, which suggests that changes in Mg^{2+} may also have regulatory functions in cell organelles. In addition to the previously mentioned Mg^{2+} influx into the SR during Ca^{2+} release, we have detected changes in the mitochondrial and cytoplasmic Mg^{2+} content in smooth muscle (Bond et al., 1984*b*), retinal rods (A. P. Somlyo and Walz, 1985), and parenchymal liver cells (see below). Mitochondrial Mg^{2+} can change in parallel with or independently of cytoplasmic Mg^{2+}. The large mitochondrial Mg^{2+} uptake observed during massive Na^+ efflux from rabbit portal vein (Broderick et al., 1986) probably reflects the very active Mg^{2+} transport system previously identified in mitochondria isolated from smooth muscle (Sloane et al., 1978).

The injection of vasopressin and glucagon into the jugular vein increased mitochondrial Mg^{2+} in rat liver to 57 mmol/kg dry wt, compared with the 43 mmol/kg dry wt in control animals injected with the diluent Krebs solution (Bond, M., G. Vadasz, A. V. Somlyo, and A. P. Somlyo, manuscript in preparation). The increase in total mitochondrial Mg^{2+} (32%) is highly significant ($P < 0.01$) and, if the relationship between free and total mitochondrial Mg^{2+} is indeed linear (Corkey et al., 1986), it also represents a 32% increase in matrix free Mg^{2+}. Such an increase in matrix [Mg^{2+}] could modulate mitochondrial enzymes and, perhaps, even be responsible for some of the effects ascribed to changes in matrix free [Ca^{2+}].

Ca^{2+} in the Endoplasmic Reticulum in Liver and Other Nonmuscle Cells

The endoplasmic reticulum (ER) can actively transport Ca^{2+} via a Ca^{2+}-ATPase (for review, see Martonosi, 1983; A. P. Somlyo, 1984) and is the organelle that plays the major role in regulating cytoplasmic free [Ca^{2+}]. Sr^{2+} uptake by endothelial ER was observed some time ago (A. V. Somlyo and Somlyo, 1971), and EPMA of cryosections localized high concentrations of Ca^{2+} in the ER of normal vertebrate retinal rods (A. P. Somlyo and Walz, 1985) and parenchymal liver cells (A. P. Somlyo et al., 1985). Compelling evidence has also implicated the ER as the "nonmitochondrial site" from which excitatory transmitters, such as vasopressin, release Ca^{2+} using $InsP_3$ as their secondary messenger (e.g., Burgess et al., 1983, 1984). Vasopressin reduced the Ca^{2+} content (measured by EPMA) of the rough ER in the liver and increased glycogen phosphorylase activity (Bond et al., 1986).

Conclusions

The subcellular movements of Ca^{2+} and other elements between cytoplasm and organelles can be followed in situ by EPMA of tissues rapidly frozen in various functional states. We have summarized here some of the applications of this method to the measurement of Ca^{2+} and counterion movements between the SR and cytoplasmic Ca-binding proteins in skeletal muscle, and to the identification of the SR as the intracellular source of activator Ca^{2+} in smooth muscle. EPMA also provided strong evidence showing that, in liver and in other nonmuscle cells, the ER, but not mitochondria, plays the predominant role in the physiological regulation of cytoplasmic free Ca^{2+}. The possible roles of Mg^{2+} uptake into mitochondria and into the SR remain to be established, but the detection, with EPMA, of subcellular Mg^{2+} movements is yet another example of obtaining new information with a new method.

The kinetics of the contractions induced by photolysis of an inert, photolabile precursor of inositol 1,4,5-trisphosphate in permeabilized muscles support the role of $InsP_3$ as a physiological messenger in smooth, but *not* in striated, muscle. However, given the possibility of a "missed modulator" or another, more active inositol phosphate, it would be premature to completely exclude this mechanism. Finally, the obvious benefits of obtaining time-resolved information from, respectively, laser-flash photolysis and rapid freezing lead us to predict that the combination of the two methods will find useful application in future studies.

Acknowledgments

This work was supported by National Institutes of Health grant HL-15835 and Training Grant HL-07499 to the Pennsylvania Muscle Institute, and by grant AM-36064 to M.B.

References

Altin, J. G., and F. L. Bygrave. 1986. Synergistic stimulation of Ca^{2+} uptake by glucagon and Ca^{2+}-mobilizing hormones in the perfused rat liver. *Biochemical Journal.* 238:653–661.

Assimacopoulos-Jeannet, F., J. G. McCormack, and B. Jeanrenaud. 1986. Vasopressin and/ or glucagon rapidly increases mitochondrial calcium and oxidative enzyme activities in the perfused rat liver. *Journal of Biological Chemistry.* 261:8799–8804.

Baron, C. B., M. Cunningham, J. F. Strauss, and R. F. Coburn. 1984. Pharmacomechanical coupling in smooth muscle may involve phosphatidylinositol metabolism. *Proceedings of the National Academy of Sciences.* 81:6899–6903.

Berridge, M. J., and R. F. Irvine. 1984. Inositol trisphosphate, a novel second messenger in cellular signal transduction. *Nature.* 312:315–321.

Bond, M., T. Kitazawa, A. P. Somlyo, and A. V. Somlyo. 1984*a*. Release and recycling of calcium by the sarcoplasmic reticulum in guinea pig portal vein smooth muscle. *Journal of Physiology.* 355:677–695.

Bond, M., H. Shuman, A. P. Somlyo, and A. V. Somlyo. 1984*b*. Total cytoplasmic calcium in relaxed and maximally contracted rabbit portal vein smooth muscle. *Journal of Physiology.* 357:185–201.

Bond, M., G. Vadasz, A. V. Somlyo, and A. P. Somylo. 1986. Electron probe analysis of Ca release from the sarcoplasmic reticulum and mitochondrial Ca content during vasopressin stimulation of rat liver in situ. *Journal of General Physiology.* 88:12*a*. (Abstr.)

Broderick, R., A. J. Wasserman, T. Fujimori, and A. P. Somlyo. 1986. Mitochondrial Ca^{2+} uptake during massive cellular Na^+ efflux and its reversibility in situ: an electron probe study. *Journal of General Physiology.* 88:13*a*. (Abstr.)

Burgess, G. M., P. O. Godfrey, J. S. McKinney, M. J. Berridge, R. F. Irvine, and J. W. Putney, Jr. 1984. Second messenger linking receptor activation to internal calcium release in liver. *Nature.* 309:63–66.

Burgess, G. M., J. S. McKinney, A. Fabiato, B. A. Leslie, and J. W. Putney, Jr. 1983. Calcium pools in saponin-permeabilized guinea pig hepatocytes. *Journal of Biological Chemistry.* 258:15336–15345.

Chance, B., and G. R. Williams. 1956. The respiratory chain and oxidative phosphorylation. *Advances in Enzymology and Related Subjects of Biochemistry.* 17:65–134.

Coll, K. E., S. K. Joseph, B. E. Corkey, and J. R. Williamson. 1982. Determination of the matrix free Ca^{2+} concentration and kinetics of Ca^{2+} efflux in liver and heart mitochondria. *Journal of Biological Chemistry.* 257:8696–8704.

Corkey, B. E., J. Duszynski, T. L. Rich, B. Matschinsky, and J. R. Williamson. 1986. Regulation of free and bound magnesium in rat hepatocytes and isolated mitochondria. *Journal of Biological Chemistry.* 261:2567–2574.

DeBold, A. J. 1986. Atrial natriuretic factor: an overview. *Federation Proceedings.* 45:2081–2085.

Denton, R. M., and J. G. McCormack. 1980. On the role of the calcium transport cycle in heart and other mammalian mitochondria. *FEBS Letters.* 119:1–8.

Devine, C. E., A. P. Somlyo, and A. V. Somlyo. 1972. Sarcoplasmic reticulum and excitation-contraction coupling in mammalian smooth muscle. *Journal of Cell Biology.* 52:690–718.

Endo, M., S. Yagi, and M. Iino. 1982. Tension-pCa relation and sarcoplasmic reticulum responses in chemically skinned smooth muscle fibers. *Federation Proceedings.* 41:2245–2250.

Fliegel, L., M. Ohnishi, M. R. Carpenter, U. K. Khanna, R. A. Reithmeier, and D. H. MacLennan. 1987. Amino acid sequence of rabbit fast-twitch skeletal muscle calsequestrin deduced from cDNA and peptide sequencing. *Proceedings of the National Academy of Sciences.* In press.

Gold, G. H. 1986. Plasma membrane calcium fluxes in intact rods are inconsistent with the "calcium hypothesis." *Proceedings of the National Academy of Sciences.* 83:1150–1154.

Goldman, Y. E., G. P. Reid, A. P. Somlyo, A. V. Somlyo, D. R. Trentham, and J. W. Walker. 1986. Activation of skinned vascular smooth muscle by photolysis of 'caged inositol trisphosphate' to inositol 1,4,5-trisphosphate ($InsP_3$). *Journal of Physiology.* 377:100P. (Abstr.)

Hall, T. A. 1971. The microprobe assay of chemical elements. *In* Physical Techniques in Biological Research. G. Oster, editor. Academic Press, Inc., New York. 1A:157–275.

Hansford, R. G. 1985. Relation between mitochondrial calcium transport and control of energy metabolism. *Review of Physiology, Biochemistry and Pharmacology.* 102:1–72.

Hansford, R. G., and F. Castro. 1982. Effect of micromolar concentrations of free calcium ions on the reduction of heart mitochondrial NAD(P) by 2-oxoglutarate. *Biochemical Journal.* 198:525–533.

Hibberd, M. G., Y. E. Goldman, and D. R. Trentham. 1984. Laser-induced photogeneration of ATP: a new approach to the study of chemical kinetics of muscle contraction. *In* Current Topics in Cellular Regulation: Mechanism of Enzyme Action. M. Deluca, H. Lardy, and R. L. Cross, editors. Academic Press, Inc., New York. 357–364.

Jamieson, T. D., and G. E. Palade. 1969. Specific granules in atrial muscle cells. *Journal of Cell Biology.* 23:151–172.

Kaplan, J. H., B. Forbush III, and J. F. Hoffman. 1978. Rapid photolytic release of adenosine 5′-triphosphate from a protected analogue: utilization by the Na:K pump of human red blood cell ghosts. *Biochemistry.* 17:1929–1935.

Kisch, B. 1956. Electron microscopy of atrium of heart; guinea pig. *Experimental Medicine and Surgery.* 14:99–112.

Kitazawa, T., H. Shuman, and A. P. Somlyo. 1982. Calcium and magnesium binding to thin and thick filaments in skinned muscle fibres: electron probe analysis. *Journal of Muscle Research and Cell Motility.* 3:437–454.

Kitazawa, T., H. Shuman, and A. P. Somlyo. 1983. Quantitative electron probe analysis: problems and solutions. *Ultramicroscopy.* 11:251–262.

Kowarski, D., H. Shuman, A. P. Somlyo, and A. V. Somlyo. 1985. Calcium release by norepinephrine from central sarcoplasmic reticulum in rabbit main pulmonary artery smooth muscle. *Journal of Physiology.* 366:153–175.

Lester, H. A., and J. M. Nerbonne. 1982. Physiological and pharmacological manipulations with light flashes. *Annual Review of Biophysics and Bioengineering.* 11:151–175.

Martonosi, A. N. 1983. The regulation of cytoplasmic Ca^{2+} concentration in muscle and nonmuscle cells. *In* Muscle and Non-Muscle Motility. A. Stracher, editor. Academic Press, Inc., New York. 1:233–358.

Shuman, H., C.-F. Chang, and A. P. Somlyo. 1986. Elemental imaging and resolution in energy-filtered conventional electron microscopy. *Ultramicroscopy.* 19:121–134.

Shuman, H., and A. P. Somlyo. 1987. Electron energy loss analysis of near trace element concentrations of calcium. *Ultramicroscopy.* 21:23–32.

Shuman, H., A. V. Somlyo, and A. P. Somlyo. 1976. Quantitative electron probe microanalysis of biological thin sections: methods and validity. *Ultramicroscopy.* 1:317–339.

Sloane, B. F., A. Scarpa, and A. P. Somlyo. 1978. Vascular smooth muscle mitochondria: magnesium content and transport. *Archives of Biochemistry and Biophysics.* 189:409–416.

Somlyo, A. P. 1984. Cellular site of calcium regulation. *Nature.* 308:516–517.

Somlyo, A. P. 1985*a*. Cell calcium measurement with electronprobe and electron energy loss analysis. *Cell Calcium.* 6:197–212.

Somlyo, A. P. 1985*b*. Excitation-contraction coupling. The messenger across the gap. *Nature.* 316:298–299.

Somlyo, A. P. 1985*c*. Excitation-contraction coupling and the ultrastructure of smooth muscle. *Circulation Research.* 57:497–507.

Somlyo, A. P., editor. 1986. Recent advances in electron and light optical imaging in biology and medicine. *Annals of the New York Academy of Sciences.* 483:1–472.

Somlyo, A. P., M. Bond, and A. V. Somlyo. 1985*a*. Calcium content of mitochondria and endoplasmic reticulum in liver frozen rapidly *in vivo*. *Nature.* 314:622–625.

Somlyo, A. P., R. Urbanics, G. Vadasz, A. G. B. Kovach, and A. V. Somlyo. 1985*b*. Mitochondrial calcium and cellular electrolytes in brain cortex frozen *in situ:* electron probe analysis. *Biochemistry and Biophysical Research Communications.* 132:1071–1078.

Somlyo, A. P., and H. Shuman. 1982. Electron probe and electron energy loss analysis in biology. *Ultramicroscopy.* 8:219–234.

Somlyo, A. P., A. V. Somlyo, and H. Shuman. 1979. Electron probe analysis of vascular smooth muscle: composition of mitochondria, nuclei and cytoplasm. *Journal of Cell Biology.* 81:316–335.

Somlyo, A. P., A. V. Somlyo, H. Shuman, and M. Endo. 1982. Calcium and monovalent ions in smooth muscle. *Federation Proceedings.* 41:2883–2890.

Somlyo, A. P., A. V. Somlyo, H. Shuman, B. Sloane, and A. Scarpa. 1978. Electron probe analysis of calcium compartments in cryo sections of smooth and striated muscles. *Annals of the New York Academy of Sciences.* 307:523–544.

Somlyo, A. P., and B. Walz. 1985. Elemental distribution in *Rana pipiens* retinal rods: quantitative electron probe analysis. *Journal of Physiology.* 358:183–195.

Somlyo, A. V., M. Bond, A. P. Somlyo, and A. Scarpa. 1985*a*. Inositol-trisphosphate ($InsP_3$) induced calcium release and contraction in vascular smooth muscle. *Proceedings of the National Academy of Sciences.* 82:5231–5235.

Somlyo, A. V., G. McClellan, H. Gonzalez-Serratos, and A. P. Somlyo. 1985*b*. Electron probe x-ray microanalysis of post tetanic Ca and Mg movements across the sarcoplasmic reticulum *in situ*. *Journal of Biological Chemistry.* 260:6801–6807.

Somlyo, A. V., H. Gonzalez-Serratos, H. Shuman, G. McClellan, and A. P. Somlyo. 1981.

Calcium release and ion changes in the sarcoplasmic reticulum of tetanized muscle: an electron probe study. *Journal of Cell Biology.* 90:577–594.

Somlyo, A. V., and A. P. Somlyo. 1968. Electromechanical and pharmacomechanical coupling in vascular smooth muscle. *Journal of Pharmacology and Experimental Therapeutics.* 159:129–145.

Somlyo, A. V., and A. P. Somlyo. 1971. Strontium accumulation by sarcoplasmic reticulum and mitochondria in vascular smooth muscle. *Science.* 174:955–958.

Sommer, J. R., and E. A. Johnson. 1980. Ultrastructure of cardiac muscle. *In* Handbook of Physiology. The Cardiovascular System. R. M. Berne, N. Sperelakis, and S. R. Geiger, editors. American Physiological Society, Washington, DC. 1:113–186.

Suematsu, E., M. Hirata, T. Hashimoto, and H. Kuriyama. 1984. Inositol 1,4,5-trisphosphate releases Ca^{2+} from intracellular store sites in skinned single cells of porcine coronary artery. *Biochemical and Biophysical Research Communications.* 120:481–485.

Vergara, J., R. Y. Tsien, and M. Delay. 1985. Inositol 1,4,5-trisphosphate: a possible chemical link in excitation-contraction coupling in muscle. *Proceedings of the National Academy of Sciences.* 82:6352–6356.

Volpe, P., F. DiVirgilio, T. Pozzan, and G. Salviati. 1986. Role of inositol 1,4,5-trisphosphate in excitation-contraction coupling in skeletal muscle. *FEBS Letters.* 197:1–4.

Volpe, P., G. Salviati, F. DiVirgillio, and T. Pozzan. 1985. Inositol 1,4,5-trisphosphate induced calcium release from sarcoplasmic reticulum of skeletal muscle. *Nature.* 316:347–349.

Walker, J. W., A. V. Somlyo, Y. E. Goldman, A. P. Somlyo, and D. R. Trentham. 1987. Inositol 1,4,5-trisphosphate ($InsP_3$) induces contractions at physiological rates in smooth but not in fast-twitch skeletal muscle. *Biophysical Journal.* 51:552*a*. (Abstr.)

Wuytack, F., L. Raeymaekers, J. Verbist, H. DeSmedt, and R. Casteels. 1984. Evidence for the presence in smooth muscle of two types of Ca^{2+} transport ATPase. *Biochemical Journal.* 224:445–451.

Yoshioka, T., and A. P. Somlyo. 1984. The calcium and magnesium contents and volume of the terminal cisternae in caffeine-treated skeletal muscle. *Journal of Cell Biology.* 99:558–568.

Yoshioka, T., and A. P. Somlyo. 1987. The effects of quinine on the calcium and magnesium content of the sarcoplasmic reticulum and the temperature-dependence of quinine contractures. *Journal of Muscle Research and Cell Motility.* In press.

Receptor-mediated Changes in Intracellular Calcium

Chapter 7

Mechanisms Involved in Receptor-mediated Changes of Intracellular Ca^{2+} in Liver

John R. Williamson, Carl A. Hansen, Arthur Verhoeven, Kathleen E. Coll, Roy Johanson, Michael T. Williamson, and Charles Filburn

Department of Biochemistry and Biophysics, University of Pennsylvania School of Medicine, Philadelphia, Pennsylvania

Introduction

It has been recognized for a number of years that changes of the intracellular free Ca^{2+} concentration caused by cell-activating agonists serve as an important signaling mechanism for the regulation of different cell functions. Ca^{2+} provides a coordinating link by causing activity changes of a variety of proteins, including protein kinases and phosphatases (Cohen, 1985) and several mitochondrial dehydrogenases (Denton and McCormack, 1985), either directly or after binding to calmodulin or other Ca^{2+}-binding proteins. Changes in the activity of these Ca^{2+}-dependent proteins are uniquely involved in eliciting cellular functions characteristic of the cell type, notably secretion in many cells and activation of metabolic pathways, such as hepatic glycogenolysis and gluconeogenesis in liver (Williamson et al., 1981; Rasmussen and Barrett, 1984). Until recently, however, the source of the Ca^{2+} and the mechanism of hormone-stimulated cellular Ca^{2+} mobilization were largely unknown.

The link between hormone or ligand interactions with cell surface receptors and the subsequent increase of cytosolic free Ca^{2+} were discovered as an extension of earlier work demonstrating increases of phosphatidylinositol turnover in a variety of tissues where intracellular signaling was thought to occur via Ca^{2+} (reviewed by Downes and Michell, 1985; Hokin, 1985). Many studies have now established that a wide range of compounds, including hormones, neurotransmitters, secretagogues, antigens, chemoattractants, and cycloxygenase and lipoxygenase products of arachidonic acid metabolism, activate a specific phosphodiesterase, termed phospholipase C, which breaks down inositol lipids in the plasma membrane. These lipids, comprised mainly of phosphatidylinositol (PI), are minor components of membranes, but are highly metabolically active compared with the structural phospholipids forming the membrane bilayer. Together with the phosphorylated derivatives phosphatidylinositol-4-phosphate (PIP) and phosphatidylinositol-4,5-bisphosphate (PIP_2), these inositol lipids serve as precursors for metabolic signal generators responsible for the mediation of the Ca^{2+}-dependent biological responses (Williamson et al., 1985).

A number of recent reviews have provided different perspectives of the role of inositol lipid metabolism in signal transduction and the separate roles of inositol 1,4,5-trisphosphate (IP_3) in causing the release of intracellular sequestered Ca^{2+} and 1,2-diacylglycerol in the activation of protein kinase C (Berridge and Irvine, 1984; Hokin, 1985; Downes and Michell, 1985; Williamson, 1986; Nishizuka, 1986; Williamson and Hansen, 1987). This chapter is concerned primarily with the Ca^{2+}-mobilizing effect of glucagon and the negative feedback effects exerted by activation of protein kinase C.

Signal Generation

Fig. 1 provides a schematic representation of the major metabolic events that occur after the addition of a Ca^{2+}-mobilizing agonist to a target cell. Typical receptors present in the hepatocyte plasma membrane that are coupled to Ca^{2+} mobilization are those for vasopressin, angiotensin II, and α_1-adrenergic agonists such as phenylephrine. A number of recent studies (see Litosch and Fain, 1986, and Williamson and Hansen, 1987, for reviews) have ascertained that receptor-mediated activation of phospholipase C is accomplished by a GTP-binding protein, denoted by G_x in

Fig. 1. Unlike the GTP-binding proteins G_s and G_i, which are involved in modulating adenylate cyclase activity (Gilman, 1984; Northup, 1985), the GTP-binding protein that is assumed to activate phospholipase C has not been isolated or characterized. In liver, receptor coupling through this putative G-protein to inositol lipid metabolism is not sensitive to inhibition by pertussis toxin (Pobiner et al., 1985; Uhing et al., 1986).

Immediately after the addition of vasopressin to hepatocytes, in which the inositol lipids have been prelabeled in the inositol ring with [^{3}H]*myo*-inositol or in

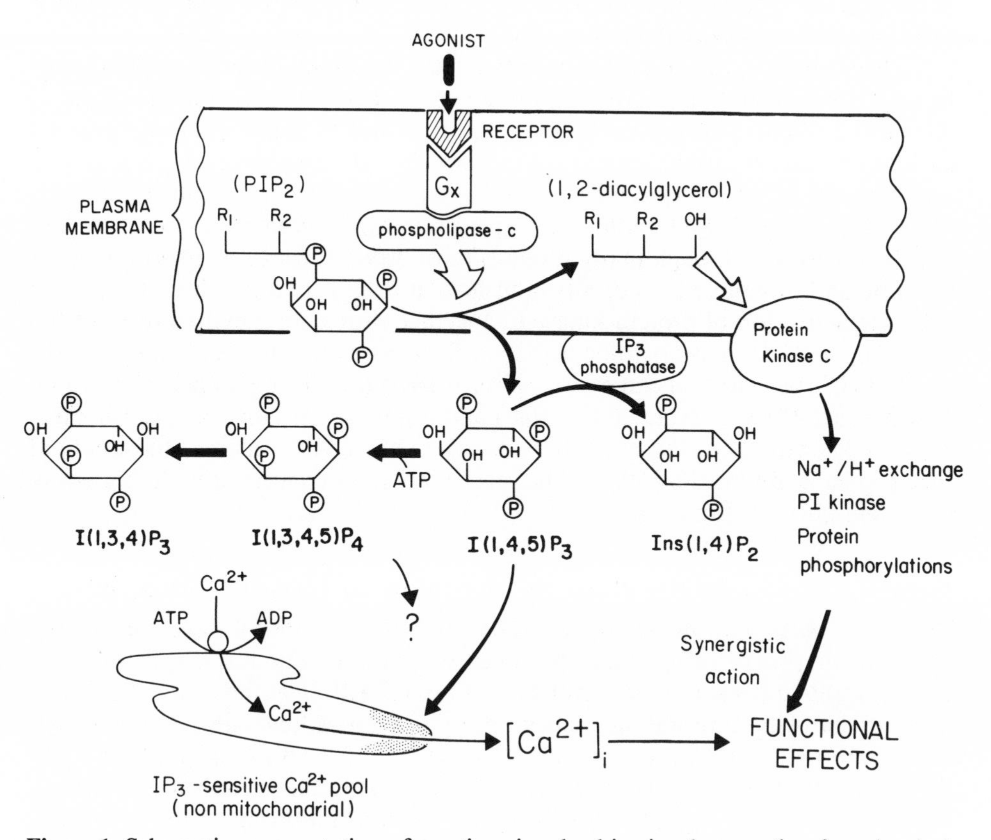

Figure 1. Schematic representation of reactions involved in signal generation from inositol lipid metabolism. G_x, postulated unidentified GTP-binding protein. R_1, fatty acid side chain (mainly stearyl). R_2, fatty acid side chain (mainly arachidonyl).

the phosphate head groups with ^{32}P inorganic phosphate, there is a rapid activation of phospholipase C, as evidenced by decreases of PIP and PIP_2 contents and increases of cellular *myo*-inositol 1,4-bisphosphate [Ins(1,4)P_2], *myo*-inositol 1,4,5-trisphosphate [Ins(1,4,5)P_3], and diacylglycerol levels (Thomas et al., 1983, 1984; Creba et al., 1983; Charest et al., 1985; Bocckino et al., 1985). Ins(1,4,5)P_3 activates a putative Ca^{2+} channel in a subpopulation of the hepatic endoplasmic reticulum (Joseph et al., 1984; Burgess et al., 1984; Joseph and Williamson, 1986), resulting in a rapid rise of the cytosolic free Ca^{2+} and activation of phosphorylase (Charest et al., 1983; Thomas et al., 1984). Ins(1,4,5)P_3 is inactivated through metabolism by two routes: by a 5-phosphatase to Ins(1,4)P_2 and by an ATP-dependent 3-kinase to *myo*-inositol 1,3,4,5-tetrakisphosphate (IP_4) (Irvine et al., 1986; Hansen et al., 1986; Cerdan et al., 1986). Degradation of IP_4 also occurs by the 5-phosphatase to a second inositol trisphosphate isomer, namely *myo*-inositol 1,3,4-trisphosphate [Ins(1,3,4)P_3] (Batty et al., 1985; Hansen et al., 1986; Irvine et al., 1986). In liver, the 5-phosphatase is mostly bound to the plasma membrane (Storey et al., 1984; Joseph and Williams, 1985), while the 3-kinase is located primarily in the soluble fraction (Hansen et al., 1986). Ins(1,3,4,5)P_4 and Ins(1,3,4)P_3 are essentially inactive in causing Ca^{2+} release from intracellular organelles and presently have no known function (Williamson and Hansen, 1987).

Diacylglycerol, the other second messenger released by the action of phospholipase C on inositol lipids, is the physiological activator of protein kinase C (Nishizuka, 1986). This protein kinase is also activated by tumor-promoting agents such as phorbol myristate acetate (PMA), which, like diacylglycerol, decrease the requirements of the enzyme for free Ca^{2+} to values found in normal resting cells (Ashendel, 1985). The physiological target proteins for protein kinase C are still largely unknown, although many proteins have been shown to be phosphorylated at serine and threonine sites in vitro (Nishizuka, 1986). In a number of secretory cells, the activation of protein kinase C, together with a sustained increase of the cytosolic free Ca^{2+}, appears to be necessary for prolonged functional effects (Nishizuka et al., 1984; Rasmussen et al., 1984). However, in liver (Cooper et al., 1985; Lynch et al., 1985; Corvera et al., 1986) and a number of other cells (Nishizuka, 1986; Williamson and Hansen, 1987), agonist-mediated activation of inositol lipid metabolism is inhibited by PMA. The present article is concerned with this aspect of protein kinase C function.

Kinetics of Hormone-induced Increases of Inositol Phosphates

Previous studies with hepatocytes prelabeled with [^{3}H]inositol have shown that after the addition of vasopressin, there was a rapid production of Ins(1,4,5)P_3 and Ins(1,3,4,5)P_4 and a slower accumulation of Ins(1,3,4)P_3 (Hansen et al., 1986). Fig. 2 shows that with a saturating concentration of the α_1-adrenergic agonist phenylephrine, the accumulations of Ins(1,4,5)P_3 and Ins(1,3,4,5)P_4 peaked after 5 s and fell within 30 s to values that were only slightly elevated above control. In contrast, the accumulation of Ins(1,3,4)P_3 was slower, reaching a maximal value after 30 s, and thereafter declined gradually. After 1 min of hormonal stimulation, Ins(1,3,4)P_3 accounted for ~90% of the increased production of IP_3 isomers. These changes were quantitatively much smaller than those induced by vasopressin (Hansen et al., 1986).

Glucagon has also been shown to cause a mobilization of intracellular Ca^{2+} (e.g., Charest et al., 1983), but has been reported not to cause an increased breakdown of inositol lipids in hepatocytes (Creba et al., 1983; Charest et al., 1985; Poggioli et al., 1986). Fig. 3 shows that 10 nM glucagon, which causes a maximal increase of cytosolic free Ca^{2+} in hepatocytes, produced a very small but statistically significant increase of Ins(1,4,5)P_3. This was associated with relatively greater percentage increases of Ins(1,3,4,5)P_4 and Ins(1,3,4)P_3. The rate of accumulation of Ins(1,3,4)P_3 was slower than that of Ins(1,3,4,5)P_4, as with phenylephrine-stimulated hepatocytes (Fig. 2). These data demonstrate, therefore, that glucagon

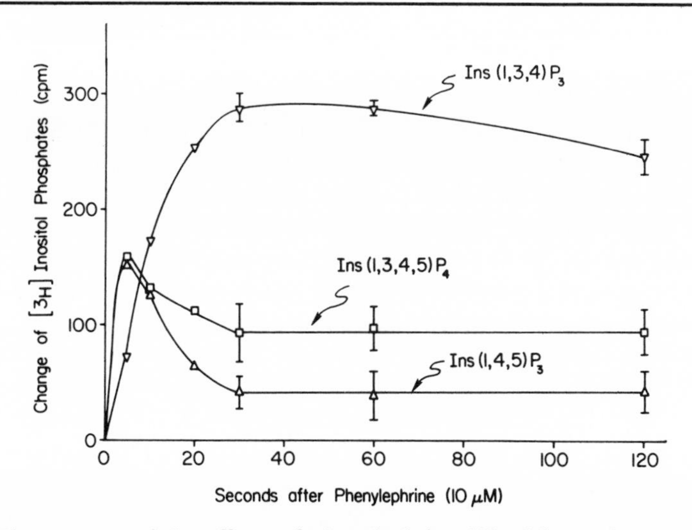

Figure 2. Time course of the effects of phenylephrine (10 μM) on the accumulation of [^{3}H]inositol phosphates in rat hepatocytes. The experimental conditions for preparation of hepatocytes, incubation with [2-^{3}H]*myo*-inositol, and separation and analysis of the inositol phosphates were the same as described by Hansen et al. (1986). Control values were: Ins(1,4,5)P_3, 288 ± 26 cpm; Ins(1,3,4,5)P_4, 54 ± 8 cpm; Ins(1,3,4)P_3, 42 ± 4 cpm (means ± SEM of four experiments).

does produce a small increase of Ins(1,4,5)P_3, presumably derived from the breakdown of PIP_2, which may account for its Ca^{2+}-mobilizing effect.

Glucagon-induced Mobilization of Intracellular Ca^{2+}

Fig. 4*A* shows the effect of different concentrations of glucagon in causing an increase of the cytosolic free Ca^{2+} in quin2-loaded hepatocytes. Half-maximum

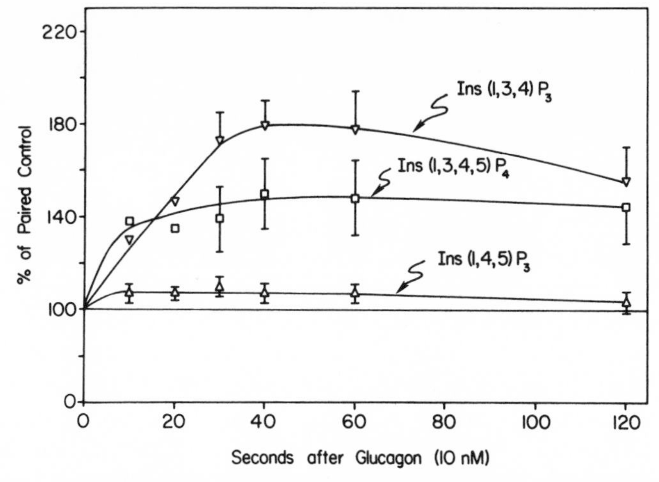

Figure 3. Time course of the effects of glucagon (10 nM) on the accumulation of [^{3}H]inositol phosphates in rat hepatocytes. The experimental conditions were the same as in Fig. 2 and the values shown are means ± SEM of nine experiments.

effects were obtained at 1 nM glucagon and maximum effects were obtained at ~10 nM, in agreement with the data of Sistare et al. (1985). Characteristically, there was a short delay of 5–15 s before the rise of cytosolic free Ca^{2+} could be observed. With increasing concentrations of glucagon, this delay diminished, while the rate of increase of cytosolic free Ca^{2+} and its peak value increased. Fig. 4*B* shows that the addition of 8-bromo–cyclic AMP (8-Br-cAMP) to quin2-loaded hepatocytes also produced a concentration-dependent increase of cytosolic free Ca^{2+}, with half-maximum and maximum effects at 20 and 100 μM, respectively. These results suggest that the Ca^{2+}-mobilizing effects of glucagon may be secondary to the production of cAMP, rather than parallel events.

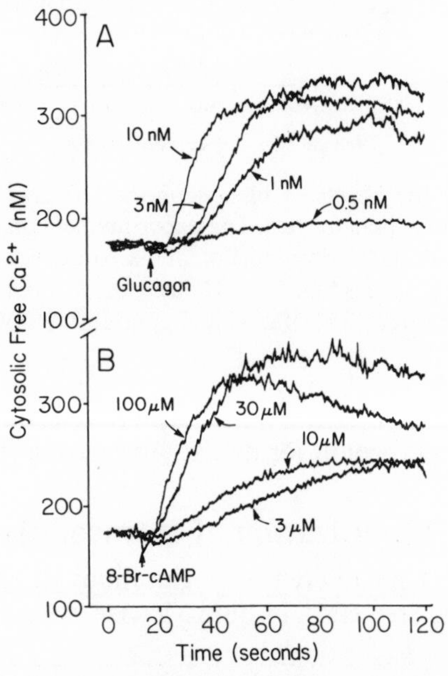

Figure 4. Effect of different concentrations of glucagon (*A*) or 8-Br-cAMP (*B*) on the cytosolic free Ca^{2+} of isolated hepatocytes. The cytosolic free Ca^{2+} in cells loaded with quin2 was measured as described by Thomas et al. (1984).

The initial phase of the increase of cytosolic free Ca^{2+} induced by 10 nM glucagon was completely dependent on the mobilization of intracellular Ca^{2+}, as illustrated in Fig. 5. A small excess addition of EGTA just before glucagon had no effect on the rate or extent of the initial rise of Ca^{2+}, but in the absence of extracellular Ca^{2+}, the cytosolic free Ca^{2+} returned to control values within 3 min rather than remaining elevated. This transient increase of cytosolic free Ca^{2+} with hepatocytes incubated in low extracellular Ca^{2+} medium is associated with a net loss of Ca^{2+} from the hepatocytes (Mauger and Claret, 1986). In common with vasopressin, angiotensin II, and α_1-adrenergic agonists (Mauger et al., 1984, 1985; Blackmore and Exton, 1985; Joseph et al., 1985), it is apparent that the glucagon-induced Ca^{2+} transient has two components, an intracellular Ca^{2+} mobilization and an influx of Ca^{2+} from the extracellular medium.

Fig. 5 also shows that a subsequent addition of phenylephrine after glucagon addition caused a further small increase of the cytosolic free Ca^{2+} in either the

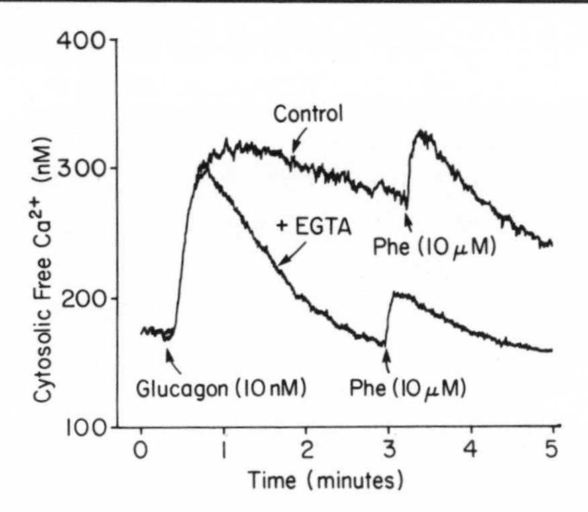

Figure 5. Effects of sequential additions of glucagon and phenylephrine (Phe) on the cytosolic free Ca^{2+} of isolated hepatocytes incubated in the presence and absence of extracellular Ca^{2+}. A low extracellular Ca^{2+} concentration was obtained by addition of 1.4 mM EGTA to medium containing 1.3 mM Ca^{2+} 30 s before hormone addition.

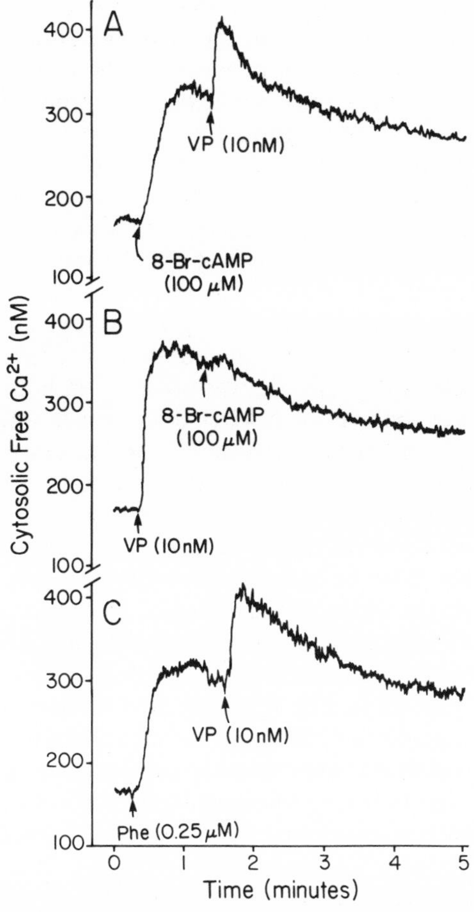

Figure 6. Effects of sequential additions of agonists on the cytosolic free Ca^{2+} of isolated hepatocytes. VP, vasopressin; Phe, phenylephrine.

presence or absence of extracellular Ca^{2+}. Under these conditions, the phenylephrine-induced increase of cytosolic free Ca^{2+} was only 20% of that obtained without prior glucagon addition, which indicates that the two hormones may be mobilizing Ca^{2+} from the same intracellular pool. This was confirmed in further experiments, which are shown in Fig. 6. When vasopressin was added at the peak of the 8-Br-cAMP–induced increase of cytosolic free Ca^{2+}, a small additional mobilization of Ca^{2+} was observed (Fig. 6*A*). On the other hand, the addition of 8-Br-cAMP at the peak of the vasopressin-induced Ca^{2+} transient showed a negligible further response (Fig. 6*B*). That the different hormones were probably mobilizing Ca^{2+} from the same intracellular Ca^{2+} pool is also illustrated by the fact that the addition of vasopressin after a submaximal concentration of phenylephrine (Fig. 6*C*) produced a further increase of cytosolic free Ca^{2+} to the same value of ~400 nM as observed when vasopressin was added as the first agonist (Fig. 6*B*). Similar results have been reported recently by Combettes et al. (1986).

As an alternative approach to determine whether glucagon and vasopressin were mobilizing Ca^{2+} from the same intracellular pool, the hormone-induced changes of cytosolic free Ca^{2+} were measured in hepatocytes after the prior addition of uncoupling agents to deplete the mitochondria of Ca^{2+}. For this purpose, it was necessary to correct for the spillover of the NADH fluorescence into the photomultiplier measuring the quin2-Ca fluorescence in the two-channel filter fluorometer (Thomas et al., 1984). The importance of this correction is illustrated by the data shown in Fig. 7 and derives from the fact that the fluorescence emission spectrum for NADH overlaps with that for the quin2-Ca complex (Tsien et al., 1982). The uncoupling agent chosen for these experiments was 1799 (Heytler, 1979), which is nonfluorescent when excited at the wavelength of 339 nm required for quin2-Ca excitation. The upper trace of Fig. 7 shows the quin2 fluorescence recorded from the output of a photomultiplier with a bandpass filter of 490–550 nm, while the middle trace shows the NADH fluorescence recorded using a bandpass filter of 420–450 nm. After the addition of 1799 to the quin2-loaded hepatocytes, a decrease of fluorescence was observed in both channels. However, after correction for the spillover of NADH fluorescence into the quin2 channel (see legend to Fig. 7), the corrected quin2 fluorescence (third trace of Fig. 7) increased upon addition of 1799, showing the expected increase of cytosolic free Ca^{2+} as a result of efflux of Ca^{2+} from the mitochondria. In further experiments reported here, only the corrected quin2 fluorescence changes are shown.

Fig. 8*A* shows the effect of vasopressin on the cytosolic free Ca^{2+} of quin2-loaded hepatocytes in the absence and presence of a prior addition of 1799. A smaller vasopressin-induced increase of the cytosolic free Ca^{2+} was observed in the presence of the uncoupler than was observed in its absence, although the peak of the Ca^{2+} increase was the same. Fig. 8*B* shows that similar qualitative results were obtained after glucagon addition, although the rate of change of the cytosolic free Ca^{2+} with glucagon was slower than with vasopressin (Charest et al., 1983). These results indicate that the hormone-sensitive intracellular Ca^{2+} pool was partially sensitive to mitochondrial uncoupling agents with both vasopressin and glucagon as agonists.

The addition of agents to intact cells or isolated organs to produce an uncoupling of oxidative phosphorylation, a collapse of the mitochondrial membrane potential, and efflux of Ca^{2+} has been used as a nondisruptive technique to

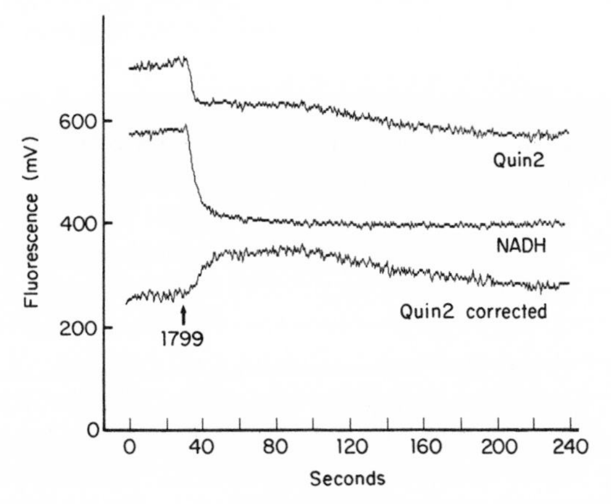

Figure 7. Effect of 1799 (5 μM), an uncoupler of oxidative phosphorylation, on quin2-Ca and NADH fluorescence changes in isolated hepatocytes. The upper trace (quin2) records the fluorescence emission between 490 and 550 nm and the middle trace (NADH) records the fluorescence emission between 420 and 450 nm, using two photomultipliers and an excitation wavelength of 399 nm with a two-channel filter fluorometer (see Thomas et al., 1984). The quin2-corrected trace (F_{Qcorr}) was obtained by correcting for NADH spillover using the formula: $F_{Qcorr} = (F_{QT} - xF_{NT})/1 - xy$, where F_{QT} is the total fluorescence in the quin2 channel, F_{NT} is the total fluorescence in the NADH channel, x is the ratio of fluorescence observed in the quin2 channel to that in the NADH channel upon addition of a standard NADH, and y is the ratio of fluorescence observed in the NADH channel to that in the quin2 channel upon addition of quin2 free acid with saturating Ca^{2+}. The constants x and y were determined by adding NADH or quin2 to hepatocytes (5 mg/ml) after permeabilization with 0.1% Nonidet, and were 0.62 and 0.18, respectively. Data were accumulated using an IBM personal computer fitted with a Data Translation DT 2801 fast A/D board and were collected every 500 ms for 4 min.

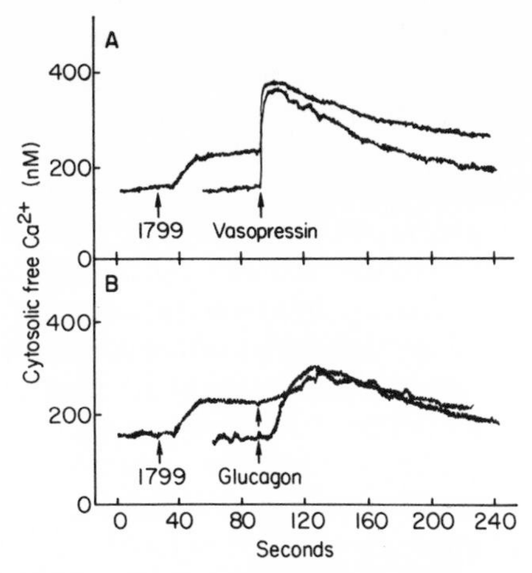

Figure 8. Effect of vasopressin (*A*) and glucagon (*B*) added after the uncoupler of mitochondrial oxidative phosphorylation 1799 (5 μM) on the cytosolic free Ca^{2+} concentration of isolated hepatocytes. The cytosolic free Ca^{2+} was measured with quin2 and the quin2-Ca fluorescence change was corrected for simultaneous NADH fluorescence changes as described in the legend to Fig. 7.

estimate the mitochondrial Ca^{2+} content of intact cells (Bellomo et al., 1982; Joseph and Williamson, 1983). Studies with a flowthrough perfused rat liver have shown that when vasopressin was infused after prior treatment with the uncoupling agent 2,4-dinitrophenol, the total amount of hormone-induced net Ca^{2+} efflux was decreased to 53 ± 6 nmol/g wet wt, compared with a value of 97 ± 1 nmol/g wet wt for vasopressin alone (Kleineke and Soling, 1985). In these studies, the values found for total net Ca^{2+} efflux after separate additions of the Ca^{2+} ionophore A23187 and 2,4-dinitrophenol were 96 ± 7 and 19 ± 2 nmol/g wet wt, respectively. These results show that the uncoupler-sensitive Ca^{2+} pool in the intact liver is small relative to the vasopressin-sensitive Ca^{2+} pool, and that part of the latter pool is sensitive to uncoupling agents, in agreement with the data shown in Fig. 8*A*.

This finding is at first sight unexpected, since it has been shown that the addition of Ins(1,4,5)P_3 to permeabilized cells causes the release of Ca^{2+} only from a nonmitochondrial intracellular Ca^{2+} pool (Joseph et al., 1984; Burgess et al., 1984; Streb et al., 1984). Consequently, Ca^{2+} mobilization elicited by vasopressin should be insensitive to mitochondrial uncoupling agents. Since this is not the case, it appears that mitochondrial uncoupling agents cannot be used as a tool to produce a selective release of Ca^{2+} from the mitochondrial Ca^{2+} compartment of the cell as they also release Ca^{2+} from part of the endoplasmic reticular Ca^{2+} pool. This conclusion probably accounts for the finding that the hormone-sensitive Ca^{2+} pool obtained in an earlier study with isolated hepatocytes, which employed uncoupling agents and A23187 to investigate the relative size of intracellular Ca^{2+} pools, was partly mitochondrial and partly endoplasmic reticular (Joseph and Williamson, 1983). It is inappropriate, therefore, to conclude from the data of Fig. 8*B* and similar experimental approaches using mitochondrial uncoupling agents to probe the nature of the intracellular source of glucagon-releasable Ca^{2+} (Kraus-Friedmann, 1986) that glucagon elicits release of Ca^{2+} from a mitochondrial Ca^{2+} pool.

It is seen from further experiments illustrated in Fig. 9 that when a comparison is made of the maximum vasopressin- and glucagon-induced changes of cytosolic free Ca^{2+} in quin2-loaded hepatocytes, the responses to the two hormones are dissimilar in several respects. With vasopressin, the peak increase of cytosolic free Ca^{2+} was achieved after 5 s. With glucagon, there was a longer lag before the cytosolic free Ca^{2+} started to increase, and the peak increase was achieved after 30 s. Also, the maximum increase of cytosolic free Ca^{2+} with glucagon was only about half that achieved with vasopressin. Phenylephrine (10 μM) gave responses that were essentially the same as those produced by vasopressin, while forskolin (10 μM), added to activate adenylate cyclase directly (Daly, 1984), gave an increase of cytosolic free Ca^{2+} similar to that produced with glucagon or 8-Br-cAMP (data not shown; Staddon and Hansford, 1986; Combettes et al., 1986). These results, together with the data shown in Figs. 5 and 6, demonstrate that glucagon, externally added cAMP derivatives, or intracellularly generated cAMP are only capable of eliciting a partial release of Ca^{2+} from a Ca^{2+} pool that is common to other Ca^{2+}-mobilizing agonists and appears to require Ins(1,4,5)P_3 as a second messenger.

Inhibitory Effects Induced by Activation of Protein Kinase C

Previous studies from this (Cooper et al., 1985) and other laboratories (Lynch et al., 1985; Corvera et al., 1986) have shown that pretreatment of hepatocytes for

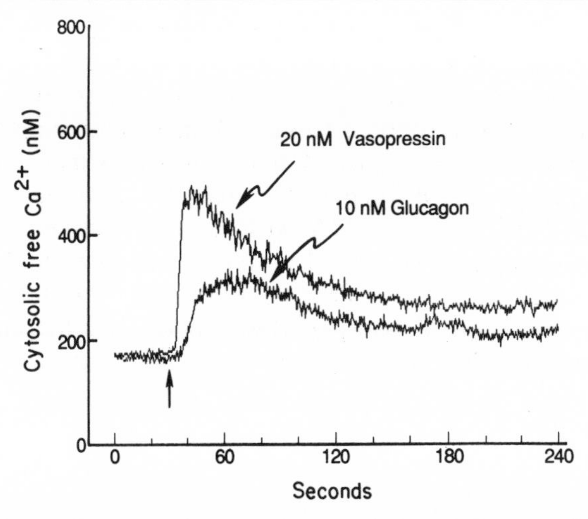

Figure 9. Comparison of vasopressin- and glucagon-induced changes of cytosolic free Ca^{2+} in isolated hepatocytes. The cytosolic free Ca^{2+} was measured with quin2 and the quin2-Ca fluorescence change was corrected for simultaneous NADH fluorescence changes as described in the legend to Fig. 7.

short times with the phorbol ester PMA caused a complete inhibition of the phenylephrine-induced increase of cytosolic free Ca^{2+}, activation of phosphorylase, and PI turnover. A similar but less complete inhibition of the glucagon-induced (Fig. 10*A*) or 8-Br-cAMP–induced (Fig. 10*B*) increase of cytosolic free Ca^{2+} is produced by pretreatment of hepatocytes for 2 min with a maximum concentration (100 ng/ml) of PMA (see also Blackmore and Exton, 1986; Staddon and Hansford,

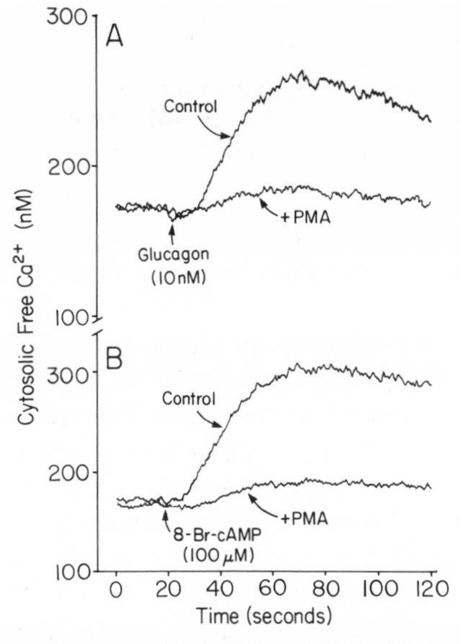

Figure 10. Effect of PMA on glucagon- (*A*) or 8-Br-cAMP– (*B*) induced increase of cytosolic free Ca^{2+} in isolated hepatocytes. PMA (100 ng/ml) was added to hepatocytes (5 mg dry wt/ml) 2 min before agonist addition.

1986). The small increase of cytosolic free Ca^{2+} observed after the addition of these agonists in the presence of PMA was also observed when the hormone was added immediately after a slight excess of EGTA, which indicates that the incomplete inhibition was not due to Ca^{2+} entry across the plasma membrane (data not shown).

Further studies (Fig. 11*A*) showed that a half-maximal effect of PMA in inhibiting the peak increase of the cytosolic free Ca^{2+} induced by 100 μM 8-Br-cAMP was obtained at 2 ng/ml of PMA, with maximum inhibition at ~10 ng/ml. A similar sensitivity was obtained for PMA inhibition of the phenylephrine-induced increase of cytosolic free Ca^{2+} (Cooper et al., 1985). Fig. 11*B* shows that PMA had no effect on the activation of phosphorylase induced by the addition of 8-Br-cAMP, which indicates that the steps involving activation of cAMP-dependent

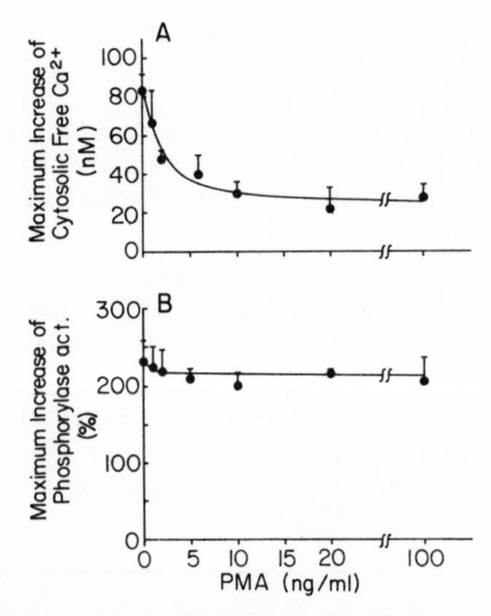

Figure 11. Concentration dependence of the effect of PMA on changes of cytosolic free Ca^{2+} (*A*) or phosphorylase activity (*B*) induced by 100 μM 8-Br-cAMP in isolated hepatocytes (5 mg dry wt/ml). Results are means ± SEM of three separate experiments.

protein kinase, its phosphorylation of phosphorylase kinase, and the further phosphorylation and activation of phosphorylase were not affected by PMA (Blackmore et al., 1986).

The effect of PMA pretreatment on phenylephrine-induced increases of inositol polyphosphates in hepatocytes prelabeled with [^{3}H]inositol is shown in Fig. 12. PMA (100 ng/ml) prevented the hormone-induced increases of Ins(1,4,5)P_3, Ins(1,3,4,5)P_4, and Ins(1,3,4)P_3 (Fig. 12), as well as the increases of inositol bis- and monophosphates (data not shown). This effect is most readily seen in the tissue content of the Ins(1,3,4)P_3 isomer, which shows the greatest percentage increase with phenylephrine. In control, nonstimulated cells, the content of [^{3}H]Ins(1,4,5)P_3 was sixfold greater than the contents of [^{3}H]Ins(1,3,4,5)P_4 and [^{3}H]Ins(1,3,4)P_3, and was decreased slightly (15%) by PMA pretreatment alone.

Fig. 13 (control panel) shows that 30 s after the addition of 100 μM 8-Br-cAMP to [^{3}H]inositol-labeled hepatocytes, there was a small increase of Ins(1,4,5)P_3, but larger and statistically significant increases of Ins(1,3,4)P_3 and Ins(1,3,4,5)P_4.

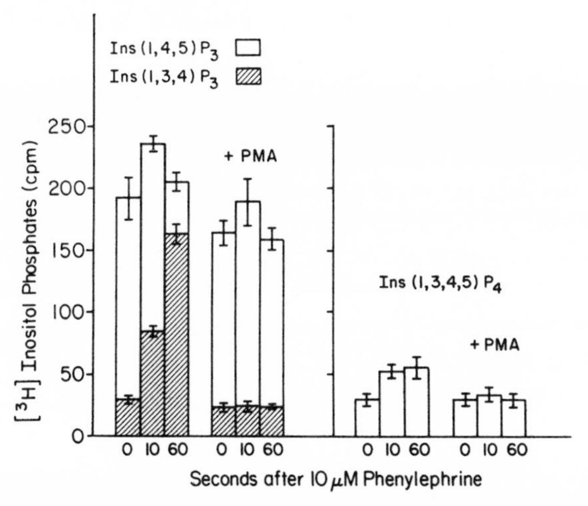

Figure 12. Effect of PMA pretreatment (100 ng/ml for 3 min) on phenylephrine-induced changes of [^{3}H]inositol phosphates in isolated hepatocytes (20 mg dry wt/ml). Results are means ± SEM of four separate experiments.

The right-hand panel of Fig. 13 shows that these changes were all abolished by a 3-min pretreatment of the hepatocytes with 100 ng/ml PMA. Similarly, the increases of inositol phosphates induced by glucagon were abolished by PMA pretreatment of hepatocytes (data not shown).

Epidermal growth factor (EGF) has recently been shown (Johnson et al., 1986) to elicit an increase of cytosolic free Ca^{2+} in hepatocytes, with this effect being inhibited by pretreatment of the cells with pertussis toxin or PMA. The EGF-induced increase of cytosolic free Ca^{2+} as measured with quin2 resembles that

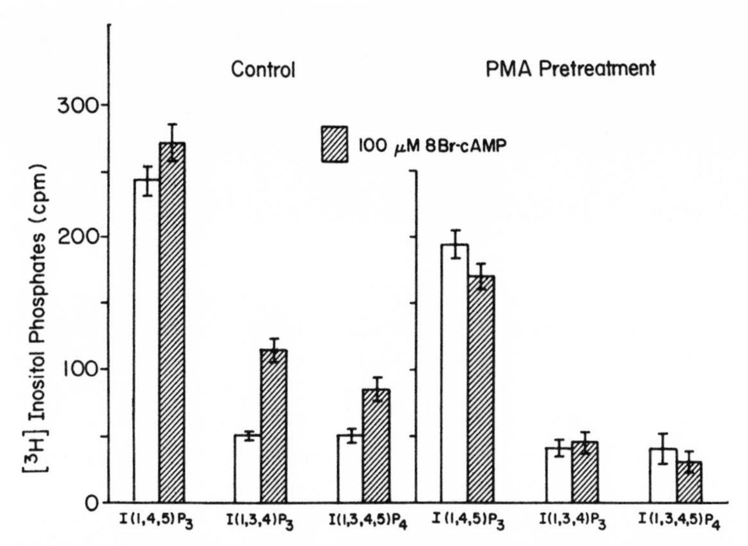

Figure 13. Effect of PMA pretreatment (100 ng/ml for 3 min) on 8-Br-cAMP–induced changes of [^{3}H]inositol phosphates in isolated hepatocytes (20 mg dry wt/ml). Results are means ± SEM of six separate experiments assayed 30 s after the addition of 100 μM 8-Br-cAMP.

elicited by glucagon, forskolin, or cAMP analogues in being smaller and of slower onset than the vasopressin-induced increase of cytosolic free Ca^{2+} (data not shown; Bosch et al., 1986). However, Bosch et al. (1986) were unable to demonstrate that EGF had any stimulatory effect on the accumulation of inositol phosphates in rat hepatocytes. Fig. 14 shows that 100 μM EGF, which gives a maximum increase of the cytosolic free Ca^{2+}, caused statistically significant increases of Ins(1,4,5)P_3, Ins(1,3,4)P_3, and Ins(1,3,4,5)P_4. This effect of EGF on inositol lipid metabolism was blocked by pretreatment of the hepatocytes for 3 min with 100 ng/ml PMA (Fig. 14). These results, with a number of receptor-dependent and -independent

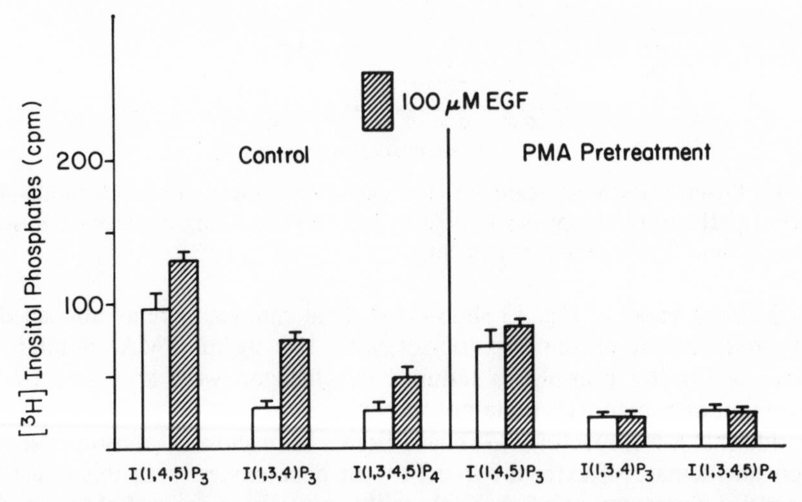

Figure 14. Effect of PMA pretreatment (100 ng/ml for 3 min) on EGF-induced changes of [^{3}H]inositol phosphates in isolated hepatocytes (20 mg dry wt/ml). Results are means ± SEM of four separate experiments assayed 30 s after the addition of 100 μM EGF.

agonists that produce an activation of phospholipase C, increased formation of Ins(1,4,5)P_3, and mobilization of intracellular Ca^{2+}, suggest that the PMA-mediated inhibition of these responses may be caused by a common mechanism.

Mechanisms of Action of Glucagon- and PMA-induced Negative Feedback

The present results are in substantial agreement with a number of other studies, which collectively show that glucagon increases the cytosolic free Ca^{2+} of hepatocytes with approximately the same sensitivity as it elevates cAMP levels and activates phosphorylase. As with other hepatic Ca^{2+}-mobilizing hormones, which include such diverse agonists as vasopressin, angiotensin II, α_1-adrenergic agents (reviewed by Blackmore and Exton, 1985; Williamson et al., 1985), EGF (Johnson et al., 1986), and opiate peptides (Leach et al., 1986), glucagon causes an initial mobilization of Ca^{2+} from intracellular storage sites and also promotes an increased flux of Ca^{2+} across the plasma membrane. This is shown by the fact that the hormone-induced increase of cytosolic free Ca^{2+} to its peak value is independent of the presence of extracellular Ca^{2+}, but Ca^{2+} is required to produce a sustained elevation of the cytosolic free Ca^{2+} above resting levels (cf. Fig. 5).

Each of the agonists appears to elicit the release of Ca^{2+} from a common intracellular Ca^{2+} pool, which is partially sensitive to the Ca^{2+}-mobilizing effect of mitochondrial uncoupling agents. However, in contrast to earlier reports (reviewed by Kraus-Friedmann, 1984), more recent evidence strongly favors the concept that treatment of hepatocytes with Ca^{2+}-mobilizing hormones leads to an increase rather than a decrease of the mitochondrial Ca^{2+} content, mainly as a secondary consequence of the elevated cytosolic free Ca^{2+} (Shears and Kirk, 1984; McCormack, 1985). Since mitochondrial uncoupling agents appear not to cause a selective release of Ca^{2+} from the mitochondria, their use for probing the nature of the intracellular organelles responsible for hormone-induced Ca^{2+} release is subject to doubt. Consequently, from the observation that uncoupling agents diminish glucagon-induced Ca^{2+} release in hepatocytes (Fig. 8) or perfused liver (Kraus-Friedmann, 1986), it is not possible to deduce that the mechanism of intracellular Ca^{2+} release induced by glucagon or cAMP is different from that of the other Ca^{2+}-mobilizing hormones.

The common mechanism of action for all these agonists appears to be an activation of phospholipase C, with production of Ins(1,4,5)P_3. The changes of Ins(1,4,5)P_3 and other inositol phosphates induced by these different agents, however, show large quantitative differences, which are not proportional to their ability to mobilize Ca^{2+} and increase the cytosolic free Ca^{2+} concentration of hepatocytes (Williamson and Hansen, 1987). A peak increase of Ins(1,4,5)P_3 levels of 45% above control values produced by phenylephrine gives the same peak increase of cytosolic free Ca^{2+} as vasopressin, which increases Ins(1,4,5)P_3 levels at least 10-fold. However, glucagon, 8-Br-cAMP, and EGF, which increase Ins(1,4,5)P_3 levels by only 5–10%, cause a both smaller and slower increase of the cytosolic free Ca^{2+}, which suggests that the Ca^{2+}-mobilizing effect of these weak agonists is limited by their rate of production of Ins(1,4,5)P_3. Prentki et al. (1985) have shown that by infusing Ins(1,4,5)P_3 at different rates into the medium containing permeabilized cells supplemented with ATP to allow Ca^{2+} uptake by the endoplasmic reticulum, different steady state levels of Ca^{2+} in the medium can be achieved. These authors suggested that suboptimal concentrations of Ins(1,4,5)P_3 modulate Ca^{2+} cycling across the endoplasmic reticulum membrane, with Ins(1,4,5)P_3-mediated Ca^{2+} efflux being balanced by ATP-dependent Ca^{2+} influx, which results in regulation of both the Ca^{2+} content of the Ins(1,4,5)P_3-sensitive Ca^{2+} pool and the cytosolic free Ca^{2+} concentration. This concept is compatible with the smaller and much less biphasic nature of the Ca^{2+} transient induced by glucagon as compared with vasopressin (Fig. 9).

The kinetics of accumulation of Ins(1,4,5)P_3 with the different agonists also show differences. With vasopressin (Hansen et al., 1986) and phenylephrine (Fig. 2), there is an initial overshoot of Ins(1,4,5)P_3 production before a steady state level is reached within the first minute of hormone action, while with glucagon (Fig. 3), 8-Br-cAMP, and EGF (data not shown), the steady state level is achieved within 10 s. During the steady state, the rate of production of Ins(1,4,5)P_3 from PIP_2 becomes equal to the sum of its rate of degradation to Ins(1,4)P_2 by the 5-phosphatase and its rate of phosphorylation to Ins(1,3,4,5)P_4 by the inositol phosphate 3-kinase. The flux through the two disposal pathways for Ins(1,4,5)P_3 is probably about equal (Williamson and Hansen, 1987), with the relatively higher activity of the phosphatase being compensated by a higher K_m (3 μM) than that for the kinase (0.2 μM). From studies with cytosolic fractions from insulin-secreting RIN m5F cells, it has been suggested that the biphasic kinetics of Ins(1,4,5)P_3 accumulation frequently observed during agonist stimulation of cells are caused by a secondary activation of the kinase (V_{max} effect) mediated by the rise of the cytosolic free Ca^{2+} (Biden and Wollheim, 1986). However, the purified Ins(1,4,5)P_3 kinase from brain shows only an inhibition of activity with Ca^{2+} concentrations of $\geq$1 μM (Johanson, R., and K. E. Coll, unpublished observations). Hence, at present,

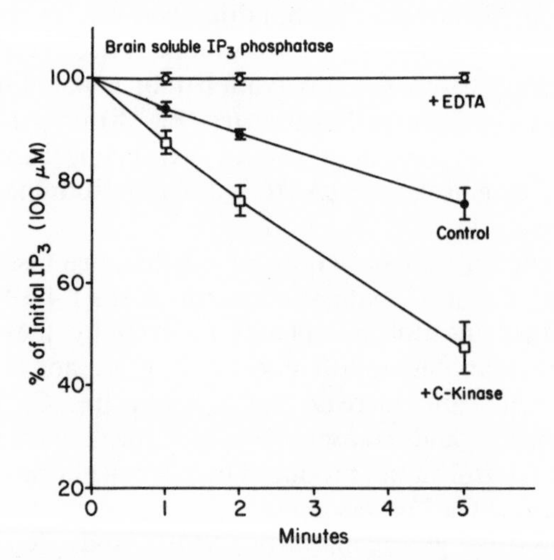

Figure 15. Effect of protein kinase C on Ins(1,4,5)P_3 5-phosphatase activity. Partially purified 5-phosphatase from rat brain was incubated for 10 min at 30°C with protein kinase C (1 nmol/min sp. act.) in a buffer containing 50 mM MOPS, pH 7.1, 200 μg/ml phosphatidylserine, 1 μM PMA, 1 mM $CaCl_2$, 3 mM $MgCl_2$, 1 mM ATP, 200 μM EGTA, 12.5 mM creatine phosphate, 2 mM dithiothreitol, and 10 U/ml creatine phosphokinase. Control conditions contained a final EGTA concentration of 2 mM. After preincubation, 45 μl was transferred to Eppendorf tubes containing 5 μl of 1 mM [^{3}H]Ins(1,4,5)P_3 and quenched at appropriate times with 100 μl of 200 mM formic acid. Subsequently, remaining Ins(1,4,5)P_3 was quantified by separation from other inositol phosphates by Dowex AG-1 $\times$ 8 chromatography. Values shown are means $\pm$ SEM of three experiments.

it is unclear whether Ca^{2+} activation of the kinase requires a cofactor protein, whether the phenomenon is dependent on the cell type, or whether a different mechanism is responsible for the decline of Ins(1,4,5)P_3 levels from their peak value (see below).

The activity of crude (Biden and Wollheim, 1986) and purified (Hansen, C. A., and J. R. Williamson, unpublished observations) Ins(1,4,5)P_3 5-phosphatase is not affected by Ca^{2+}, but a purified enzyme isolated from platelets has been shown to be phosphorylated and activated by protein kinase C (Connolly et al., 1986). Similar results have been obtained with purified brain 5-phosphatase, as illustrated in Fig. 15. The addition of purified protein kinase C to the Ins(1,4,5)P_3 5-phosphatase assay medium resulted in a doubling of the rate of Ins(1,4,5)P_3 degradation when the substrate was added at a saturating concentration of 100 μM. The phosphatase activity was obligatorily dependent on an excess of free Mg^{2+}, as shown by the lack of Ins(1,4,5)P_3 degradation in the presence of excess EDTA. The addition of excess EGTA in the presence of activated protein kinase C decreased the phosphatase activity to control rates (data not shown).

In accordance with the above studies showing activation of the 5-phosphatase by protein kinase C in vitro, pretreatment of human platelets with diacylglycerol to activate protein kinase C in situ enhanced the rate of hydrolysis of Ins(1,4,5)P_3 when added to subsequently permeabilized cells (Molina y Vedia and Lapetina, 1986). An alternative mechanism to a Ca^{2+}-mediated activation of the Ins(1,4,5)P_3 kinase to account for the biphasic Ins(1,4,5)P_3 kinetics after agonist stimulation, therefore, is a diacylglycerol-mediated activation of protein kinase C and enhancement of Ins(1,4,5)$_3$ degradation by the 5-phosphatase. Clearly, this mechanism will be more effective with strong than with weak agonists, since the former type will generate more diacylglycerol and hence produce a greater degree of activation of protein kinase C. As previously mentioned, a biphasic accumulation of Ins(1,4,5)P_3 appears to be dependent on the degree of agonist-induced activation of inositol lipid breakdown. Hence, a modulatory negative feedback effect on the disposal of Ins(1,4,5)P_3 by the diacylglycerol branch rather than the Ca^{2+} branch of the inositol lipid signaling pathway appears to be more in accordance with the overall results obtained with a variety of different agonists in hepatocytes. However, it should be stressed that in liver, at least, this potential mechanism for the modulation of Ins(1,4,5)P_3 levels does not necessarily attenuate the Ca^{2+}-signaling effect of the agonist, since the addition of phenylephrine after the peak of the vasopressin-induced Ca^{2+} transient, or vice versa, does not augment Ca^{2+} mobilization (Reinhart et al., 1984; Joseph et al., 1985), which indicates that the steady state Ins(1,4,5)P_3 concentration is more than sufficient to elicit a maximal release of Ca^{2+} from the hormone-sensitive Ca^{2+} pool (see also Williamson, 1986).

In contrast, when protein kinase C is fully activated by the addition of PMA or suitable exogenous diacylglycerols to hepatocytes, the Ca^{2+}-mobilizing effects of phenylephrine, glucagon, and EGF and the increases of inositol phosphates induced by these agonists are abolished or greatly attenuated. Vasopressin-induced Ca^{2+} mobilization in hepatocytes is only inhibited by PMA pretreatment at very low agonist concentrations, even though the accumulation of inositol phosphates is decreased at all vasopressin concentrations (Lynch et al., 1985). These results reflect the fact that, with this agonist, the negative feedback effect of PMA is not sufficiently powerful to diminish the Ins(1,4,5)P_3 level to below its effective concentration for eliciting Ca^{2+} release, except at suboptimal vasopressin concentrations. It is plausible that a protein kinase C–mediated activation of the inositol phosphate 5-phosphatase contributes to the PMA-induced inhibition of agonist-induced IP_3 mobilization. However, as illustrated in Fig. 16, interactions at other sites, notably the receptor itself or the GTP-binding coupling protein, resulting in an inhibition of the agonist-induced activation of phospholipase C, also appear to be involved in the general phenomenon of protein kinase C–mediated negative signaling (reviewed by Williamson and Hansen, 1987). The extent to which negative feedback by agonist-induced activation of protein kinase C in different systems can account for receptor desensitization under physiological conditions and the nature of the molecular mechanisms involved nevertheless remain speculative (for review, see Sibley and Lefkowitz, 1985).

The mechanism by which glucagon causes an activation of phospholipase C also cannot be defined unambiguously. In a recent study with isolated hepatocytes, Wakelam et al. (1986) showed that glucagon over the concentration range from 0.1 to 10 nM gave a biphasic increase of inositol phosphates, while a glucagon analogue, (1-*N*-α-trinitrophenylhistidine-12-homoarginine) glucagon, which is inactive in stimulatory adenylate cyclase, produced a normal dose-response curve for stimulation of inositol phosphate production. These authors also found that the sensitivity of glucagon to stimulate inositol phosphate formation (K_a = 0.25 nM) was considerably higher than that for stimulation of adenylate cyclase (6.3 nM), and suggested that hepatocytes possess two distinct glucagon receptors, one coupled to phospholipase C and the other to adenylate cyclase, with mutual inhibitory interactions by the separately generated signaling systems. Cyclic AMP does not

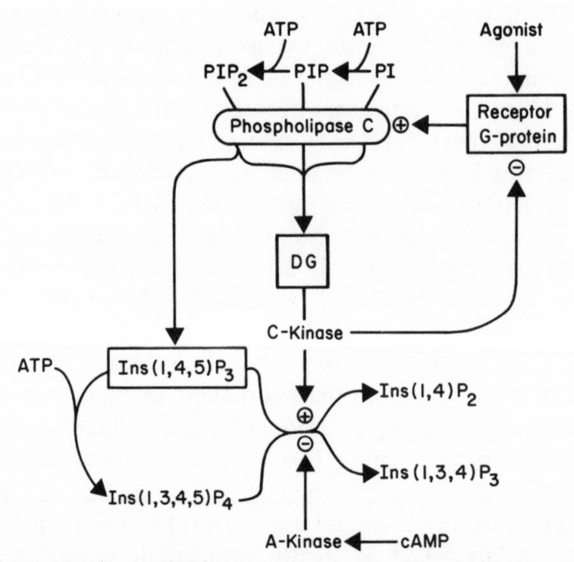

Figure 16. Scheme illustrating potential sites for negative feedback signaling for agonist-stimulated inositol lipid metabolism by diacylglycerol (DG) activation of protein kinase C (C-kinase). Also illustrated is the possibility that the inositol phosphate 5-phosphatase may be inhibited by cAMP-dependent protein kinase phosphorylation.

inhibit inositol lipid breakdown, as shown by the present data, which demonstrate cAMP's positive signaling role in Ca^{2+} mobilization. Also, a biphasic glucagon dose-response curve for Ins(1,4,5)P_3 production does not seem compatible with the normal monotonic dose-response curve for the increase of cytosolic free Ca^{2+}. Furthermore, only a single class of glucagon-binding sites has been described in rat hepatocytes, although they apparently exist in interconvertible high- and low-affinity sites (Horwitz et al., 1985, 1986). Although a direct coupling of the glucagon receptor (or a particular state of the receptor) through a GTP-binding protein to phospholipase C cannot be excluded, cAMP appears to interact with the inositol lipid signaling pathway to induce Ca^{2+} mobilization independently of the glucagon receptor. One possible site of action could be at the level of the G-protein α-subunit, with promotion of dissociation of the $\beta\gamma$-subunits from the inactive heterotrimer or stabilization of the active GTP-bound α-subunit upon phosphorylation of the protein. An alternative possibility, illustrated in Fig. 16, is a possible inhibition of the inositol phosphate 5-phosphatase by cAMP-dependent protein kinase. A negative interaction at this site implies a basal activity of phospholipase C and would be consistent with the small increases of Ins(1,4,5)P_3 and its metabolic products, as observed with the rise of cAMP in hepatocytes. Phospholipase C itself has not been reported to be phosphorylated or altered by either cAMP-dependent protein kinase or protein kinase C. If the site of action of these two kinases is on the same protein, then phosphorylation by the different kinases would have the interesting effect of modulating its activity in opposite directions. Further studies are required to determine whether the glucagon receptor can couple via a G-protein to phospholipase C or whether glucagon effects are exerted entirely as a secondary consequence of cAMP production.

Summary

Interactions between the different signaling roles of *myo*-inositol 1,4,5-trisphosphate and 1,2-diacylglycerol, the products of agonist-stimulated phosphatidylinositol 4,5-bisphosphate breakdown, are assessed in isolated rat hepatocytes. Measurements of the kinetics of accumulation of individual [^{3}H]inositol phosphates after the addition of different Ca^{2+}-mobilizing agonists in general support the role of inositol 1,4,5-trisphosphate as the second messenger responsible for release of sequestered intracellular Ca^{2+}. Various agonists, when added at maximal concentrations, however, produce qualitatively and quantitatively different responses, which reflect varying abilities of the agonists to activate phospholipase C. Qualitative differences are revealed by a pronounced biphasic pattern to the Ins(1,4,5)P_3 accumulation after vasopressin and phenylephrine stimulation, which is indicative of negative feedback. It is suggested that this effect is mediated by a partial diacylglycerol activation of protein kinase C, which in vitro causes an activation of inositol phosphate 5-phosphatase and hence promotes removal of Ins(1,4,5)P_3 to Ins(1,4)P_2. An alternative mechanism proposed by Biden and Wollheim (1986) of a secondary Ca^{2+} activation of Ins(1,4,5)P_3 3-kinase is considered less likely as a general mechanism, since highly purified kinase prepared from rat brain shows only an inhibition by Ca^{2+}. Glucagon, 8-Br-cAMP, and EGF induce small increases of Ins(1,4,5)P_3 in hepatocytes, together with slower and smaller increases of cytosolic free Ca^{2+} than those produced by vasopressin or phenylephrine, with Ca^{2+} being mobilized from the same intracellular pools with each of the agonists. The Ca^{2+}-mobilizing effect of glucagon, therefore, may be entirely due to a cAMP-dependent process, although a direct receptor-mediated activation of phospholipase C, as suggested by Wakelam et al. (1986), remains a possibility. The EGF receptor appears to be coupled to phospholipase C, presumably via a G-protein. It is speculated that the mechanism by which cAMP increases Ins(1,4,5)P_3 levels in hepatocytes could either be by phosphorylation and inhibition of inositol phosphate 5-phosphatase or by phosphorylation and facilitation of the coupling between the G-protein and phospholipase C. When protein kinase C is maximally activated by pretreatment of hepatocytes with PMA, the stimulatory effects of phenylephrine, glucagon, 8-Br-cAMP, and EGF on the accumulation of inositol phosphates and increase of cytosolic free Ca^{2+} are largely inhibited. It is doubtful that a protein kinase C–mediated stimulation of the inositol phosphate 5-phosphatase is sufficient to account for this effect since the products of this enzyme step, namely Ins(1,4)P_2 and Ins(1,3,4)P_3, do not accumulate, even transiently. PMA-induced negative feedback of agonist responses probably also involves inhibitory effects at the level of the receptor (Leeb-Lundberg et al., 1985) or the G-protein responsible for coupling receptors to phospholipase C.

Acknowledgments

This work was supported by National Institutes of Health grants DK-15120 and HL-14461.

References

Ashendel, C. L. 1985. The phorbol ester receptor: a phospholipid-regulated protein kinase. *Biochimica et Biophysica Acta.* 822:219–242.

Batty, I. R., S. R. Nahorski, and R. F. Irvine. 1985. Rapid formation of inositol 1,3,4,5-tetrakisphosphate following muscarinic receptor stimulation of rat cerebral cortical slices. *Biochemical Journal.* 232:211–215.

Bellomo, G., S. A. Jewell, H. Thor, and S. Orrenius. 1982. Regulation of intracellular calcium compartmentation: studies with isolated hepatocytes and t-butyl hydroperoxide. *Proceedings of the National Academy of Sciences.* 79:6842–6846.

Berridge, M. J., and R. F. Irvine. 1984. Inositol trisphosphate: a novel second messenger in signal transduction. *Nature.* 312:315–321.

Biden, T. J., and C. B. Wollheim. 1986. Ca^{2+} regulates the inositol tris/tetrakisphosphate pathway in intact and broken preparations of insulin-secreting RIN m5F cells. *Journal of Biological Chemistry.* 261:11931–11934.

Blackmore, P. F., and J. H. Exton. 1985. Mechanisms involved in the actions of calcium-dependent hormones. *In* Biochemical Actions of Hormones. G. Litwack, editor. Academic Press, Inc., New York. 12:215–235.

Blackmore, P. F., and J. H. Exton. 1986. Studies on the hepatic calcium-mobilizing activity of aluminum fluoride and glucagon. *Journal of Biological Chemistry.* 261:11056–11063.

Blackmore, P. F., W. G. Strickland, S. B. Bocckino, and J. H. Exton. 1986. Mechanism of hepatic glucagon synthase inactivation induced by Ca^{2+}-mobilizing hormones. *Biochemical Journal.* 237:235–242.

Bocckino, S. B., P. F. Blackmore, and J. H. Exton. 1985. Stimulation of 1,2-diacylglycerol accumulation in hepatocytes by vasopressin, epinephrine and angiotensin II. *Journal of Biological Chemistry.* 260:14201–14207.

Bosch, F., B. Bouscarel, J. Slaton, P. F. Blackmore, and J. H. Exton. 1986. Epidermal growth factor mimics insulin effects in rat hepatocytes. *Biochemical Journal.* 239:523–530.

Burgess, G. M., P. P. Godfrey, J. S. McKinney, M. J. Berridge, and J. W. Putney, Jr. 1984. The second messenger linking receptor activation to internal Ca^{2+} release in liver. *Nature.* 309:63–66.

Cerdan, S., C. A. Hansen, R. Johanson, T. Inubushi, and J. R. Williamson. 1986. Nuclear magnetic resonance spectroscopic analysis of *myo*inositol 1,3,4,5-tetrakisphosphate. *Journal of Biological Chemistry.* 261:14676–14680.

Charest, R., P. F. Blackmore, B. Berthon, and J. H. Exton. 1983. Changes in free cytosolic

Ca^{2+} in hepatocytes following α_1-adrenergic stimulation: studies on Quin 2-loaded hepatocytes. *Journal of Biological Chemistry.* 258:8769–8773.

Charest, R., V. Prpic, J. H. Exton, and P. F. Blackmore. 1985. Stimulation of inositol trisphosphate formation in hepatocytes by vasopressin, adrenaline and angiotensin II and its relationship to changes in cytosolic free Ca^{2+}. *Biochemical Journal.* 227:79–90.

Cohen, P. 1985. The role of protein phosphorylation in the hormonal control of enzyme activity. *European Journal of Biochemistry.* 151:439–448.

Combettes, L., B. Berthon, A. Binet, and M. Claret. 1986. Glucagon and vasopressin interactions on Ca^{2+} movements in isolated hepatocytes. *Biochemical Journal.* 237:675–683.

Connolly, T. M., W. J. Lawing, Jr., and P. W. Majerus. 1986. Protein kinase C phosphorylates human platelet inositol trisphosphate 5′-phosphomonoesterase, increasing the phosphatase activity. *Cell.* 49:951–958.

Cooper, R. H., K. E. Coll, and J. R. Williamson. 1985. Differential effects of phorbol ester on phenylephrine and vasopressin-induced Ca^{2+} mobilization in isolated hepatocytes. *Journal of Biological Chemistry.* 260:3281–3288.

Corvera, S., K. R. Schwarz, R. M. Graham, and J. A. Garcia-Sainz. 1986. Phorbol esters inhibit α_1-adrenergic effects and decrease the affinity of liver cell α_1-adrenergic receptors for (−)epinephrine. *Journal of Biological Chemistry.* 261:520–526.

Creba, J. A., C. P. Downes, P. T. Hawkins, G. Brewster, R. H. Michell, and C. J. Kirk. 1983. Rapid breakdown of phosphatidylinositol 4-phosphate and phosphatidylinositol 4,5-bisphosphate in rat hepatocytes stimulated by vasopressin and other Ca^{2+} mobilizing hormones. *Biochemical Journal.* 212:733–747.

Daly, J. W. 1984. Forskolin, adenylate cyclase, and cell physiology: an overview. *Advances in Cyclic Nucleotide and Protein Phosphorylation Research.* 17:81–89.

Denton, R. M., and J. G. McCormack. 1985. Ca^{2+} transport by mammalian mitochondria and its role in hormone action. *American Journal of Physiology.* 249:E543–E554.

Downes, P., and R. Michell. 1985. Inositol phospholipid breakdown as a receptor controlled generator of second messengers. *In* Molecular Mechanisms of Transmembrane Signalling. P. Cohen and M. D. Houslay, editors. Elsevier/North-Holland, Amsterdam. 3–56.

Gilman, A. G. 1984. G-proteins and dual control of adenylate cyclase. *Cell.* 36:577–579.

Hansen, C. A., S. Mah, and J. R. Williamson. 1986. Formation and metabolism of inositol 1,3,4,5-tetrakisphosphate in liver. *Journal of Biological Chemistry.* 261:8100–8103.

Heytler, P. 1979. Uncouplers of oxidative phosphorylation. *Methods in Enzymology.* 55:462–472.

Hokin, L. E. 1985. Receptors and phosphoinositide-generated second messengers. *Annual Review of Biochemistry.* 54:205–235.

Horwitz, E. M., W. T. Jenkins, N. M. Hoosein, and R. S. Gurd. 1985. Kinetic identification of a two-state glucagon receptor system in isolated hepatocytes: interconversion of homogeneous receptors. *Journal of Biological Chemistry.* 260:9307–9315.

Horwitz, E. M., R. J. Wyborski, and R. S. Gurd. 1986. Partial agonism in the glucagon receptor system is a consequence of the two-state rat hepatic receptor. *Journal of Biological Chemistry.* 261:13670–13676.

Irvine, R. F., A. J. Letcher, J. P. Heslop, and M. J. Berridge. 1986. The inositol tris/

tetrakisphosphate pathway—demonstration of Ins(1,4,5)P_3 3-kinase activity in animal tissues. *Nature.* 320:631–634.

Johnson, R. M., P. A. Connelly, R. B. Sisk, B. F. Pobiner, E. L. Hewlett, and J. C. Garrison. 1986. Pertussis toxin or phorbol 12-myristate 13 acetate can distinguish between epidermal growth factor and angiotensin-stimulated signals in hepatocytes. *Proceedings of the National Academy of Sciences.* 83:2032–2036.

Joseph, S. K., K. E. Coll, A. P. Thomas, R. Rubin, and J. R. Williamson. 1985. The role of extracellular Ca^{2+} in the response of the hepatocyte to Ca^{2+}-dependent hormones. *Journal of Biological Chemistry.* 260:12508–12515.

Joseph, S. K., A. P. Thomas, R. J. Williams, R. F. Irvine, and J. R. Williamson. 1984. *Myo*-inositol 1,4,5-trisphosphate: a second messenger for the hormonal mobilization of intracellular Ca^{2+} in liver. *Journal of Biological Chemistry.* 259:3077–3081.

Joseph, S. K., and R. J. Williams. 1985. Subcellular localization and properties of the enzymes hydrolysing inositol polyphosphates in rat liver. *FEBS Letters.* 180:150–154.

Joseph, S. K., and J. R. Williamson. 1983. The origin, quantitation, and kinetics of intracellular calcium mobilization by vasopressin and phenylephrine in hepatocytes. *Journal of Biological Chemistry.* 258:10425–10432.

Joseph, S. K., and J. R. Williamson. 1986. Characteristics of inositol trisphosphate-mediated Ca^{2+} release from permeabilized hepatocytes. *Journal of Biological Chemistry.* 261:14658–14664.

Kleineke, J., and H.-D. Soling. 1985. Mitochondrial and extramitochondrial Ca^{2+} pools in the perfused rat liver: mitochondria are not the origin of calcium mobilized by vasopressin. *Journal of Biological Chemistry.* 260:1040–1045.

Kraus-Friedmann, N. 1984. Hormonal regulation of gluconeogenesis. *Physiological Reviews.* 64:170–259.

Kraus-Friedmann, N. 1986. Effects of glucagon and vasopressin on hepatic Ca^{2+} release. *Proceedings of the National Academy of Sciences.* 83:8943–8946.

Leach, R. P., S. B. Shears, C. J. Kirk, and M. A. Titheradge. 1986. Changes in free cytosolic calcium and accumulation of inositol phosphates in isolated hepatocytes by [Leu]enkephalin. *Biochemical Journal.* 238:537–542.

Leeb-Lundberg, L. M. F., S. Cotecchia, J. W. Lomasney, J. F. DeBernardis, R. J. Lefkowitz, and M. G. Caron. 1985. Phorbol esters promote α_1-adrenergic receptor phosphorylation and receptor uncoupling from inositol phospholipid metabolism. *Proceedings of the National Academy of Sciences.* 82:5651–5655.

Litosch, I., and J. N. Fain. 1986. Regulation of phosphoinositide breakdown by guanine nucleotides. *Life Sciences.* 39:187–194.

Lynch, C. J., R. Charest, S. B. Bocckino, J. H. Exton, and P. F. Blackmore. 1985. Inhibition of hepatic α_1-adrenergic effects and binding by phorbol myristate acetate. *Journal of Biological Chemistry.* 260:2844–2851.

Mauger, J.-P., and M. Claret. 1986. Mobilization of intracellular calcium by glucagon and cyclic AMP analogues in isolated rat hepatocytes. *FEBS Letters.* 195:106–110.

Mauger, J.-P., J. Poggioli, and M. Claret. 1985. Synergistic stimulation of the Ca^{2+} influx in rat hepatocytes by glucagon and the Ca^{2+}-linked hormones vasopressin and angiotensin II. *Journal of Biological Chemistry.* 260:11635–11642.

Mauger, J.-P., J. Poggioli, F. Guesdon, and M. Claret. 1984. Noradrenaline, vasopressin and angiotensin increase Ca^{2+} influx by opening a common pool of Ca^{2+} channels in isolated rat liver cells. *Biochemical Journal.* 221:121–127.

McCormack, J. G. 1985. Studies on the activation of rat liver pyruvate dehydrogenase and 2-oxoglutarate dehydrogenase by adrenaline and glucagon. *Biochemical Journal.* 231:597–608.

Molina y Vedia, L. M., and E. G. Lapetina. 1986. Phorbol 12,13-dibutyrate and 1-oleoyl-2-acetyldiglycerol stimulate inositol trisphosphate dephosphorylation in human platelets. *Journal of Biological Chemistry.* 261:10493–10495.

Nishizuka, Y. 1986. Studies and perspectives of protein kinase C. *Science.* 233:305–312.

Nishizuka, Y., Y. Takai, A. Kishimoto, V. Kikkawa, and K. Kaibuchi. 1984. Phospholipid turnover in hormone action. *Recent Progress in Hormone Research.* 40:301–341.

Northup, J. K. 1985. Overview of the guanine nucleotide regulatory protein systems, N_s and N_i, which control adenylate cyclase activity in plasma membranes. *In* Molecular Mechanisms of Transmembrane Signalling. P. Cohen and M. D. Houslay, editors. Elsevier/North-Holland, Amsterdam. 91–116.

Pobiner, B. F., E. L. Hewlitt, and J. C. Garrison. 1985. Role of N_i in coupling angiotensin receptors to inhibition of adenylate cyclase in hepatocytes. *Journal of Biological Chemistry.* 260:16200–16209.

Poggioli, J., J.-P. Mauger, and M. Claret. 1986. Effect of cyclic AMP-dependent hormones and Ca^{2+}-mobilizing hormones on the Ca^{2+} influx and polyphospho-inositide metabolism in isolated rat hepatocytes. *Biochemical Journal.* 235:663–669.

Prentki, M., B. E. Corkey, and F. M. Matschinsky. 1985. Inositol 1,4,5-trisphosphate and the endoplasmic reticulum Ca^{2+} cycle of a rat insulinoma cell line. *Journal of Biological Chemistry.* 260:9185–9190.

Rasmussen, H., and P. Q. Barrett. 1984. Calcium messenger system: an integrated view. *Physiological Reviews.* 64:938–984.

Rasmussen, H. K., K. Kojima, W. Kojima, W. Zawalich, and W. Apfeldorf. 1984. Calcium as intracellular messenger: sensitivity modulation, C-kinase pathway, and sustained cellular response. *Advances in Cyclic Nucleotide Research.* 18:159–193.

Reinhart, P. H., W. M. Taylor, and F. L. Bygrave. 1984. The role of calcium ions in the mechanism of action of α-adrenergic agonists in liver. *Biochemical Journal.* 223:1–13.

Shears, S. B., and C. J. Kirk. 1984. Determination of mitochondrial calcium content in hepatocytes by a rapid cellular fractionation technique. *Biochemical Journal.* 220:417–421.

Sibley, D. R., and R. J. Lefkowitz. 1985. Molecular mechanisms of receptor desensitization using the β-adrenergic receptor-coupled adenylate cyclase system as a model. *Nature.* 317:124–129.

Sistare, F. D., R. A. Pickining, and R. C. Haynes, Jr. 1985. Sensitivity of the response of cytosolic calcium in Quin2-loaded rat hepatocytes to glucagon, adenine nucleosides and adenine nucleotides. *Journal of Biological Chemistry.* 260:12744–12747.

Staddon, J. M., and R. G. Hansford. 1986. 4β-Phorbol 12-myristate 13-acetate attenuates the glucagon-induced increase in cytoplasmic free Ca^{2+} concentration in isolated rat hepatocytes. *Biochemical Journal.* 238:737–743.

Storey, D. J., S. B. Shears, C. J. Kirk, and R. H. Michell. 1984. Stepwise enzymatic dephosphorylation of inositol 1,4,5-trisphosphate to inositol in liver. *Nature.* 312:374–376.

Streb, H., E. Bayerdorffer, W. Haase, R. F. Irvine, and I. Schulz. 1984. Effect of inositol

1,4,5-trisphosphate on isolated subcellular fractions of rat pancreas. *Journal of Membrane Biology.* 81:241–253.

Thomas, A. P., J. Alexander, and J. R. Williamson. 1984. Relationship between inositol polyphosphate production and the increase of cytosolic free Ca^{2+} induced by vasopressin in isolated hepatocytes. *Journal of Biological Chemistry.* 259:5574–5584.

Thomas, A. P., J. S. Marks, K. E. Coll, and J. R. Williamson. 1983. Quantitation and early kinetics of inositol lipid changes induced by vasopressin in isolated and cultured hepatocytes. *Journal of Biological Chemistry.* 258:5716–5725.

Tsien, R. Y., T. Pozzan, and T. J. Rink. 1982. Calcium homeostasis in intact lymphocytes: cytoplasmic free calcium monitored with a new, intracellularly trapped fluorescent indicator. *Journal of Cell Biology.* 94:325–334.

Uhing, R. J., V. Prpic, H. Jiang, and J. H. Exton. 1986. Hormone-stimulated polyphosphoinositide breakdown in rat liver plasma membranes. *Journal of Biological Chemistry.* 261:2140–2146.

Wakelam, M. J. O., G. J. Murphy, V. J. Hruby, and M. D. Houslay. 1986. Activation of two signal-transduction systems in hepatocytes by glucagon. *Nature.* 323:68–71.

Williamson, J. R. 1986. Role of inositol lipid breakdown in the generation of intracellular signals. *Hypertension.* 8:II140–II156.

Williamson, J. R., R. H. Cooper, and J. B. Hoek. 1981. The role of calcium in the hormonal regulation of liver metabolism. *Biochimica et Biophysica Acta.* 639:243–295.

Williamson, J. R., R. H. Cooper, S. K. Joseph, and A. P. Thomas. 1985. Inositol trisphosphate and diacylglycerol as intracellular second messengers in liver. *American Journal of Physiology.* 248:C203–C216.

Williamson, J. R., and C. A. Hansen. 1987. Signalling systems in stimulus-response coupling. *In* Biochemical Actions of Hormones. G. Litwack, editor. Academic Press, Inc., New York. In press.

Chapter 8

The Role of Phosphatidylinositides in Stimulus-Secretion Coupling in the Exocrine Pancreas

I. Schulz, S. Schnefel, H. Banfić, F. Thévenod, T. Kemmer, and L. Eckhardt

Max-Planck-Institut für Biophysik, Frankfurt/Main, Federal Republic of Germany

Introduction

Cellular activation of a wide variety of cell types is accompanied by enhanced metabolism of phosphatidylinositides. Both the hydrolysis products diacylglycerol (DG) and inositol 1,4,5-trisphosphate (IP_3) are intracellular messengers for cellular responses such as secretion, metabolism, contraction, phototransduction, and proliferation (Berridge and Irvine, 1984).

While IP_3 leads to Ca^{2+} release from intracellular stores (Streb et al., 1983) and a consequent rise in the cytosolic free Ca^{2+} concentration by which Ca^{2+} and Ca^{2+}-binding, protein-dependent protein kinases are activated, DG activates protein kinase C. In a further step in this cascade of events, protein kinases phosphorylate target proteins involved in the ultimate cell response.

In acinar cells of the exocrine pancreas, one important pathway by which enzyme$^-$, $NaCl^-$, and fluid secretion are stimulated involves the activation of phospholipase C by muscarinic and peptidergic receptors, which leads to a breakdown of phosphatidylinositol 4,5-bisphosphate (PIP_2). Another pathway that leads to enzyme secretion is governed by the stimulation of adenylate cyclase, a rise of cyclic AMP, and the activation of protein kinase A in the same type of cell, whereas in duct cells this pathway is coupled to $NaHCO_3$ and fluid secretion.

In this chapter, some steps in intracellular signal transduction will be discussed and illustrated by the pancreatic acinar cell. These include receptor coupling to phospholipase C, generation of IP_3, Ca^{2+} release from intracellular Ca^{2+} pools, Ca^{2+} uptake mechanisms into Ca^{2+} stores, and direct triggering of exocytosis by activators of protein kinases, such as Ca^{2+} and phorbol ester for protein kinase C and cAMP for protein kinase A. Experiments were done on isolated acinar cells and acini obtained by treatment of pancreatic tissue with collagenase (Kimura et al., 1986). In some experiments, cells were permeabilized with digitonin (Kimura et al., 1986) or by washing in Ca^{2+}-free media (Streb and Schulz, 1983). Ca^{2+} uptake and Ca^{2+} release from permeabilized cells was measured by monitoring free Ca^{2+} concentrations in the incubation media using a Ca^{2+} electrode (Streb and Schulz, 1983).

IP_3 production was estimated after prelabeling of cells with [^{3}H]*myo*-inositol and separation of inositides released into the incubation medium by means of anion-exchange columns (Streb et al., 1985).

Is a GTP-binding Protein Involved in the Activation of Phospholipase C?

Guanine nucleotide–binding proteins (N- or G-proteins) play an important role in regulatory processes such as hormonal activation of adenylate cyclase (N_s and N_i) or in light-catalyzed activation of cGMP phosphodiesterase in rod disk membranes (transducin) (Rodbell, 1980; Manning and Gilman, 1983; Levitzki, 1984). Recently, evidence has been obtained that GTP-binding proteins are also involved in the coupling of Ca^{2+}-mobilizing receptors with phospholipase C, which leads to hydrolysis of PIP_2 (Cockcroft and Gomperts, 1985). The weakly hydrolyzable GTP analogue GTPγS has been shown to activate phospholipase C in many systems (Gomperts, 1983; Haslam and Davidson, 1984; Cockcroft and Gomperts, 1985; Litosch et al., 1985; Wallace and Fain, 1985; Uhing et al., 1986). In permeabilized pancreatic acinar cells, the effect of cholecystokinin-pancreozymin octapeptide (CCK-OP) on Ca^{2+} release is increased when GTPγS is added to cells together with the hormone (Fig. 1). As shown in Fig. 2, the addition of GTPγS to leaky acinar cells by itself increases IP_3 production. Ca^{2+} uptake is not affected by GTPγS. However, when steady state is reached, Ca^{2+} release occurs (Fig. 2). The following CCK-OP effect is reduced, probably by an alteration of the GTP-binding protein in the presence of GTP or its analogues that leads to a decrease in the receptor affinity to the agonist (Sokolovsky et al., 1980; Koo et al., 1983; Levitzki, 1984).

Fluoride in the presence of Al^{3+} has been reported to activate N_s, N_i, and transducin (Sternweis and Gilman, 1982; Kanaho et al., 1985). Recent studies on the GTP-binding subunit of transducin suggests that AlF_4^- can substitute for the γ-phosphate of GTP because of structural similarities between AlF_4^- and PO_4^{3-} (Bigay et al., 1985). It has been shown in isolated hepatocytes that AlF_4^- is able to mimic the effects of the Ca^{2+}-mobilizing hormones on PIP_2 hydrolysis, IP_3 formation, and

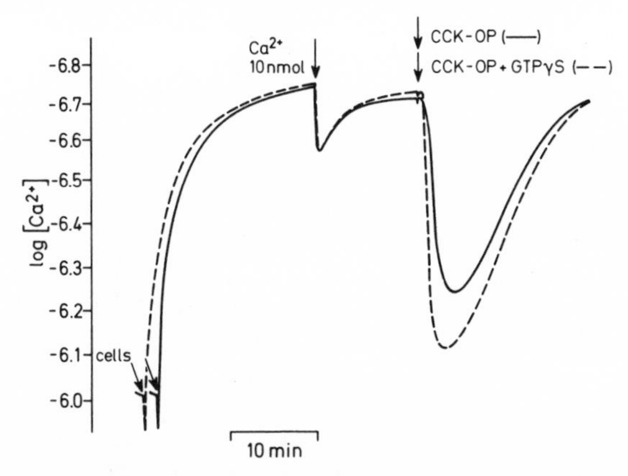

Figure 1. Effects of CCK-OP (0.3 μM) and GTPγS (50 μM) on Ca^{2+} release from isolated permeabilized pancreatic acinar cells. The plasma membranes of isolated cells were permeabilized by washing cells with a nominally Ca^{2+}-free buffer containing (mM): 135 KCl, 10 HEPES, 1 $MgCl_2$, 0.1 mg/ml trypsin inhibitor, pH 7.4 adjusted with KOH. Permeabilized cells were incubated in a buffer containing (mM): 110 KCl, 7 $MgCl_2$, 5 K_2ATP, 10 creatine phosphate, 5 K_2-succinate, 5 K-pyruvate, 8 U/ml creatine kinase, 25 HEPES, pH 7.4, at 25°C. The free Ca^{2+} concentration was recorded with a Ca^{2+}-specific macroelectrode as described (Streb and Schulz, 1983). A Ca^{2+} calibration pulse of 10 nmol (final concentration, 3.3 μM) was added where indicated.

Ca^{2+} mobilization (Blackmore et al., 1985). When fluoride (10 mmol/liter) is added to permeabilized pancreatic acinar cells, a similar increase in IP_3 production and Ca^{2+} release is seen, as with GTPγS (Fig. 3).

GTP-binding proteins of the adenylate cyclase system are affected by cholera and pertussis toxin. Cholera toxin ADP-ribosylates the α-subunit of N_s and thereby leads to its dissociation from the heterotrimer (αβγ) so that it can interact with the catalytic unit of the enzyme. Pertussis toxin inhibits N_i activation by ADP-ribosylating α_i, thereby inhibiting the exchange of GDP for GTP and consequently the activation of N_i (Levitzki, 1984). It has been shown that pertussis toxin also inhibits phospholipase C activation in leukocytes (Bradford and Rubin, 1985; Nakamura and Ui, 1985; Verghese et al., 1986) and mast cells (Nakamura and Ui, 1985), whereas in epithelial cells, inhibition of phospholipase C by pertussis toxin has not been observed. Cholera toxin was shown to ADP-ribosylate the same N-protein that is also substrate for pertussis-mediated ADP ribosylation in leukocytes (Verghese et al., 1986) and to inhibit agonist-activated phospholipase C by a cAMP-independent mechanism (Okajima and Ui, 1984). In permeabilized pancreatic acinar cells, preincubation with activated cholera toxin for 30 min inhibits subsequent hormone effects on both IP_3 production and Ca^{2+} release (Fig. 4). This effect of cholera toxin does not appear to be mediated by cAMP, which is increased because of activation of adenylate cyclase by cholera toxin, since the addition of either cAMP or 8-Br-cAMP to permeabilized cells does not mimic the effect of cholera toxin on IP_3 production or on Ca^{2+} release (Schnefel, S., H. Banfić, G.

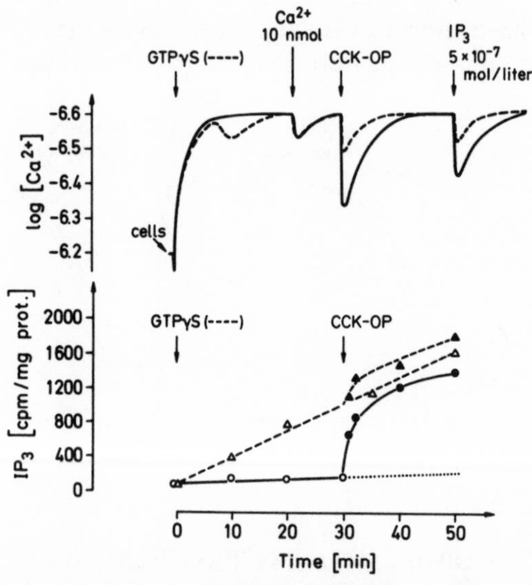

Figure 2. Effect of GTPγS (50 μM), CCK-OP (0.3 μM), and IP_3 (0.5 μM) on Ca^{2+} release (upper panel) and of GTPγS and CCK-OP on IP_3 production (lower panel). For determination of IP_3 production, inositol phospholipids were prelabeled by addition of 100 μCi/ml (~6 μM) [2-^{3}H]*myo*-inositol during the last hour of collagenase digestion. Cells were then washed three times with the "Ca^{2+}-free buffer" supplemented with 1 mM unlabeled *myo*-inositol and once again without *myo*-inositol. At the given times, aliquots of the cell suspension were removed from the incubation medium and mixed with the same volume of ice-cold 20% wt/vol trichloroacetic acid, and the samples were centrifuged for 4 min at 12,000 rpm. The supernatant was analyzed for inositol phosphates, using anion-exchange columns as described previously (Streb et al., 1985).

Schultz, and I. Schulz, manuscript submitted for publication). We therefore assume that the effect of cholera toxin on IP_3 production is due to a direct effect of cholera toxin on an N-protein that couples the hormone receptor to phospholipase C. Another possibility is an effect of N_s on adenylate cyclase and direct inhibition of phospholipase C by α_s. Inhibition of phospholipase C by the $\beta\gamma$ complex of the N_s-protein that is liberated from N_s by cholera toxin-induced ADP ribosylation of the α_s-subunit is another, but less likely, possibility to explain the data obtained.

IP_3-mediated Ca^{2+} Release

IP_3, when added to permeabilized cells (Streb et al., 1983) or isolated endoplasmic reticulum (ER) (Streb et al., 1984), releases Ca^{2+}. Since only 30–40% of the total

Ca^{2+} that is taken up by the ER can be released by IP_3, it is evident that only part of the ER is IP_3 sensitive. It is also possible that this part is a subcompartment or does not belong to the ER. So far, this IP_3-sensitive organelle has not been clearly distinguished from the ER, but it does not appear to be mitochondria, zymogen granules, or plasma membranes (Streb et al., 1984).

The mechanism of IP_3 action is not yet clear. High-affinity binding sites for IP_3 have been identified in microsomes from different cells (Baukal et al., 1985;

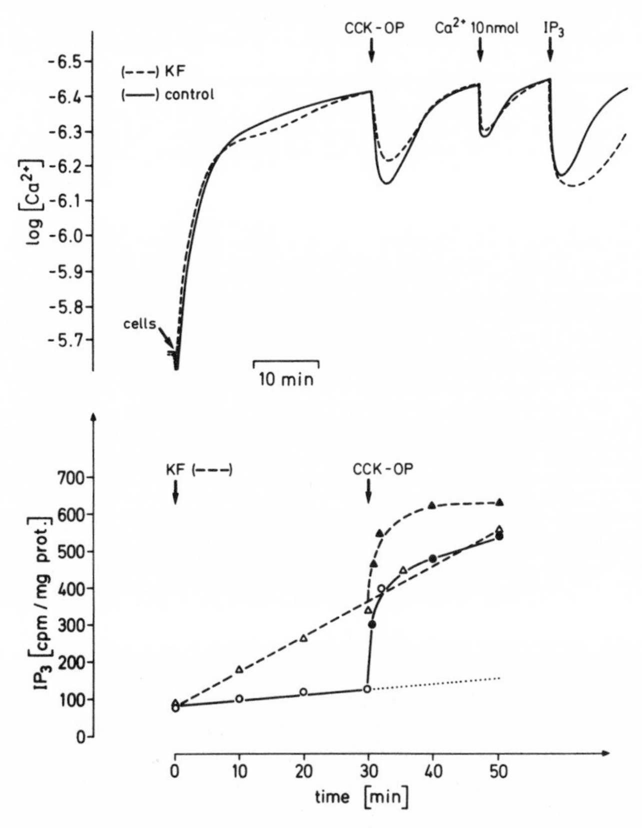

Figure 3. Effects of KF, CCK-OP, and IP_3 on Ca^{2+} release and of KF and CCK-OP on IP_3 production in isolated permeabilized pancreatic acinar cells. Permeabilized cells were incubated in a standard medium in the absence (continuous lines, open circles) or presence (dashed lines, open triangles) of KF (10 mM). Where indicated, $CaCl_2$ (10 nmol/3 ml to a final concentration of 3.3 μmol/liter), CCK-OP (0.3 μM), or IP_3 (0.5 μM) was added.

Spät et al., 1986*a*, *b*) and it is assumed that IP_3 opens a Ca^{2+} channel and that electrogenic Ca^{2+} release can take place only if a counterion is present (Muallem et al., 1985). Recent evidence suggests that GTP hydrolysis might be involved in the mechanism of IP_3 action, since in some tissues the IP_3 effect on Ca^{2+} release can be only seen in the presence of GTP, and not in the presence of GTP analogues that are weakly hydrolyzable, such as GTPγS and guanylylimidodiphosphate (GPPNHP) (Dawson, 1985; Dawson et al., 1986; Gill et al., 1986; Ueda et al., 1986). As shown in Fig. 2, in our preparation, the effect of exogeneously added IP_3 on Ca^{2+} release was inhibited even in the presence of GTPγS. It is therefore not very likely that an N-protein is involved in the action of IP_3, and GTP-dependent phosphorylation of a regulatory protein seems to be more likely.

Ca^{2+} Uptake and Ca^{2+} Extrusion Mechanisms in the ER and the Plasma Membrane

After Ca^{2+} release from the ER, the cytosolic free Ca^{2+} concentration rises and then declines to a lower level within minutes. If Ca^{2+} is present in the extracellular medium, Ca^{2+} influx into cells takes place (Kondo and Schulz, 1976) and the cytosolic Ca^{2+} concentration stays at a higher level as compared with the absence

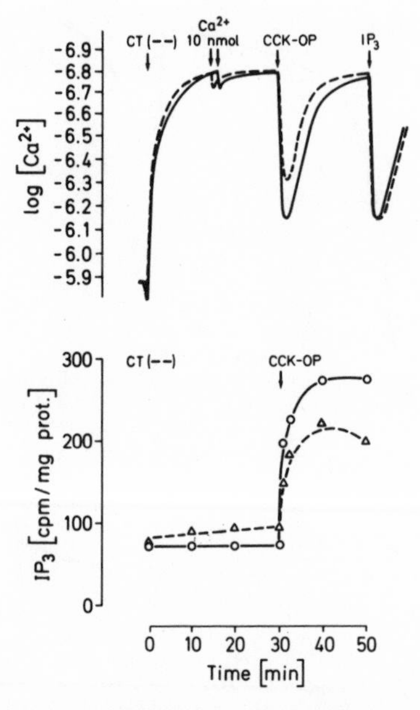

Figure 4. Effects of cholera toxin (CT) (40 μg/ml) on CCK-OP– and IP_3–induced Ca^{2+} release and on IP_3 production in permeabilized pancreatic acinar cells. Permeabilized cells were incubated in the absence (circles) or presence (triangles) of cholera toxin that had been preactivated with 20 mM dithiothreitol for 30 min at 37°C. In addition, the incubation medium contained 1 mM NAD^+. Where indicated, $CaCl_2$ (3.3 μmol/liter added as 10 nmol/3 ml), CCK-OP (0.3 μM), or IP_3 (0.5 μM) was added.

of extracellular Ca^{2+} (Ochs et al., 1983; Schulz et al., 1985). This observation agrees with the previous conclusion that the initial event of secretagogue action is Ca^{2+} release from an intracellular store that leads to a burst of secretion, which ceases if Ca^{2+} is absent in the extracellular medium. However, if Ca^{2+} is present, secretagogue-induced Ca^{2+} influx leads to sustained secretion (Petersen and Ueda, 1976). The cytosolic free Ca^{2+} concentration is regulated by intracellular Ca^{2+} uptake mechanisms as well as by Ca^{2+} extrusion over the plasma membrane. Although mitochondria can take up Ca^{2+}, they seem to act as an emergency Ca^{2+} pool if the cell has an overflow of Ca^{2+}, whereas at physiological conditions, mitochondrial Ca^{2+} is rather low (Somlyo et al., 1985). Fine regulation to ~100 nmol/liter free cytosolic Ca^{2+} concentration is achieved by the ER and the plasma membrane but not by mitochondria (Streb and Schulz, 1983). In the presence of mitochondrial inhibitors, Ca^{2+} uptake is slower, as compared with the control; however, a steady state free Ca^{2+} concentration of ~2×10^{-7} mol/liter is reached (Streb and Schulz, 1983).

If nonmitochondrial Ca^{2+} uptake is inhibited by the ATPase inhibitor vanadate, the initial rate of Ca^{2+} uptake is the same as in control cells; however, the steady state free Ca^{2+} concentration is not reached. Ca^{2+} uptake into the nonmitochondrial Ca^{2+} pool was further investigated using cellular subfractions isolated by differential centrifugation of the cell homogenate. We found that Ca^{2+} uptake into isolated ER had the same properties as that into permeabilized cells concerning the dependences on Ca^{2+}, Mg^{2+}, cation, anion, pH, and ATP (Bayerdörffer et al.,

Table I
Transport Characteristics of the (Ca^{2+} + Mg^{2+})-ATPases in Vesicles from Plasma Membrane and Rough ER*

For Ca^{2+} uptake	Plasma membrane	Rough ER
Ca^{2+}		
Maximum	10	2
K_m	0.88	0.54‡
Mg^{2+}		
Maximum	3,000	200
K_m	30	8
ATP		
Maximum	5,000	1,000
K_m	2,000	10
Optimum pH	6.5–7.0	6.5–7.0
Substrate specificity	ATP	ATP ≫ GTP > UTP > ITP > CTP
Specificity for Mg^{2+}	$Mg^{2+} > Mn^{2+} \gg Zn^{2+}$	$Mg^{2+} \gg Mn^{2+}$
Cation dependence	$K^+ > Rb^+ > Na^+ > Li^+ >$ choline$^+$	$Rb^+ \geq K^+ \geq Na^+ > Li^+ >$ choline$^+$
Anion dependence	$Cl^- \geq Br^- \geq I^- > SCN^- > NO_3^- >$ isethionate$^- >$ gluconate$^- >$ cyclamate$^- > SO_4^{2-} \geq$ glutarate^{2-}	$Cl^- > Br^- >$ gluconate$^- > SO_4^{2-} \geq NO_3^- > I^- >$ cyclamate$^- \geq SCN^-$
Oxalate dependence	No	Yes

* Summarized from the results of Bayerdörffer et al. (1985).
‡ Ca^{2+} uptake decreases above 2 μmol/liter Ca^{2+}.

1984). Furthermore, a Ca^{2+}-ATPase was localized in the ER that also showed Ca^{2+}, cation, and anion dependences in its phosphorylation-dephosphorylation cycle similar to Ca^{2+} uptake, which indicates that this Ca^{2+}-ATPase is the Ca^{2+} pump that adjusts the cytosolic free Ca^{2+} concentration to a low cytosolic level (Imamura and Schulz, 1985). However, in intact cells, the final adjustment of the cytosolic free Ca^{2+} concentration is performed by Ca^{2+}-extrusion mechanisms in the plasma membrane. We have therefore also studied both MgATP-dependent Ca^{2+} transport and an electrochemical gradient-coupled Na^+/Ca^{2+} exchange mechanism in isolated plasma membranes of pancreatic acinar cells (Bayerdörffer et al., 1985*a*, *b*). As shown in Table I, MgATP-driven Ca^{2+} transport in vesicles of plasma membranes differs in many respects from that in the ER. Most striking is the anion dependence. Whereas the lipophilic anion SCN^- augments MgATP-driven Ca^{2+}

uptake into plasma membrane vesicles, Ca^{2+} uptake into endoplasmic vesicles is inhibited.

In the presence of weakly permeant anions such as gluconate, cyclamate, and SO_4^{2-}, MgATP-driven Ca^{2+} uptake into vesicles from the plasma membrane is inhibited. This indicates that Ca^{2+} transport in the plasma membrane is electro-

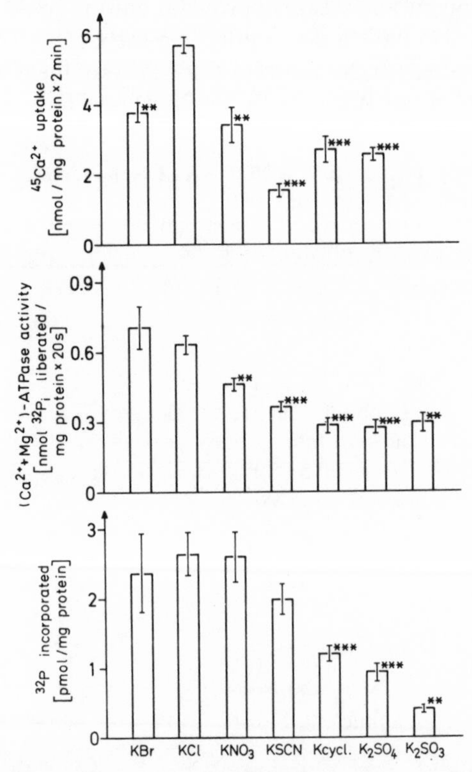

Figure 5. Effect of different anions on $^{45}Ca^{2+}$ uptake, (Ca^{2+} + Mg^{2+})-ATPase activity, and ^{32}P incorporation of the (Ca^{2+} + Mg^{2+})-ATPase intermediate in vesicles from pancreatic ER. A fraction with ER had been obtained by differential centrifugation at 27,000 *g* for 15 min. Ca^{2+} uptake was measured in a buffer containing 130 mM K^+ salt with anions as indicated, 30 mM HEPES buffer (pH 7 adjusted with Tris) in the presence of 0.01 mM antimycin A, and 0.05 mM oligomycin using $^{45}Ca^{2+}$. Ca^{2+} uptake was initiated by adding Tris-ATP (1 mmol/liter). At given times, samples were filtered through cellulose nitrate filters (pore size, 0.65 μm). The values for ATP-driven Ca^{2+} transport into vesicles were calculated as the difference between Ca^{2+} content in the presence and absence of ATP as described (Bayerdörffer et al., 1984). (Ca^{2+} + Mg^{2+})-ATPase activity and ^{32}P incorporation into the 100-kD (Ca^{2+} + Mg^{2+})-ATPase intermediate were measured in the presence of 130 mM K^+ salt of the indicated anions in the presence of [^{32}P]ATP (5 μmol/liter). The reaction was stopped with 10% trichloroacetic acid and the sample was centrifuged at 2,250 *g* for 5 min. The pellet was dissolved with lithium dodecyl sulfate (LDS) and submitted to LDS-polyacrylamide gel electrophoresis. The phosphorylated 100-kD protein was excised from the gel and radioactivity was counted as described (Imamura and Schulz, 1985). The supernatant was mixed with charcoal and centrifuged at 2,500 *g* for 10 min, and the $^{32}P_i$ liberated was counted and determined in the supernatant by liquid scintillation counting. (Ca^{2+} + Mg^{2+})-ATPase activity was estimated by subtracting the ATPase activity in the absence of Ca^{2+} from that in the presence of Ca^{2+} as described (Imamura and Schulz, 1985).

genic, the charge of transported Ca^{2+} being compensated by lipophilic anions. Indeed, under conditions at which Ca^{2+} transport was studied in the presence of electrochemical K^+ gradients, the electrogenicity of Ca^{2+} transport in plasma membranes has been revealed (Bayerdörffer et al., 1985*a*).

In the ER, a clear distinction between electrogenic and neutral Ca^{2+} uptake in the presence of permeant and weakly permeant anions could not be made since it appears that anions also have a direct effect on the Ca^{2+}-ATPase molecule. Thus, Cl ions stimulate, whereas SCN^- inhibits, Ca^{2+} uptake, the (Ca^{2+} + Mg^{2+})-ATPase, and phosphorylation of the 100-kD (Ca^{2+} + Mg^{2+})-ATPase intermediate (Fig. 5).

Involvement of H Ions in Ca^{2+} Uptake into the ER

An unexpected observation on the effect of inhibitors of Ca^{2+} uptake in the ER was the finding that protonophores such as nigericin and carbonyl cyanide *m*-chlorophenyl-hydrazone (CCCP), as well as the H^+-ATPase inhibitors *N*-ethylmaleimide (NEM) , dio-9, and 7-chloro-4-nitro-2,1,3-benzoxadiazole (NBD-Cl) inhib-

Table II
Effects of Ionophores and Inhibitors on MgATP-driven $^{45}Ca^{2+}$ Uptake, (Ca^{2+} + Mg^{2+})-ATPase Activity, and H^+ Transport into Isolated Vesicles from Pancreatic Rough ER

Inhibitor	$^{45}Ca^{2+}$ uptake	(Ca^{2+} + Mg^{2+})-ATPase activity	H^+ transport
	%	%	%
Control	100	100	100
CCCP (10^{-5} M)	78±7 (5)	124±7 (3)	0 (5)
Nigericin (10^{-5} M)	76±5 (3)	103±8 (3)	0 (29)
A23187 (2 × 10^{-5} M)	4±1 (4)	153±45 (3)	0 (1)
NBD-Cl (10^{-5} M)	65 (2)	125±6 (2)	0 (2)
NEM (5 × 10^{-4} M)	55±17 (3)	30±3 (4)	0 (2)
Dio 9 (0.25 mg/ml)	3±1 (4)	138±12 (4)	21±10 (3)
Vanadate (2 × 10^{-3} M)	10±2 (10)	15±4 (3)	98±25 (4)

$^{45}Ca^{2+}$ uptake, (Ca^{2+} + Mg^{2+})-ATPase activity, and H^+ transport were determined as described in the legends to Figs. 5 and 6.

ited Ca^{2+} uptake into the ER (Table II). These results led us to assume that an inside-to-outside–directed H^+ gradient over the endoplasmic membrane might be involved in Ca^{2+} uptake. We therefore looked for the presence of an H^+ pump in endoplasmic membrane vesicles. As shown in Fig. 6, H^+ uptake, as visualized by the acridine orange method (Rottenberg and Lee, 1975), takes place in the presence of MgATP. The H^+ gradient formed can be dissipated by protonophores. Inhibitors of H^+-ATPase and H^+ uptake decreased intravesicular acidification to various degrees, NBD-Cl being the most effective. Vanadate, which blocks Ca^{2+}-ATPase and Ca^{2+} uptake into the ER (Streb and Schulz, 1983; Imamura and Schulz, 1985), had no effect on H^+ transport (Table II). Since Ca^{2+}-ATPase is stimulated by the protonophore CCCP, whereas Ca^{2+} uptake is inhibited (Table II), the easiest assumption is that there is a pH-regulated Ca^{2+} channel present in the ER membrane. At a pH that can be generated by the H^+ pump (~pH 6) (Thévenod, F., and I. Schulz, manuscript in preparation), the Ca^{2+} channel is closed, whereas at higher pH, it opens up. Because of a Ca^{2+} leak at alkaline pH inside the vesicle, the Ca^{2+} pump is not as efficient as at a lower pH and Ca^{2+} uptake is decreased. However, under these conditions of dissipated Ca^{2+} gradient, Ca^{2+}-ATPase activity, as measured in closed vesicles, is increased. Evidence that both Ca^{2+} and H^+ uptake occur in the same pool that is also sensitive to IP_3 comes from the observation that the effect of IP_3 on Ca^{2+} release is diminished after protonophore addition (Thévenod, F., and I. Schulz, manuscript in preparation). These data indicate that Ca^{2+} is released from the same Ca^{2+} pool into which H ions are taken up and regulate a Ca^{2+} leak pathway. Whether this pathway is the same as that by which IP_3-induced Ca^{2+} release also occurs remains to be determined.

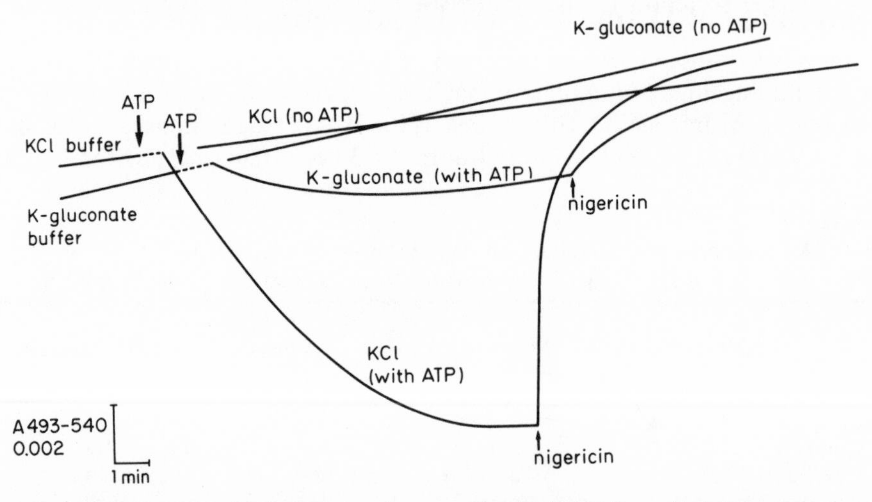

Figure 6. MgATP-driven H^+ transport into rough ER (27,000-*g* pellet) from rat pancreas as measured by absorbance changes due to acridine orange uptake. The incubation buffer contained (mM): 100 mannitol, 100 KCl or K-gluconate, 5 HEPES, 5 $MgSO_4$ (pH 7.0 adjusted with Tris), 0.006 acridine orange. K_2ATP (1.5 mM) was added to the medium where indicated; the pH gradient was dissipated by addition of the protonophore CCCP (10^{-5} M). The difference in absorbance at 493–540 nm was measured in a dual-wavelength spectrophotometer (Thévenod, F., and I. Schulz, manuscript in preparation).

Involvement of Ca^{2+} and Protein Kinases in Secretion from the Exocrine Pancreas

Both hydrolysis products of phosphatidylinositol-4,5-trisphosphate, IP_3 and DG, are important intracellular messengers in the triggering of exocytosis. IP_3-induced Ca^{2+} release from the ER leads to a transient rise in cytosolic free Ca^{2+}, which initiates secretion; for sustained secretion, Ca^{2+} influx into the cell is necessary (Petersen and Ueda, 1976) to keep the cytosolic free Ca^{2+} at a slightly elevated level (Ochs et al., 1983; Schulz et al., 1985). In this second phase, the rise of DG and the activation of protein kinase C are probably the dominant pathways involved in secretion (Rasmussen and Barrett, 1984). If the Ca^{2+} ionophore A23187 is added to isolated pancreatic acinar cells, which elevates cytosolic free Ca^{2+} to the levels obtained with secretagogues, a short burst of secretion occurs that ceases within minutes (Fig. 7). When phorbol ester, a protein kinase C activator, is added, a slow secretory response can be elicited. However, if both A23187 and phorbol ester are added together, a response is obtained that mimics the effect of secretagogues. Initially, secretion increases at a rate similar to that obtained in the presence of A23187. After 10–20 min, the rate decreases and secretion continues at a lower rate, which is similar to that obtained in the presence of phorbol 12-myristate 13-acetate (TPA) (Fig. 7).

These observations indicate that both Ca^{2+} and the activation of protein kinase C are involved in the triggering of exocytosis by secretagogues that act via phospholipase C. The other pathway mediated by cAMP is also dependent on Ca^{2+}. In permeabilized acinar cells, an increase in the free Ca^{2+} concentration in the surrounding medium increases enzyme secretion very little. In the presence of TPA, enzyme secretion is further increased. Similarly, cAMP also shows Ca^{2+}-

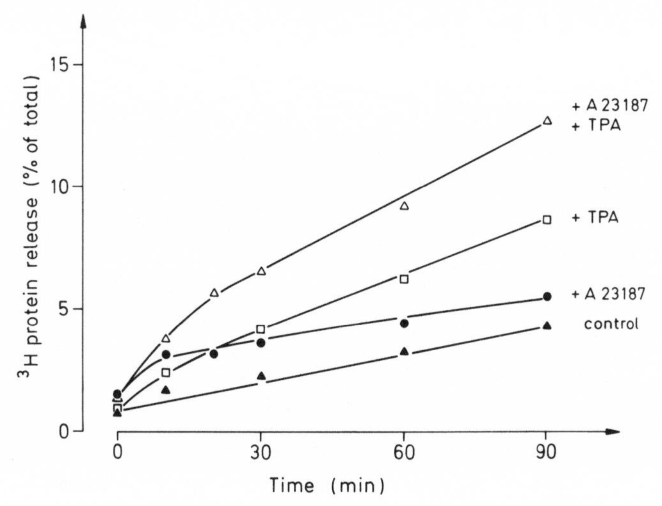

Figure 7. Effects of Ca^{2+} ionophore A23187 (10^{-5} mol/liter), TPA (10^{-7} mol/liter), and A23187 plus TPA on enzyme (^{3}H protein) release with time in isolated pancreatic acini. Rat pancreatic tissue was prelabeled with [^{3}H]leucine. Acini were isolated by collagenase treatment (30 min) and preincubated in the absence of added Ca^{2+} (10 min) with or without A23187 as indicated in a buffer containing (mmol/liter): 145 NaCl, 4.7 KCl, 1.2 $MgCl_2$, 1.2 KH_2PO_4, 10 HEPES, adjusted with NaOH to pH 7.4 at 37°C and gassed with 100% O_2 in a shaking water bath. At zero time, Ca^{2+} (1 mmol/liter) was added to the incubation media, and TPA (10^{-7} mol/liter) was added where indicated. Samples were taken immediately after these additions (time zero) and after 10, 30, 60, and 90 min of incubation. Radioactivity was counted and [^{3}H]protein released by cells is expressed as the percent of the total trichloroacetic acid–precipitable ^{3}H-labeled protein content present in the cells.

dependent stimulation. If both TPA and cAMP are added together, a marked increase in enzyme release, with a shift in the activation curve to lower free Ca^{2+} concentrations, is seen (Fig. 8).

We interpret our data to mean that Ca^{2+} is necessary for stimulation but it is not the only intracellular messenger. The small stimulation by Ca^{2+} could mean activation of Ca^{2+}/calmodulin-dependent or other Ca^{2+}-binding, protein-dependent protein kinases. TPA activates protein kinase C, which is also Ca^{2+} dependent. Although cAMP-dependent protein kinase A is not dependent on Ca^{2+} (Burnham and Williams, 1984), the stimulation curve in Fig. 8 shows that cAMP-induced enzyme secretion is dependent on Ca^{2+}. This could mean that Ca^{2+}/calmodulin-, cAMP-, and DG-dependent protein kinases act synergistically on some step involved in secretion. Protein phosphorylation by these protein kinases is probably critically involved in the fusion of zymogen granules with the luminal plasma membrane (Burnham and Williams, 1984). It is also possible that GTP-binding proteins are involved in the last step of exocytosis, as has been described for other secretory cells (Cockcroft and Gomperts, 1985; Knight and Baker, 1985; Gomperts et al., 1986). The exact location of these functions and their interrelation remain to be determined.

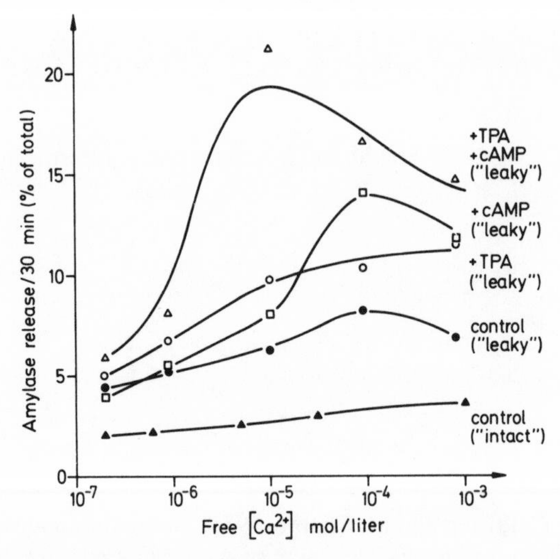

Figure 8. Effect of TPA (10^{-7} M) and cAMP (2×10^{-3} M) on enzyme secretion from isolated intact or permeabilized pancreatic acini, which had been treated with digitonin as described (Kimura et al., 1986). The free Ca^{2+} concentration was adjusted with EGTA. Amylase was determined using the Phadebas amylase test kit (Pharmacia, Freiburg, Federal Republic of Germany) and release was expressed as the percentage of total amylase activity present in cells before stimulation (Kimura et al., 1986).

Summary

Cell activation of different cell types is accompanied by receptor-mediated stimulation of phospholipase C and a consequent breakdown of phosphatidylinositol 4,5-bisphosphate. Evidence suggests that GTP-binding proteins are involved in this signal transduction mechanism, which couples receptors to phospholipase C. Both the hydrolysis products diacylglycerol (DG) and inositol 1,4,5-trisphosphate (IP_3) are intracellular messengers for cellular responses such as secretion, as illustrated by the pancreatic acinar cell. IP_3 releases Ca^{2+} from a nonmitochondrial Ca^{2+} pool likely to be the endoplasmic reticulum (ER). This Ca^{2+} release leads to a transient rise in the cytosolic free Ca^{2+} concentration from ~100 to ~800 nmol/liter, by which enzyme secretion is initiated. For sustained secretion, Ca^{2+} influx into the cell is necessary to keep the cytosolic free Ca^{2+} concentration at a slightly elevated level. Activation of protein kinase C by DG and Ca^{2+} seems to play a major role in the second, sustained phase of secretion. Ca^{2+} reuptake into the ER and Ca^{2+} extrusion from the cell are achieved by (Ca^{2+} + Mg^{2+})-ATPase in both the ER and the plasma membrane as well as by an Na^+/Ca^{2+} exchange in the latter. In the final step of exocytosis, protein phosphorylation by Ca^{2+}-, DG-, and cAMP-dependent protein kinases is probably involved.

Acknowledgments

The authors wish to thank Prof. K. J. Ullrich for helpful discussions.

This work was supported in part by Deutsche Forschungsgemeinschaft grant no. Schu 429/2-2.

References

Baukal, A. J., G. Guillemette, R. P. Rubin, A. Spät, and K. J. Catt. 1985. Binding sites for inositol trisphosphate in the bovine adrenal cortex. *Biochemical and Biophysical Research Communications.* 133:532–538.

Bayerdörffer, E., L. Eckhardt, W. Haase, and I. Schulz. 1985*a*. Electrogenic calcium transport in plasma membrane of rat pancreatic acinar cells. *Journal of Membrane Biology.* 84:45–60.

Bayerdörffer, E., W. Haase, and I. Schulz. 1985*b*. Na^+/Ca^{2+} countertransport in plasma membrane of rat pancreatic acinar cells. *Journal of Membrane Biology.* 87:107–119.

Bayerdörffer, E., H. Streb, L. Eckhardt, W. Haase, and I. Schulz. 1984. Characterization of calcium uptake into rough endoplasmic reticulum of rat pancreas. *Journal of Membrane Biology.* 81:69–82.

Berridge, M. J., and R. F. Irvine. 1984. Inositol trisphosphate, a novel second messenger in cellular signal transduction. *Nature.* 312:315–321.

Bigay, J., P. Deterre, C. Pfister, and M. Chabre. 1985. Fluoroaluminates activate transducin-GDP by mimicking the γ-phosphate of GTP in its binding site. *FEBS Letters.* 191:181–185.

Blackmore, P. F., S. B. Bocckino, L. E. Waynick, and J. H. Exton. 1985. Role of a guanine nucleotide-binding regulatory protein in the hydrolysis of hepatocyte phosphatidylinositol 4,5-bisphosphate by calcium-mobilizing hormones and the control of cell calcium. Studies utilizing aluminum fluoride. *Journal of Biological Chemistry.* 260:14477–14483.

Bradford, P. G., and R. P. Rubin. 1985. Pertussis toxin inhibits chemotactic factor-induced phospholipase C stimulation and lysosomal enzyme secretion in rabbit neutrophils. *FEBS Letters.* 183:317–320.

Burnham, D. B., and J. A. Williams. 1984. Activation of protein kinase activity in pancreatic acini by calcium and cAMP. *American Journal of Physiology.* 246:G500–G508.

Cockcroft, S., and B. D. Gomperts. 1985. Role of guanine nucleotide binding protein in the activation of polyphosphoinositide phosphodiesterase. *Nature.* 314:534–536.

Dawson, A. P. 1985. GTP enhances inositol trisphosphate-stimulated Ca^{2+} release from rat liver microsomes. *FEBS Letters.* 185:147–150.

Dawson, A. P., J. G. Comerford, and D. V. Fulton. 1986. The effect of GTP on inositol 1,4,5-trisphosphate-stimulated Ca^{2+} efflux from a rat liver microsomal fraction: is a GTP-dependent protein phosphorylation involved? *Biochemical Journal.* 234:311–315.

Gill, D. L., T. Ueda, S.-H. Chuek, and M. W. Noel. 1986. Ca^{2+} release from endoplasmic reticulum is mediated by a guanine nucleotide regulatory mechanism. *Nature.* 320:461–464.

Gomperts, B. D. 1983. Involvement of guanine nucleotide-binding protein in the gating of Ca^{2+} by receptors. *Nature.* 306:64–66.

Gomperts, B. D., M. M. Barrowman, and S. Cockcroft. 1986. Dual role for guanine nucleotides in stimulus-secretion coupling. *Federation Proceedings.* 45:2156–2161.

Haslam, R. J., and M. M. L. Davidson. 1984. Receptor-induced diacylglycerol formation in permeabilized platelets; possible role for a GTP-binding protein. *Journal of Receptor Research.* 4:605–629.

Imamura, K., and I. Schulz. 1985. Phosphorylated intermediate of (Ca^{2+} + K^{+})-stimulated Mg^{2+}-dependent transport ATPase in endoplasmic reticulum from rat pancreatic acinar cells. *Journal of Biological Chemistry.* 260:11339–11347.

Kanaho, Y., J. Moss, and M. Vaughan. 1985. Mechanism of inhibition of transducin GTPase activity by fluoride and aluminum. *Journal of Biological Chemistry.* 260:11493–11497.

Kimura, T., K. Imamura, L. Eckhardt, and I. Schulz. 1986. Ca^{2+}-, phorbol ester-, and cAMP-stimulated enzyme secretion from permeabilized rat pancreatic acini. *American Journal of Physiology.* 250:G698–G708.

Knight, D. E., and P. F. Baker. 1985. Guanine nucleotides and Ca-dependent exocytosis. Studies on two adrenal cell preparations. *FEBS Letters.* 189:345–349.

Kondo, S., and I. Schulz. 1976. Calcium ion uptake in isolated pancreas cells induced by secretagogues. *Biochimica et Biophysica Acta.* 419:76–92.

Koo, C., R. J. Lefkowitz, and R. Snyderman. 1983. Guanine nucleotides modulate the binding affinity of the oligopeptide chemoattractant receptor on human polymorphonuclear leukocytes. *Journal of Clinical Investigation.* 72:748–753.

Levitzki, A. 1984. Receptor to effector coupling in the receptor-dependent adenylate cyclase system. *Journal of Receptor Research.* 4:399–409.

Litosch, I., C. Wallis, and J. N. Fain. 1985. 5-Hydroxytryptamine stimulates inositol phosphate production in a cell-free system from blowfly salivary glands. Evidence for a role of GTP in coupling receptor activation to phosphoinositide breakdown. *Journal of Biological Chemistry.* 260:5464–5471.

Manning, D. R., and A. G. Gilman. 1983. The regulatory components of adenylate cyclase and transducin: a family of structurally homologous guanine nucleotide binding proteins. *Journal of Biological Chemistry.* 258:7059–7063.

Muallem, S., M. Schoeffield, S. Pandol, and G. Sachs. 1985. Inositol trisphosphate modification of ion transport in rough endoplasmic reticulum. *Proceedings of the National Academy of Sciences.* 82:4433–4437.

Nakamura, T., and M. Ui. 1985. Simultaneous inhibitions of inositol phospholipid breakdown, arachidonic acid release, and histamine secretion in mast cells by islet-activating protein, pertussis toxin. A possible involvement of the toxin-specific substrate in the Ca^{2+}-mobilizing receptor-mediated biosignaling system. *Journal of Biological Chemistry.* 260: 3584–3593.

Ochs, D. L., J. I. Korenbrot, and J. A. Williams. 1983. Intracellular free calcium concentrations in isolated pancreatic acini; effects of secretagogues. *Biochemical and Biophysical Research Communications.* 117:122–128.

Okajima, F., and M. Ui. 1984. ADP-ribosylation of the specific membrane protein by islet-activating protein, pertussis toxin, associated with inhibition of a chemotactic peptide-induced arachidonate release in neutrophils. A possible role of the toxin substrate in Ca^{2+}-mobilizing biosignaling. *Journal of Biological Chemistry.* 259:13863–13871.

Petersen, O. H., and N. Ueda. 1976. Pancreatic acinar cells: the role of calcium in stimulus-secretion coupling. *Journal of Physiology.* 254:583–606.

Rasmussen, H., and P. Q. Barrett. 1984. Calcium messenger system: an integrated view. *Physiological Reviews.* 64:938–984.

Rodbell, M. 1980. The role of hormone receptors and GTP-regulatory proteins in membrane transduction. *Nature.* 284:17–22.

Rottenberg, H., and C.-P. Lee. 1975. Energy dependent hydrogen ion accumulation in submitochondrial particles. *Biochemistry.* 14:2675–2680.

Schulz, I., H. Streb, E. Bayerdörffer, and K. Imamura. 1985. Hormonal and neurotransmitter regulation of Ca^{2+} movements in pancreatic acinar cells. *In* Hormones and Cell Regulation. J. E. Dumont, B. Hamprecht, and J. Nunez, editors. Elsevier/North-Holland, Amsterdam. 9:325–342.

Sokolovsky, M., D. Gurwitz, and R. Galron. 1980. Muscarinic receptor binding in mouse brain: regulation by guanine nucleotides. *Biochemical and Biophysical Research Communications.* 94:487–492.

Somlyo, A. P., M. Bond, and A. V. Somlyo. 1985. Calcium content of mitochondria and endoplasmic reticulum in liver frozen rapidly *in vivo. Nature.* 314:622–625.

Spät, A., P. G. Bradford, J. S. McKinney, R. P. Rubin, and J. W. Putney, Jr. 1986*a*. A saturable receptor for [^{32}P]inositol-(1,4,5)trisphosphate in guinea-pig hepatocytes and rabbit neutrophils. *Nature.* 319:514–516.

Spät, A., A. Fabiato, and R. P. Rubin. 1986*b*. Binding of inositol trisphosphate by a liver microsomal fraction. *Biochemical Journal.* 233:929–932.

Sternweis, P. C., and A. G. Gilman. 1982. Aluminum: a requirement for activation of the regulatory component of adenylate cyclase by fluoride. *Proceedings of the National Academy of Sciences.* 79:4888–4891.

Streb, H., E. Bayerdörffer, W. Haase, R. F. Irvine, and I. Schulz. 1984. Effect of inositol-1,4,5-trisphosphate on isolated subcellular fractions of rat pancreas. *Journal of Membrane Biology.* 81:241–253.

Streb, H., J. P. Heslop, R. F. Irvine, I. Schulz, and M. J. Berridge. 1985. Relationship between secretagogue-induced Ca^{2+} release and inositol polyphosphate production in permeabilized pancreatic acinar cells. *Journal of Biological Chemistry.* 260:7309–7315.

Streb, H., R. F. Irvine, M. J. Berridge, and I. Schulz. 1983. Release of Ca^{2+} from a nonmitochondrial intracellular store in pancreatic acinar cells by inositol-1,4,5-trisphosphate. *Nature.* 306:67–69.

Streb, H., and I. Schulz. 1983. Regulation of cytosolic free Ca^{2+} concentration in acinar cells of rat pancreas. *American Journal of Physiology.* 245:G347–G357.

Ueda, T., S.-H. Chueh, M. W. Noel, and D. L. Gill. 1986. Influence of inositol 1,4,5-trisphosphate and guanine nucleotides on intracellular calcium release within the N1E-115 neuronal cell line. *Journal of Biological Chemistry.* 261:3184–3192.

Uhing, R. J., V. Prpic, H. Jiang, and J. H. Exton. 1986. Hormone-stimulated polyphosphoinositide breakdown in rat liver plasma membranes. Roles of guanine nucleotides and calcium. *Journal of Biological Chemistry.* 261:2140–2146.

Verghese, M., R. J. Uhing, and R. Snyderman. 1986. A pertussis/cholera toxin-sensitive N protein may mediate chemoattractant receptor signal transduction. *Biochemical and Biophysical Research Communications.* 138:887–894.

Wallace, M. A., and J. N. Fain. 1985. Guanosine 5′-O-thiotriphosphate stimulates phospholipase C activity in plasma membranes of rat hepatocytes. *Journal of Biological Chemistry.* 260:9527–9530.

Chapter 9

A Chemical Link in Excitation-Contraction Coupling in Skeletal Muscle

Julio Vergara, Kamlesh Asotra, and Michael Delay

Department of Physiology, University of California, Los Angeles, California

Introduction

It is currently widely accepted that, in skeletal muscle fibers, the depolarization of the transverse (T-) tubular system membranes is uniquely responsible for the Ca release from the terminal cisternae of the sarcoplasmic reticulum (SR) (Huxley and Taylor, 1958; Hodgkin and Horowicz, 1960; Ebashi et al., 1969). In all likelihood, the transfer of information occurs across well-defined structures called triads or T-SR junctions (Peachey, 1965; Franzini-Armstrong, 1970) at which the two membranes (T-tubule and SR) are separated by a gap of 100–200 Å. It has been reported recently (Vergara and Tsien, 1985; Vergara et al., 1985; Volpe et al., 1985) that inositol 1,4,5-trisphosphate ($InsP_3$) may play a key role as a chemical link in excitation-contraction (EC) coupling in skeletal muscle. We proposed (Vergara et al., 1985) that the T-tubular depolarization induces the release of $InsP_3$, which diffuses through the very narrow gap separating the T-tubule from the SR and binds to a specific receptor of a chemically activatable Ca channel at the SR.

Fig. 1 illustrates the proposed chemical sequence for the involvement of inositol phosphates and phosphoinositides in muscle EC coupling; it is taken almost literally from the scheme proposed by Berridge and Irvine (1984) involving $InsP_3$ in the receptor-mediated release of Ca from the endoplasmic reticulum, which has been widely verified to occur in several cell and tissue preparations. There are, however, two aspects of this scheme that have been specifically modified in our proposal for muscle. (*a*) We define a specific morphological site at which these chemical reactions occur, namely the T-SR junctions, where the membranes from the T-tubules and the SR come in contact with each other, leaving a well-defined gap of 100–200 Å. (*b*) We propose that in muscle, the T-tubular depolarization, rather than agonist-receptor binding, may be the initial signal responsible for the activation of phospholipase C, leading to the release of $InsP_3$ and diacylglycerol (DG) from the hydrolysis of phosphatidylinositol 4,5-bisphosphate (PIP_2). The $InsP_3$ released at the triadic junction would diffuse to the terminal cisternae of the SR and bind to a receptor, inducing the release of Ca.

Briefly, the evidence supporting our proposal is as follows (Vergara et al., 1985). (*a*) Electrical stimulation of skeletal muscle released $InsP_3$. (*b*) $InsP_3$ induced the release of Ca from the SR. This was verified in microinjection experiments by observing local contractures in intact and skinned muscle fibers and by measuring tension in skinned fiber experiments. (*c*) Blockers of the inositol trisphosphatase enhanced the Ca release from the SR in skinned fibers. (*d*) Blockers of the phospholipase C–dependent hydrolysis of the polyphosphoinositide PIP_2, such as neomycin and other polyamines, blocked EC coupling in isolated muscle fibers under physiological conditions.

Many of these observations have been either collectively or individually reproduced in different laboratories throughout the world. These are especially important in skinned fiber preparations, in which there has been additional evidence demonstrating that $InsP_3$ is capable of inducing contractures (Donaldson et al., 1986; Nocek et al., 1986), in some cases, at extremely low concentrations (Donaldson et al., 1986). There are, however, a few reports of negative results in skinned muscle fibers (Lea et al., 1986) and in isolated SR preparations (Scherer and Ferguson, 1985; Movsesian et al., 1985), which need to be seriously considered as a warning about the projected specific role of $InsP_3$ in the EC coupling process.

These results may point toward the possibility that an inositol phosphate(s) other than $InsP_3$ may be the actual transmitter(s) in muscle. Alternatively, they may be explained in terms of a conspiracy between a restricted diffusional space at the T-SR junction, with very strategically located degradative enzymes where the biochemical reactions may occur, and a likely modulation of the chemical processes by cofactors that may be lost in the skinning or SR membrane isolation processes.

We present here new evidence in support of a chemical coupling process at the level of the triads in skeletal muscle in which the inositol phosphate/phospho-

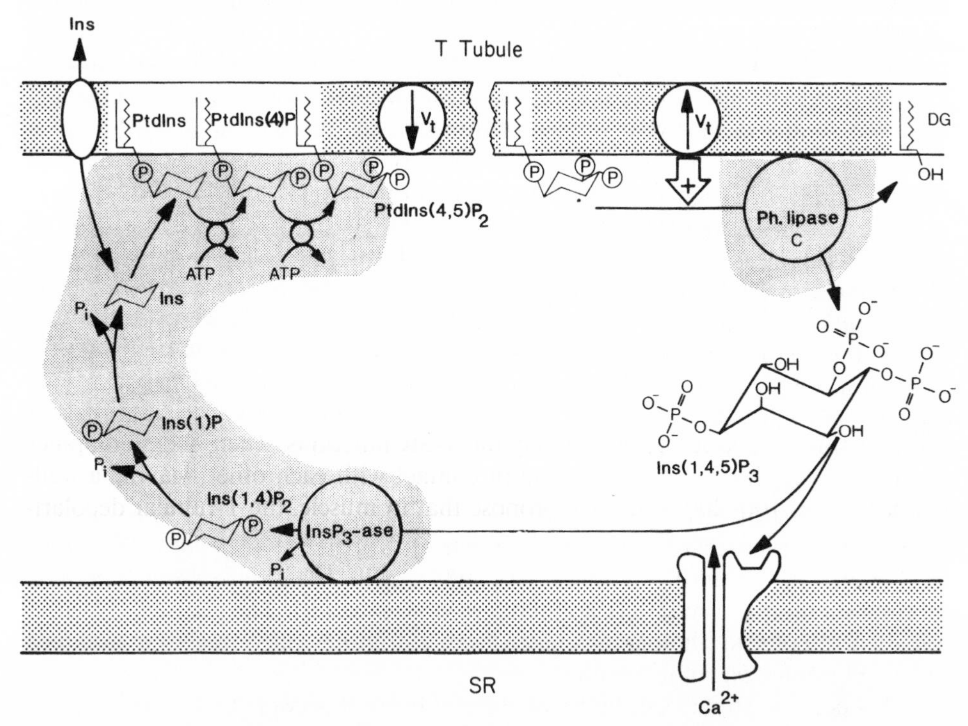

Figure 1. The proposed role of $InsP_3$ as a chemical link in EC coupling at the T-SR junction. T-tubular membrane depolarization (V_t) stimulates the hydrolysis of PI_2 by phospholipase C to form DG and $InsP_3$. The latter binds to a specific receptor on the SR to release Ca. $InsP_3$ is rapidly degraded by $InsP_3$-ase to form $InsP_2$. The soluble cycle ends with *myo*-inositol (Ins), which is readily transported across the membrane. Ins is also rapidly incorporated into the lipid phase to form phosphatidylinositol (PI), which is phosphorylated twice to form PIP_2.

inositide metabolic pathways may play a key role. Specifically, we will attempt to answer two global questions in this regard: (*a*) Does the timing of the coupling process at the triads allow for a chemical process to occur, and related to that, is the $InsP_3$-induced release of Ca fast enough to mediate the physiological coupling? (*b*) Are the critical pathways supporting the $InsP_3$ metabolism present in skeletal muscle?

Methods

The physiological experiments were performed in single semitendinosus muscle fibers of *Rana catesbeiana.* They were mounted in a Vaseline-gap chamber (Hille and Campbell, 1976) modified for optical recording (Vergara et al., 1978; Palade and Vergara, 1982; Heiny and Vergara, 1982), and were stimulated with 0.5-ms current pulses to elicit membrane action potentials from the fiber segment, ~100–150 μm long, in pool A of the chamber. The fiber ends in pools E and C of the Vaseline-gap chamber were cut in internal solution (relaxing solution) containing 100 mM K-aspartate, 20 mM K-MOPS, 5 mM phosphocreatine, 3 mM ATP, 3 mM $MgSO_4$, 0.2 mM EGTA, and adjusted to pH 7.1, pCa 7.0; the osmolality was adjusted to 245 mosmol. In experiments to measure the T-tubular signals, the fiber segment in pool A was initially stained with a Ringer's solution containing 0.3–0.5 mg/ml of the dye NK2367 (Nippon Kankoh-Shikiso Kenkyusho Co., Okayama, Japan) for 15 min and later was continuously perfused with 0.02 mg/ml dye.

Potential changes across the T-system membranes were monitored with the membrane-impermeable potentiometric dye NK2367 using nonpolarized light at 670 nm. The dye-related absorbance changes recorded under these conditions arise predominantly from the T-system and represent a "weighted average" of changes in the T-tubular membrane potential (Heiny and Vergara, 1982, 1984; Ashcroft et al., 1985). Ca release from the SR was monitored with the Ca-indicating dyes azo1 or antipyrylazo III (ApIII) diffused into the myoplasm from the cut ends of the muscle fibers (Vergara and Delay, 1985; Delay et al., 1986). The former dye was used in the majority of the experiments, mainly because it is most sensitive to Ca at 480 nm, a wavelength at which the potentiometric dye signal is negligible; hence, Ca transients at 480 nm and T-system potential signals at 670 nm do not contaminate each other in fibers simultaneously stained with both dyes (Vergara and Delay, 1986).

Microinjection experiments were performed in chemically skinned muscle fibers using the procedures described elsewhere (Vergara et al., 1985). The fibers were bathed in relaxing solution plus 0.1 mM $CdCl_2$. The micropipette contained relaxing solution, 1.5 mM $InsP_3$, and 0.5 mM arsenazo III. The pictures shown in Fig. 4 correspond to photographs taken from still-frame video images reproduced on a high-resolution color monitor.

Biochemical studies on the production of inositol phosphates and the characterization of phosphoinositides were conducted in intact pairs of sartorius and semitendinosus muscles of the frog *Rana catesbeiana* with wet weights of 1–2 g. A muscle pair was incubated for 3 h at room temperature in Ringer's solution containing either [2-^{3}H]*myo*-inositol (Amersham Corp., Arlington Heights, IL; 14 Ci/mmol sp. act.) at a total concentration of 12 μM, or carrier-free ^{32}P (Amersham Corp.; 60 μCi/ml of Ringer's solution). In both cases, the muscles were washed extensively in isotope-free Ringer's solution containing 10^{-4} M curare for up to 1 h before the experiments were started. Experimental muscles were electrically stimulated at tetanic frequencies (50 Hz) for variable periods of time with 0.5-ms current pulses applied directly to the muscles between two Pt wires located close to the tendons. The muscle temperature was equilibrated for several minutes with that of Ringer's solution in a temperature-controlled bath. A fraction of a second before the end of a stimulation period, the table supporting the temperature-controlled bath and the liquid N_2 containers cooling two copper hammers was dropped and the hammers were freed to travel toward the muscle. The smashing force that the hammers applied to flatten the muscles was ~200 lb. The thickness of the "muscle wafer" after smashing was ~0.5 mm and was adjusted by interposing metal stoppers between the hammers. The freezing times (20 to −10°C) in this system have been measured to be on the order of tens of milliseconds. Resting control muscles were immersed in Ringer's with 10^{-7} M tetrodotoxin (TTX), mounted in the smasher apparatus, and smashed without stimulation. The fast-frozen muscle wafers in either case were ground to a fine powder in liquid N_2 and homogenized in a mixture of chloroform/methanol/concentrated HCl (100:200:5, vol/vol) followed by chloroform and deionized water. Phospholipids containing phosphoinositides in the lower organic phase and inositol phosphates in the upper aqueous phase were separated by a single partitioning step after centrifugation for 15 min.

Our studies required the removal of nucleotides without the loss of inositol phosphates. This was achieved by treating the muscle extracts with activated charcoal, which completely removed nucleotides (monitored by A_{254}) but spared all the inositol phosphates (Meek, 1986). The steps followed in the separation of inositol phosphates and phosphoinositides (Asotra and Vergara, 1986) consisted mainly of the utilization of anion-exchange high-pressure liquid chromatography (HPLC) to separate inositol phosphates (Binder et al., 1985), glycophase G-bound neomycin affinity-column chromatography for the separation of phosphoinositides (Schacht, 1978), thin-layer chromatography (TLC) or HETLC, and autoradiography for further identification of phosphoinositides and other phospholipids (Emilsson and Sundler, 1984), and high-sensitivity methods for the determination of organic phosphorus (Nakamura, 1952; Hess and Derr, 1975). The anion-exchange HPLC has also been used for monitoring changes in the levels of nucleotides in resting and stimulated muscle extracts and to assay the activity of inositol (1,4,5)P_3 5-phosphatase ($InsP_3$-ase), inositol (1,4)P_2 4-phosphatase ($InsP_2$-ase), and inositol (1)P 1-phosphatase (InsP-ase) quantitatively.

Results and Discussion

Kinetics of Ca Release from the SR

As a background to the first question proposed in the Introduction, we must remember the morphological properties of the membrane systems involved in EC coupling in skeletal muscle. The T-tubule penetrates radially into the muscle fiber and, just beneath the surface membrane, is immediately apposed by the terminal cisternae from the SR, forming the typical T-SR junctions (Peachey, 1965; Franzini-Armstrong, 1970). This contact exists virtually along the entire length of the T-tubule (Zampighi et al., 1975), giving the basis for considering the triad as a specific organelle of skeletal muscle. This organelle shows a peculiar morphology, including the so-called "feet" and a well-defined gap between two membrane compartments (Franzini-Armstrong, 1970). The morphology of the T-SR junction is reminiscent of that of another junction important in muscle: the neuromuscular junction. However, it should be emphasized that the latter is used for electrical communication between two different cells separated by a gap of extracellular space, while the former responds to the functional need of a communication between two membrane compartments within the same cell separated by a gap of intracellular space. Nonetheless, we may use the analogy between these two organelles and test whether specific features of the coupling at the neuromuscular junction are observed at the T-SR junction as well. One of the reinforcements of the concept of chemical coupling at the neuromuscular junction came from experiments (Katz and Miledi, 1965*a*) showing a transmission delay compatible with an underlying chemical reaction scheme. In relation to the EC coupling process, this question has been more difficult to answer because none of the membrane compartments involved are amenable to direct study by electrical methods. However, with the aid of optical methodology, we have recently obtained evidence demonstrating that the coupling process at the T-SR junctions in skeletal muscle fibers displays a junctional latency (the "triadic delay") analogous to the synaptic delay of the neuromuscular junction (Vergara and Delay, 1986).

Fig. 2 shows a family of traces obtained from a fiber stained with both azo1 and NK2367 dyes. Trace *a* corresponds to the electrically recorded membrane action potential, trace *b* to the optically recorded average T-tubular action potential, and trace *c* to the Ca absorbance signal. A prolonged time interval of ~2 ms occurs between the upswing of the membrane action potential and the onset of the Ca transient; we call this time interval the "total coupling lag." The potentiometric

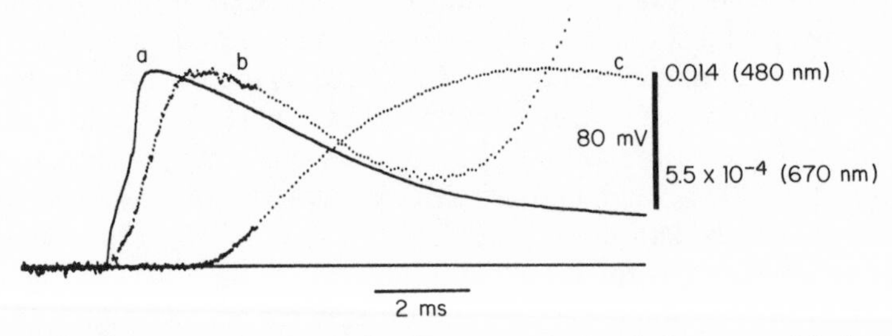

Figure 2. Surface membrane action potential (trace *a*), T-tubular potential signal (trace *b*), and Ca signal (trace *c*). The optical traces were obtained in successive recordings at illuminating wavelengths of 670 nm (16 sweeps averaged) and 480 nm (2 sweeps averaged), respectively, from a cut fiber preparation stained with NK2367 and 200 μM azo1. Temperature, 13°C. The NK2367 transient represents an increase in absorbance, and the azo1 transient represents a decrease. The late falling phase of the T-system signals always shows interference from fiber movement subsequent to the Ca release. (Modified from Vergara and Delay, 1986.)

signal has no significant lag, but shows a relatively slow rising phase, with a time to peak of ~1.8 ms. Since the depolarization of the T-tubular system is clearly interposed between the activation of the surface membrane and the release of Ca, part of the observed total coupling lag may be due to kinetic limitations of this membrane compartment. By analogy with the experiments of synaptic delay, we define the onset of the Ca release as the first perceptible systematic departure of the Ca signal from baseline. We developed a method to derive the total coupling lag with improved statistical significance by which the early phase of the record was fitted by a least-squares method to a mathematical lag function; the onset of the Ca signal was defined as the time when the fit function first departed perceptibly from zero. The total coupling lag thus obtained for trace *a* in Fig. 2 is 2.08 ms. In experiments with several Ca-indicating dyes, we verified that the lag is an intrinsic property of the Ca-release process, independent of the dyes used (Vergara and Delay, 1986).

The total coupling lag is the sum of a genuine "triadic delay" and the lag caused by the kinetic limitations of the T-system, called the "propagational lag." It is necessary to exclude the latter component in order to calculate the time required to couple the T-tubule voltage to the Ca release. A first-order estimate could be made by measuring the time elapsed between stimulation and the time when the average T-system membrane potential reached a threshold for Ca release. Appropriate numbers for this threshold can be obtained from published values of experiments monitoring tension development or Ca transients in response to steady state or prolonged depolarizations. These values range between −60 and −40 mV (Hodgkin and Horowicz, 1960; Heistracher and Hunt, 1969; Adrian et al., 1969;

Table I
Values for Kinetic Parameters Taken from Several Fibers under Different Experimental Conditions

	Ca dye	Temperature	Total coupling lag	Propagational lag	Triadic delay
		°C	*ms*	*ms*	*ms*
Fiber 1	Azo1	7	3.90	0.67	3.23
	"	12	2.38	0.57	1.81
	"	17	1.46	0.50	0.96
	"	20	1.30	0.46	0.84
	"	25.6	0.96	0.40	0.56
Fiber 2	Azo1	13	2.08	0.56	1.52
	ApIII	13	1.96	0.56	1.40
	Azo1	4.5	4.48	0.70	3.78
	ApIII	4.5	4.56	0.70	3.86
Fiber 3	Azo1	6	3.43	0.80	2.63
	"	11	2.51	0.68	1.83
	"	19	1.71	0.58	1.13
Fiber 4	Azo1	3	3.56	0.76	2.80
	"	13	1.78	0.64	1.14
Fiber 5	Azo1	4	3.46	0.77	2.69
	"	18	3.34	0.65	0.89

Vergara and Caputo, 1983); they would predict, for the experiment in Fig. 2, propagational lags ranging between 0.5 and 0.7 ms. A better method of estimating the true propagational lag from the T-tubular depolarization signals was described by Vergara and Delay (1986). Values of the propagational delay calculated by these authors, from fibers under various experimental conditions, are presented in Table I, column 5. The values of the triadic delay computed by subtracting the propagational lag from the total coupling lag (column 4) are included in the last column of Table I. It should be noted that these values appear to be sufficiently long to allow for a chemical reaction.

Temperature Dependence of the Coupling Process

The suggestion of a chemical coupling is corroborated by experiments at different temperatures. Fig. 3 shows a family of Ca transients recorded with azo1 at several temperatures in a fiber stimulated at the time indicated by the arrow. The transients illustrate the strong temperature dependence of the total coupling lag. The corresponding data on the temperature dependence of the triadic delay are given in Table I, including the values for the propagational lag at each of these temperatures. Although there is some variation in the range of triadic delay observed from fiber to fiber, the dependence on temperature found for the individual fibers is consistent. Assuming a single Arrhenius type of activation process, these data correspond to an activation energy of 16.4 kcal/mol, which is equivalent to a Q_{10} of 2.7 between 10 and 20°C. Our results demonstrate that the process of Ca release from the SR follows the T-system depolarization with a measurable delay ranging from ~0.5 ms at 25.6°C to ~4 ms at 4°C, which is compatible with the possibility that a temperature-dependent chemical coupling process occurs across the T-SR junction in skeletal muscle fibers. Even the small triadic delay at high temperatures (nearly 0.5 ms) allows ample time for the occurrence of a chemical process, considering that delays across the neuromuscular junction fall within the same range. The large Q_{10} for this process even suggests that a sequence of chemical steps may be involved. The temperature dependence obtained for Ca release from the SR is similar to that

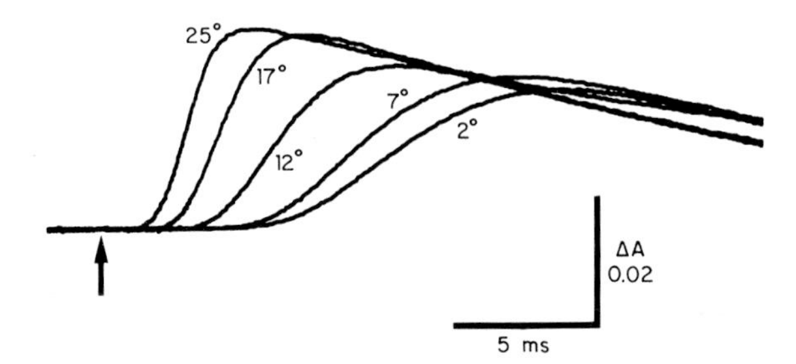

Figure 3. A series of azo1 Ca transients recorded from a single muscle fiber at the indicated temperatures; stimulation began at the arrow. Single-sweep recordings.

obtained by Katz and Miledi (1965*b*) for the neuromuscular junction, plausibly because such a similarity arises from the involvement of multiple chemical steps in both junctions.

Time Course of Ca Release in Response to $InsP_3$

The existence of a latency in the coupling process at the T-SR junctions, although suggestive of the involvement of a chemical link, does not provide any direct information about the role of inositol phosphates, $InsP_3$ in particular, in this process. Moreover, it poses the problem that the Ca release from the SR in response to these chemical transmitters has to be fast enough to be compatible with the relatively short coupling delays. The exact solution to this problem remains to be established in the future. However, we have obtained some circumstancial evidence of this effect from microinjection experiments like the one illustrated in Fig. 4. The figure shows a sequence of video images separated by 33-ms intervals before (frame 1) and during (frames 2–5) the application of an iontophoretic "pulse" of $InsP_3$ close to the edge of a skinned skeletal muscle fiber. The onset of a cathodal pulse (inside of the pipette negative with respect to the bath) was associated with a blurring of the fiber's image caused by the release of a negatively charged coloring dye (arsenazo III) together with $InsP_3$ (frame 2). Frame 3 shows the first indication of a mechanical response (shortening of sarcomeres), which becomes more pronounced at later times (frames 4 and 5). Control experiments, performed in the same fiber, demonstrated that pulses of the reverse polarity and pulses without $InsP_3$ in the micropipette failed to elicit a significant mechanical response. The

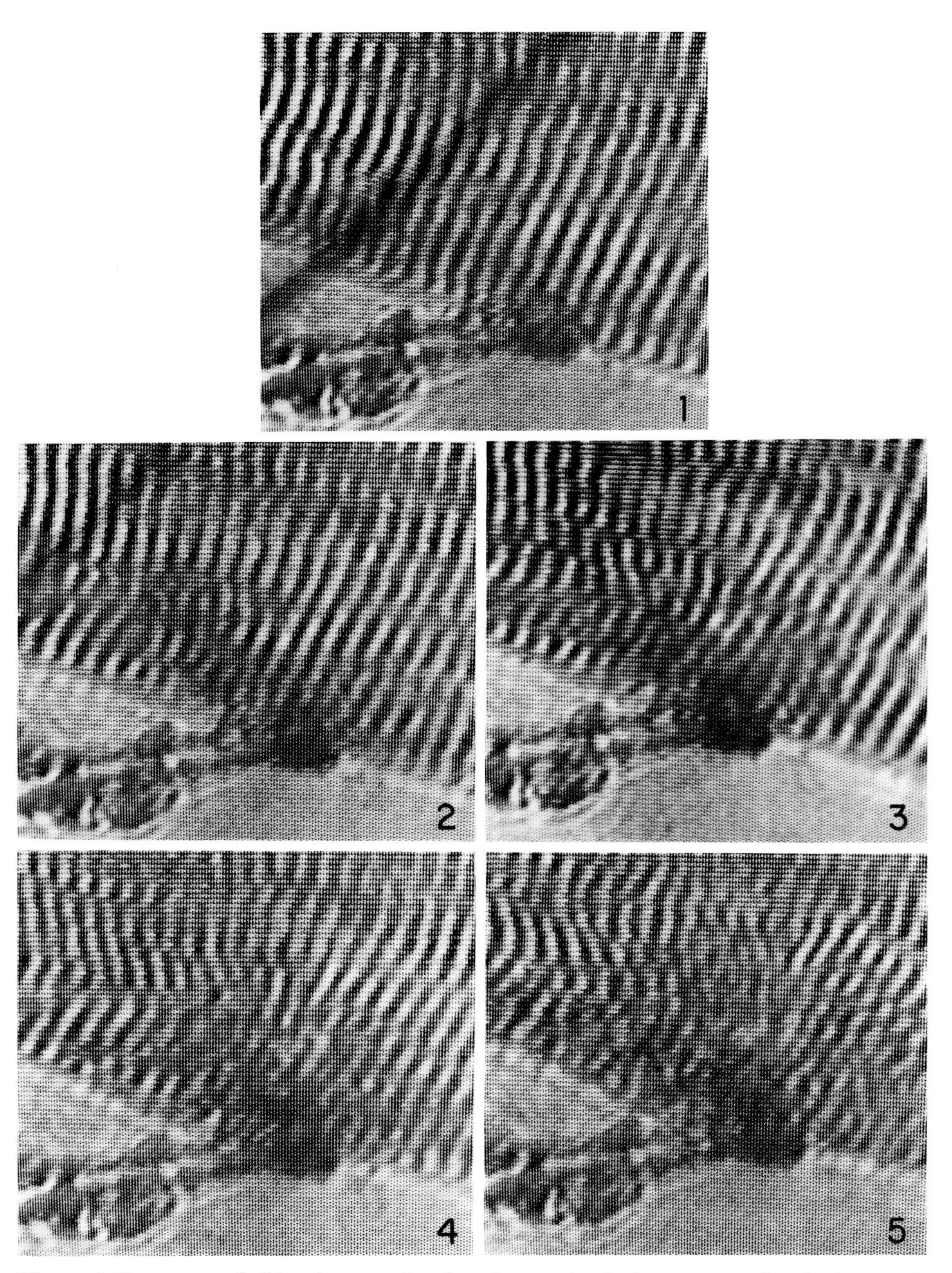

Figure 4. Sequence of video images showing the mechanical response of a single muscle fiber to iontophoretic application of $InsP_3$. The tip of the micropipette delivering $InsP_3$ is shown beneath the muscle fiber. The contractile response is evidenced by a shortening of the sarcomere spacing and eventually a loss of the striated appearance of the fiber.

important information obtained from these experiments demonstrates not only that $InsP_3$ is able to release Ca from the SR, but also that the response time by the SR has to be shorter than 33 ms. Admittedly, this interval is still too long to be compatible with the physiological delays described above. However, the visual response monitored in these experiments includes the response of the contractile machinery to Ca release, which is probably the rate-limiting process. Also, the experiment is technically limited because the video system did not provide an adequate time resolution to test a possibly faster process.

Biochemical Support for a Chemical Link in EC Coupling

An immediate consequence of the chemical coupling hypothesis is that there may be a cascade of enzymatic reactions ultimately providing the chemical link between the depolarization of the T-tubule and the release of Ca from the SR. There are two general approaches that we have followed in the initial stages of the investigation of these biochemical pathways: (*a*) we performed in vivo experiments in which electrically stimulated muscles were rapidly frozen at different levels of activation, followed by a detailed analysis of the biochemical balance of the relevant metabolites, and (*b*) we characterized biochemically the phosphoinositide metabolism in whole muscle and its subcellular fractions, which may participate in the EC coupling process.

Release of $InsP_3$ in Skeletal Muscle

One of the most important pieces of evidence supporting the role of $InsP_3$ in EC coupling came from the experiments showing that increased amounts of radiolabeled inositol phosphates, separated with anion-exchange chromatography, were detected in soluble extracts from stimulated as compared with control muscles (Vergara et al., 1985). These experiments were extended and improved in order to attempt the quantitation of inositol phosphates in muscle at rest to study the changes induced by electrical stimulation. For this purpose, we improved the methodology of fast-freezing of muscles and their subsequent chemical analysis (Mommaerts and Schilling, 1964; Asotra and Vergara, 1986). Anion-exchange HPLC provided a clear advantage over the anion-exchange chromatography in that it could quantitatively resolve the low levels of various inositol phosphates (nanomoles) as well as the high levels of adenosine nucleotides and phosphocreatine (micromoles) present in muscle extracts (Homsher et al., 1981). Fig. 5 shows the HPLC separation and quantitative phosphorus determination of the inositol phosphates extracted from a sartorius muscle stimulated for 10 s at 20°C and 50 Hz compared with that of an unstimulated contralateral control muscle. It is clear that the levels of $InsP_3$, $InsP_2$, and InsP increased in the stimulated muscle. The levels of $InsP_3$ in muscles stimulated for 5, 10, or 20 s at 20°C were maintained at a constantly high level (8 nmol/g wet wt) compared with a negligible amount in control muscles (<0.3 nmol/g wet wt). The levels of $InsP_2$ increased with the duration of electrical stimulation (reaching ~80 nmol/g wet wt after a 20-s tetanus), which suggests that the $InsP_3$ formed during muscle activation is also dynamically degraded to form $InsP_2$. In experiments where muscles were stimulated at low temperatures (1–5°C), high levels of $InsP_3$ with relatively low levels of $InsP_2$ were usually observed, which shows that the degradation of $InsP_3$ is also a temperature-dependent process. While these data on the synthesis of $InsP_3$ and $InsP_2$ can be

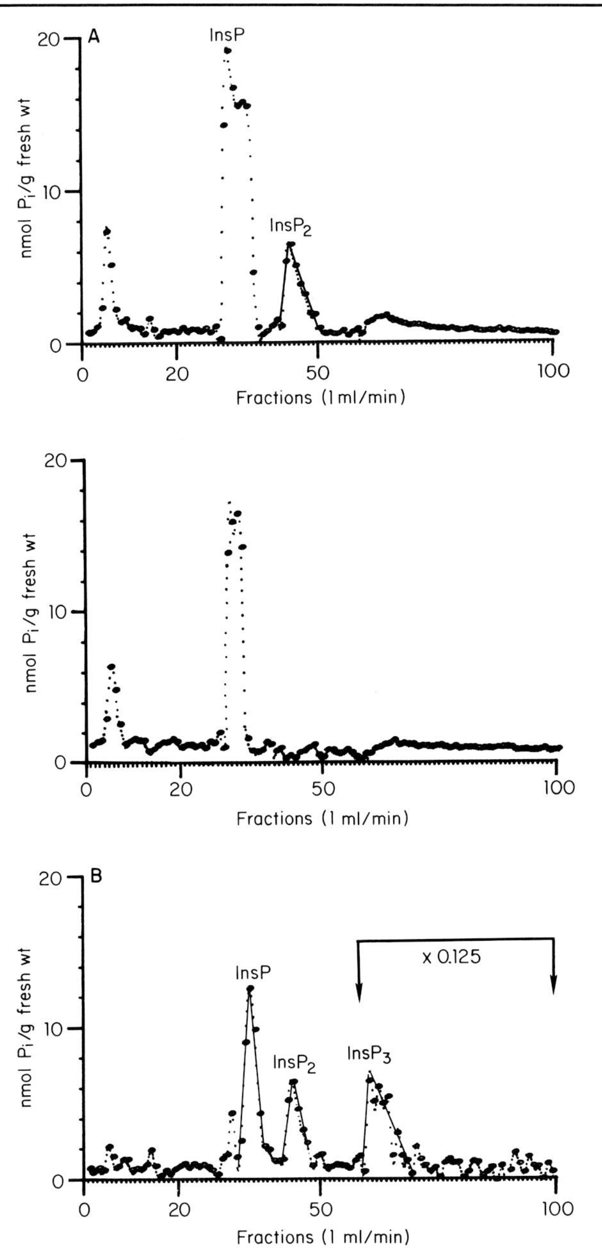

Figure 5. (*A*) Phosphorus contents of inositol phosphates extracted from resting muscles (lower panel) and muscles stimulated for 10 s (upper panel) at 20°C after separation with a uBondapak NH_2 anion-exchange HPLC column (Waters Associates, Milford, MA). Inositol phosphates were separated from charcoal-treated, nucleotide-free muscle extracts isocratically for the initial 20 min with 60 mM ammonium acetate/acetic acid, pH 4.0, and by applying a linear gradient up to 2 M ammonium acetate/acetic acid, pH 4.0, over the next 100 min. (*B*) Net increase in the phosphorus contents of InsP, $InsP_2$, and $InsP_3$ with 10 s stimulation was obtained by subtracting the phosphorus contents of resting muscle (*A*, bottom) from those of the stimulated muscle (*A*, top); the $InsP_3$ content is shown at eight-times-higher gain than the scale for InsP and $InsP_2$.

taken as an adequate support of the proposed view of the physiological significance of inositol phosphates during EC coupling, they prompted us also to investigate whether the increased levels of $InsP_3$ were maintained in the muscles, which were first stimulated and then briefly relaxed before fast-freezing. If, indeed, $InsP_3$ (and possibly other inositol phosphates) were a chemical link in EC coupling, one would expect not only that its levels would increase in muscle upon stimulation, but also that they would revert to a negligible concentration upon brief relaxation after a tetanus. Fig. 6 shows the levels of inositol phosphates in a semitendinosus muscle

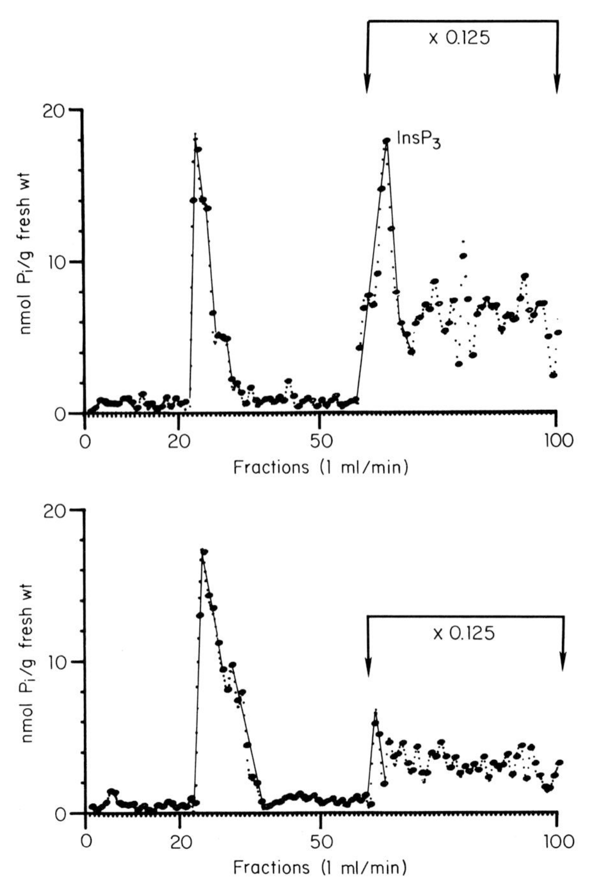

Figure 6. Levels of $InsP_3$ in the HPLC eluates of charcoal-treated aqueous extracts of muscles stimulated for 12 s (top) and equally stimulated muscles relaxed for 1 s (bottom) at 5°C. See the legend to Fig. 5 for other conditions.

that was stimulated at 5°C for 12 s and allowed to relax for 1 s before fast-freezing compared with its contralateral muscle, which was stimulated for the same time. In several such experiments, a relaxation period of 1 s after a tetanus resulted in the disappearance of $InsP_3$ and a reduction of $InsP_2$ to one-half maximal.

As shown in Fig. 1, the release of $InsP_3$ after muscle depolarization is proposed to be generated from the precursor phospholipid PIP_2 by the activation of a membrane-bound phospholipase C. An important step in the validation of this hypothesis requires the identification of this lipid in muscle. Fig. 7 shows that the autoradiogram of a TLC separation of lipids extracted from ^{32}P-labeled frog skeletal muscle indeed contains intense spots of radioactivity associated with PIP and PIP_2. The identity of both muscle phosphoinositides has been verified by coelution with standards, comparison with published R_f values, phosphorus determination, and glycophase G–bound neomycin affinity-column chromatography (Emilsson and Sundler, 1984; Schacht, 1976, 1978). The determination of phosphorus by a sensitive malachite green method (Hess and Derr, 1975) and neomycin affinity-column chromatography has allowed us to quantitate the amount of PIP_2 in resting muscles to be in the range of 4–8 nmol/g wet wt. Our yields of total phospholipids extracted are ~1 μmol/g fresh wt. At this point, it seems that the steady state PIP_2 pool in muscle is barely adequate to support the synthesis of physiologically relevant levels of $InsP_3$ during muscle contraction. Nevertheless, Fig. 8 shows that dynamic

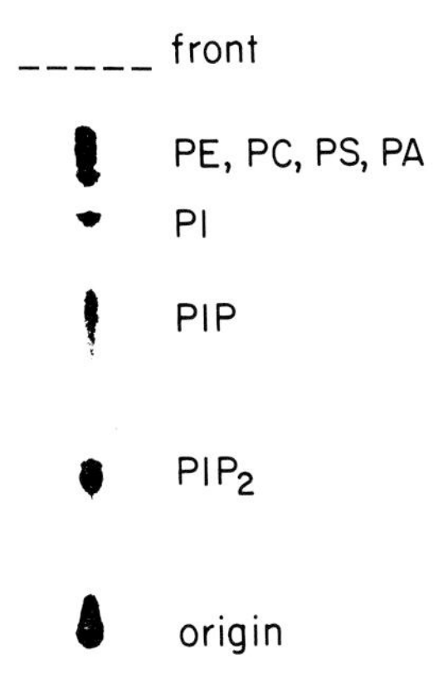

Figure 7. Autoradiograph of one-dimensional TLC of ^{32}P-labeled muscle lipids. Lipids were extracted under slightly acidic conditions and TLC plates were developed with chloroform/methanol/ammonia (22:22:7). PE, phosphatidylethanolamine; PC, phosphatidylcholine; PS, phosphatidylserine; PA, phosphatidic acid; PI, phosphatidylinositol; PIP, phosphatidylinositol 4-phosphate; PIP_2, phosphatidylinositol 4,5-bisphosphate.

changes occur in the phosphoinositide pool upon muscle stimulation. The three phosphoinositides of resting and stimulated muscles labeled with [2-^{3}H]*myo*-inositol were separated by neomycin affinity-column chromatography. In resting muscles, the amount of ^{3}H-label incorporated is maximal in phosphatidylinositol (PI) (eluted with 150 mM ammonium acetate), followed by that in phosphatidylinositol 4-phosphate (PIP) (eluted with 350 mM ammonium acetate) and in PIP_2 (eluted with 12 N HCl). The phosphoinositides of the electrically stimulated muscle exhibit a distinct decrease in their radiolabel, which demonstrates that these lipids are readily mobilized during muscle activation. We have obtained similar results for the stimulation-related loss of radiophosphorus from the phosphoinositides of skeletal muscle labeled with ^{32}P (not shown). Whether the loss of PIP upon stimulation represents exclusively its phosphorylation into PIP_2 or also includes a fraction that may be degraded directly into another product (such as $InsP_2$) is not yet clear and calls for further investigation. Additional information on the presence of phosphoinositides and their role in EC coupling has recently been provided in studies on the activities of PI-kinase and PIP-kinase in the T-tubule membranes isolated from skeletal muscle (Hidalgo et al., 1986; Varsanyi et al., 1986). These kinases have already been described in other preparations in which phosphoinositides are known to play an important regulatory role in cell function (Harwood and Hawthorne, 1969; Downes and Michell, 1982; Dale, 1985). However, their activity in muscle T-tubule membranes seems to be especially high (Hidalgo et al., 1986), and consequently this preparation may not only be relevant to the study of EC coupling but may also be useful as a general tool in membrane enzyme research. However, the formation of $InsP_2$ and InsP from the direct degradation of PIP and PI cannot be ruled out and must also be investigated. It is logical that absolute amounts of $InsP_2$ and InsP may play a role in some sort of feedback mechanism exerting an overall control on the generation of $InsP_3$ and PIP_2.

The quantitation of the levels of $InsP_3$, $InsP_2$, and InsP and of their respective precursor phosphoinositides in resting and electrically stimulated muscles, as described above, provides an insight into the rates of synthesis of the second

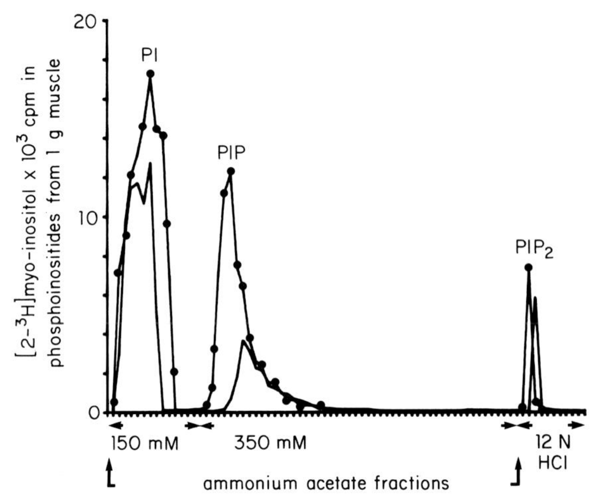

Figure 8. Neomycin affinity-column chromatography of phosphoinositides from [2-^{3}H]*myo*-inositol–labeled, resting, and stimulated muscles. PE, PC, PS, and PA are not retained by immobilized neomycin. PI, PIP, and PIP_2 are eluted with 150 mM ammonium acetate, 350 mM ammonium acetate, and 12 N HCl, respectively, in chloroform/methanol/water (3:6:1, vol/vol). Temperature, 20°C. Solid line with circles, resting muscle; solid line without symbols, stimulated for 10 s.

messenger molecule(s) during muscle contraction. However, to understand the overall involvement of the inositol phosphates in EC coupling, it is imperative that we also know about their rates of degradation. To this end, we studied the activities of three enzymes responsible for the catabolism of the inositol phosphates, viz., $InsP_3$-ase, $InsP_2$-ase, and InsP-ase, in vitro. We describe here only the results of $InsP_3$-ase. Fig. 9 shows the results of a typical experiment to determine $InsP_3$-ase activity in crude muscle homogenates that were incubated with [2-^{3}H]*myo*-inositol in the presence of appropriate cofactors such as high ATP, Mg, and EGTA (to adjust free Ca). The reaction was started by the addition of homogenate, allowed to proceed at room temperature (20°C) for various periods of time, and terminated by the addition of 10% trichloroacetic acid (TCA). $InsP_3$-ase activity was measured in terms of the radioactivity appearing in $InsP_2$ or that lost from the initial $InsP_3$ after HPLC separation. Similar experiments done with isolated SR microsomes showed that only a part of the crude muscle homogenate $InsP_3$-ase activity was present in the microsomes. These results show that skeletal muscle has $InsP_3$-ase activities comparable to those of nonmuscle preparations in which $InsP_3$ is accepted to be an important second messenger (Downes et al., 1982; Sefred et al., 1984; Storey et al., 1984; Connolly et al., 1985; Streb et al., 1985; Rana et al., 1986). However, further quantitative studies are needed to determine whether skeletal muscle $InsP_3$-ase activities are sufficient to degrade the amounts of $InsP_3$ formed in stimulated muscle, within short periods of relaxation.

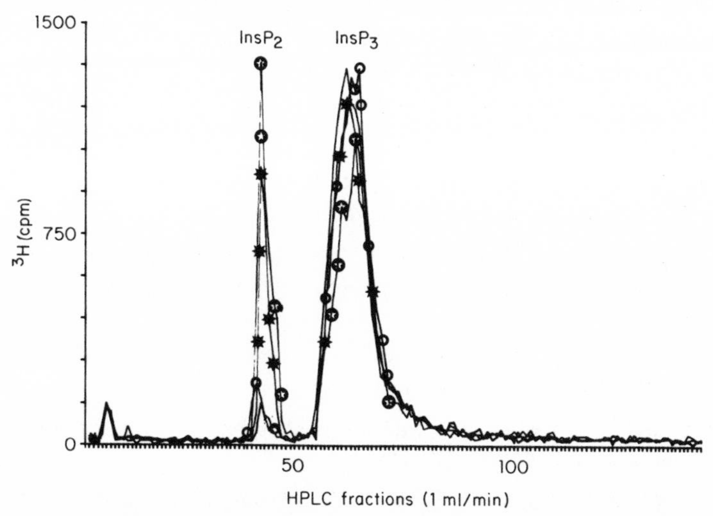

Figure 9. HPLC chromatographic separation of inositol phosphates illustrating $InsP_3$-ase activity of crude muscle homogenates. A bundle of semitendinosus muscle was homogenized in relaxing solution of pCa 7 (see Methods). An aliquot of homogenate containing 70 mg of muscle weight was added to 1 ml of the relaxing solution, which contained 5 nCi/ml [2-^{3}H]*myo*-inositol, and was incubated at 20°C for various periods of time before terminating the reaction by addition of 10% TCA. Incubation: solid line with no symbols, 30 s; open circles, 1 min; filled stars, 5 min; open stars, 10 min.

Summary

The process of EC coupling at the T-SR junction in skeletal muscle shows the characteristic features observed in other chemical junctions; namely, there is a considerable temperature-dependent latency between prejunctional and postjunctional events. This evidence gives further support to the possible involvement of inositol phosphates, $InsP_3$ in particular, in the aforementioned process. Although the kinetics of Ca release in response to $InsP_3$ microinjections have not been appropriately studied yet, our preliminary evidence does not rule out the proposed role of $InsP_3$ as the chemical transmitter. The biochemical analysis of inositol phosphates and phosphoinositides in control and stimulated muscles demonstrates significant changes in the concentration of the key moieties proposed to be relevant in several other preparations. Apparently, the enzymatic machinery for the relevant biochemical reactions of the inositol phosphate/phosphoinositide cycle is present at high levels in muscle. It remains to be investigated whether some aspects of this cycle may have been selectively adapted for the fast process of Ca release in muscle.

Acknowledgments

We thank Dr. Roger Y. Tsien for the generous gift of azo1 and for helpful comments throughout this work; Mr. A. Chertock for technical help in the isolation and characterization of muscle phospholipids; and Mr. P. Ricchiuti for technical help in the design and manufacture of the muscle-smashing apparatus.

M.D. was supported by a Lievre Award and K.A. by a postdoctoral fellowship, both from the American Heart Association, Greater Los Angeles Affiliate. This work was supported by grants from the U.S. Public Health Service (AM-25201) and the Muscular Dystrophy Association, and by a grant-in-aid to J.V. from the Jerry Lewis Neuromuscular Research Center.

References

Adrian, R. H., W. K. Chandler, and A. L. Hodgkin. 1969. The kinetics of mechanical activation in frog muscle. *Journal of Physiology.* 204:207–230.

Ashcroft, F. M., J. A. Heiny, and J. Vergara. 1985. Inward rectification in the transverse tubular system of frog skeletal muscle studied with potentiometric dyes. *Journal of Physiology.* 359:269–291.

Asotra, K., and J. Vergara. 1986. Levels of inositol phosphates in stimulated and relaxed muscles. *Biophysical Journal.* 49:190*a*. (Abstr.)

Berridge, M. J., and R. F. Irvine. 1984. Inositol trisphosphate, a novel second messenger in cellular signal transduction. *Nature.* 312:315–321.

Binder, H., P. C. Weber, and W. Siess. 1985. Separation of inositol phosphates and glycerphosphoinositol phosphates by high-performance liquid chromatography. *Analytical Biochemistry.* 48:220–227.

Connolly, M., T. E. Bross, and P. W. Majerus. 1985. Isolation of a phosphomonoesterase from human platelets that specifically hydrolyzes the 5-phosphate of inositol 1,4,5-trisphosphate. *Journal of Biological Chemistry.* 260:7868–7874.

Dale, G. L. 1985. Phosphatidylinositol 4-phosphate kinase is associated with the membrane skeleton of human erythrocytes. *Biochemical and Biophysical Research Communications.* 133:189–194.

Delay, M., B. Ribalet, and J. Vergara. 1986. Caffeine potentiation of calcium release in frog skeletal muscle fibres. *Journal of Physiology.* 375:535–559.

Donaldson, S. K., N. D. Goldberg, T. F. Walseth, and D. A. Huetteman. 1986. Inositol 1,4,5-trisphosphate induces force generation of peeled skeletal muscle fibers at 10^{-3} M free Mg^{2+}. *Biophysical Journal.* 49:191*a*. (Abstr.)

Downes, P., and R. H. Michell. 1982. Phosphatidylinositol 4-phosphate and phosphatidylinositol 4,5-bisphosphate: lipids in search of a function. *Cell Calcium.* 3:467–502.

Downes, C. P., M. C. Mussat, and R. H. Michell. 1982. The inositol trisphosphate phosphomonoesterase of the human erythrocyte membrane. *Biochemical Journal.* 203:169–177.

Ebashi, S., M. Endo, and I. Ohtsuki. 1969. Control of muscle contraction. *Quarterly Review of Biophysics.* 2:351–384.

Emilsson, A., and R. Sundler. 1984. Differential activation of phosphatidylinositol deacylation and a pathway via diphosphoinositide in macrophages responding to zymosan and ionophore A23187. *Journal of Biological Chemistry.* 259:3111–3116.

Franzini-Armstrong, C. 1970. Studies of the triad. I. Structure of the junction in frog twitch fibers. *Journal of Cell Biology.* 47:488–499.

Harwood, J. L., and J. N. Hawthorne. 1969. The properties and subcellular distribution of phosphatidylinositol kinase in mammalian tissues. *Biochimica et Biophysica Acta.* 171:75–88.

Heiny, J. A., and J. Vergara. 1982. Optical signals from surface and T-system membranes in skeletal muscle fibers. Experiments with the potentiometric dye NK2367. *Journal of General Physiology.* 80:203–230.

Heiny, J. A., and J. Vergara. 1984. Dichroic behavior of the absorbance signals from dyes NK2367 and WW375 in skeletal muscle fibers. *Journal of General Physiology.* 84:805–837.

Heistracher, P., and C. C. Hunt. 1969. The relation of membrane changes to contraction in twitch muscle fibers. *Journal of Physiology.* 201:589–611.

Hess, H. H., and J. E. Derr. 1975. Assay of inorganic and organic phosphorus in the 0.1–5 nanomole range. *Analytical Biochemistry.* 63:607–613.

Hidalgo, C., M. A. Carrasco, K. Magendo, and E. Jaimovich. 1986. Phosphorylation of phosphatidylinositol by transverse tubule vesicles and its possible role in excitation contraction coupling. *FEBS Letters.* 202:69–73.

Hille, B., and D. T. Campbell. 1976. An improved vaseline gap voltage clamp for skeletal muscle fibers. *Journal of General Physiology.* 67:265–293.

Hodgkin, A. L., and P. Horowicz. 1960. The effect of sudden changes in ionic concentrations of membrane potential of single muscle fibres. *Journal of Physiology.* 153:370–385.

Homsher, E., M. Irving, and A. Wallner. 1981. High-energy phosphate metabolism and energy liberation associated with rapid shortening in frog skeletal muscle. *Journal of Physiology.* 321:423–436.

Huxley, A. F., and R. E. Taylor. 1958. Local activation of striated muscle fibres. *Journal of Physiology.* 144:426–451.

Katz, B., and R. Miledi. 1965*a*. The measurement of synaptic delay, and the time course of acetylcholine release at the neuromuscular junction. *Proceedings of the Royal Society of London, Series B.* 161:483–495.

Katz, B., and R. Miledi. 1965*b*. The effects of temperature on the synaptic delay at the neuromuscular junction. *Journal of Physiology.* 181:656–670.

Lea, T. J., P. J. Griffiths, R. T. Tregear, and C. C. Ashley. 1986. An examination of the ability of inositol 1,4,5-trisphosphate to induce calcium release and tension development in skinned skeletal muscle fibres of frog and crustacea. *FEBS Letters.* 207:153–161.

Meek, J. L. 1986. Inositol bis-, tris-, and tetrakis(phosphate)s: analysis of tissues by HPLC. *Proceedings of the National Academy of Sciences.* 83:4162–4166.

Mommaerts, W. F. H. M., and M. O. Schilling. 1964. The rapid freezing method for the interruption of muscular contraction. *In* Rapid Mixing and Sampling Techniques in Biochemistry. B. Chance, R. H. Eisenhradt, Q. Gibson, and K. K. Lonberg-Holm, editors. Academic Press, Inc., New York. 239–254.

Movsesian, M. A., A. P. Thomas, M. Selak, and J. R. Williamson. 1985. Inositol trisphosphate does not release Ca^{2+} from permeabilized cardiac myocytes and sarcoplasmic reticulum. *FEBS Letters.* 185:329–332.

Nakamura, G. R. 1952. Microdetermination of phosphorus. *Analytical Chemistry.* 24:1372.

Nocek, T. M., M. F. Williams, S. T. Zeigler, and R. E. Godt. 1986. Inositol trisphosphate enhances calcium release in skinned cardiac and skeletal muscle. *American Journal of Physiology.* 250:C807–C811.

Palade, P., and J. Vergara. 1982. Arsenazo III and antipyrylazo III calcium transient in single skeletal muscle fibers. *Journal of General Physiology.* 79:679–707.

Peachey, L. D. 1965. The sarcoplasmic reticulum and transverse tubules of the frog's sartorius. *Journal of Cell Biology.* 25:209–231.

Ramon, F., and G. Zampighi. 1980. On the electronic coupling mechanism of crayfish septate axons: temperature dependence of junctional conductance. *Journal of Membrane Biology.* 54:165–171.

Rana, R. S., M. C. Sekar, L. E. Hokin, and M. J. MacDonald. 1986. A possible role for glucose metabolites in the regulation of inositol-1,4,5-trisphosphate 5-phosphomonoesterase activity in pancreatic islets. *Journal of Biological Chemistry.* 261:5237–5240.

Schacht, J. 1976. Inhibition by neomycin of polyphosphoinositide turnover in subcellular fractions by guinea pig cerebral cortex in vitro. *Journal of Neurochemistry.* 27:1119–1124.

Schacht, J. 1978. Purification of polyphosphoinositides by chromatography on immobilized neomycin. *Journal of Lipid Research.* 19:1063–1067.

Scherer, N. M., and J. E. Fergusson. 1985. Inositol 1,4,5-trisphosphate is not effective in releasing calcium from skeletal sarcoplasmic reticulum microsomes. *Biochemical and Biophysical Research Communications.* 128:1064–1070.

Sefred, M. A., L. E. Farrell, and W. W. Wells. 1984. Characterization of D-*myo*-inositol 1,4,5-trisphosphate phosphatase in rat liver plasma membranes. *Journal of Biological Chemistry.* 259:13204–13208.

Storey, D. J., S. B. Shears, C. J. Kirk, and R. H. Michell. 1984. Stepwise enzymatic dephosphorylation of inositol 1,4,5-trisphosphate to inositol in liver. *Nature.* 312:374–376.

Streb, H., J. P. Heslop, R. F. Irvine, I. Schulz, and M. J. Berridge. 1985. Relationship between secretagogue-induced Ca^{2+} release and inositol polyphosphate production in permeabilized pancreatic acinar cells. *Journal of Biological Chemistry.* 260:7309–7315.

Varsanyi, M., M. Messer, N. Brandt, and L. M. G. Heilmeyer, Jr. 1986. Phosphatidylinositol 4,5-bisphosphate formation in rabbit skeletal and heart muscle membranes. *Biochemical and Biophysical Research Communications.* 138:1395–1404.

Vergara, J., F. Bezanilla, and B. M. Salzberg. 1978. Nile blue fluorescence signals from cut single muscle fibers under voltage or current clamp conditions. *Journal of General Physiology.* 72:775–800.

Vergara, J., and C. Caputo. 1983. Effects of tetracaine on charge movements and calcium signals in frog skeletal muscle fibers. *Proceedings of the National Academy of Sciences.* 80:1477–1481.

Vergara, J., and M. Delay. 1985. The use of metallochromic Ca indicators in skeletal muscle. *Cell Calcium.* 6:119–132.

Vergara, J., and M. Delay. 1986. A transmission delay and the effect of temperature at the triadic junction of skeletal muscle. *Proceedings of the Royal Society of London, Series B.* 229:97–110.

Vergara, J., and R. Y. Tsien. 1985. Inositol trisphosphate induced contractures in frog skeletal muscle fibers. *Biophysical Journal.* 47:351*a*. (Abstr.)

Vergara, J., R. Y. Tsien, and M. Delay. 1985. Inositol (1,4,5) trisphosphate: possible chemical link in excitation-contraction coupling in muscle. *Proceedings of the National Academy of Sciences.* 82:6352–6356.

Volpe, P., G. Salviati, F. Di Virgilio, and T. Pozzan. 1985. Inositol 1,4,5-trisphosphate induces calcium release from sarcoplasmic reticulum of skeletal muscle. *Nature.* 316:347–349.

Zampighi, G., F. Ramon, and J. Vergara. 1975. On the connection between the transverse tubules and the plasma membrane in frog semitendinosus skeletal muscle. *Journal of Cell Biology.* 64:734–740.

Modulation of Membrane Transport by Intracellular Calcium

Chapter 10

Characterization of Voltage-gated Calcium Channels in *Xenopus* Oocytes after Injection of RNA from Electrically Excitable Tissues

Terry P. Snutch, John P. Leonard, Joël Nargeot, Hermann Lübbert, Norman Davidson, and Henry A. Lester

Divisions of Biology and Chemistry, California Institute of Technology, Pasadena, California

Introduction

The entry of Ca^{2+} through voltage-gated channels has two major functions. First, Ca^{2+} fluxes elevate the intracellular concentration of this important second messenger. Second, Ca^{2+} currents directly influence membrane potential, thereby contributing to impulse patterns. Ca^{2+} channels serve these functions in a variety of nerve, muscle, and endocrine cells (Reuter, 1983; Tsien, 1983). As might be expected from their wide distribution and their role in regulating a number of cellular functions, voltage-gated Ca^{2+} channels form a heterogenous family. Diverse types of Ca^{2+} channels can be distinguished with regard to gating kinetics, pharmacology, and permeability (Fox and Krasne, 1981; Armstrong and Matteson, 1985; Nowycky et al., 1985). For the purpose of characterizing the various forms of voltage-gated Ca^{2+} channels, it would be desirable to study them in a similar membrane environment.

The utilization of *Xenopus* oocytes as an expression system for the study of exogenous voltage-gated ion channels and neurotransmitter receptors was initially demonstrated by Barnard, Miledi, and co-workers (Barnard et al., 1982; Miledi et al., 1982; Gundersen et al., 1983). While these studies used mRNA isolated from a number of tissues and organisms, it has also been demonstrated that RNA synthesized in vitro from the cDNAs of the cloned nicotinic acetylcholine receptor and the voltage-gated Na^{+} channel are translated in *Xenopus* oocytes to give functional transmitter and voltage-gated channels, respectively (Mishina et al., 1985; White et al., 1985; Noda et al., 1986). This chapter describes a means of studying several aspects of the diversity, molecular biology, and physiology of voltage-sensitive Ca^{2+} channels using *Xenopus* oocytes that have been injected with mRNA isolated from electrically excitable tissues.

Experimental Procedures

The study of ion channels synthesized from exogenous mRNA in *Xenopus* oocytes requires both a steady supply of healthy *Xenopus laevis* frogs and a source of high-quality RNA. A number of companies, including Xenopus One (Ann Arbor, MI) and Nasco Biologicals, Inc. (Fort Atkinson, WI), supply mature females that have previously been injected with human chorionic gonadotropin to show that they contain ripe oocytes and to induce a fresh round of oocyte development (noninjected females are also available). The ovaries of such frogs usually contain stage 5 and stage 6 oocytes (Dumont, 1972) capable of translating exogenous mRNA (Gurdon et al., 1971). The care and maintenance of *Xenopus* frogs in the laboratory is relatively simple. Several discussions on this subject are available (Nieuwkoop and Faber, 1967; Brown, 1970).

RNA suitable for injection can be isolated from the tissue or cell line of interest using any procedure that results in a good yield of high-molecular-weight RNA (the rat brain Na channel, for example, is encoded by an ~9-kb mRNA). In selecting an RNA-extraction method, one must consider the nature of the source tissue. In our hands, good yields of high-molecular-weight, translatable RNA can be obtained from the brain using a modification of the LiCl-urea procedure of Auffray and Rougeon (1980). The isolation of RNA of similar quality from tissues such as the heart and skeletal muscle, however, requires the use of a more effective

RNAse inhibitor. For these purposes, we used a modification of the guanidine-HCl procedure of Chirgwin et al. (1979).

The preparation and injection of oocytes is straightforward and can easily be accomplished in a single day. Frogs are anesthetized with 0.15% tricaine and several ovarian lobes are excised. The ovarian tissue and follicular cells surrounding the oocytes can be removed either manually or enzymatically. Both procedures apparently cause some transient damage to the oocyte membrane; however, this does not seem to affect electrophysiological experiments performed 2–4 d after mRNA injection (Dascal, 1986). We do, however, allow the oocytes to recover from the defolliculation procedure for several hours before the injection of mRNA. Injection is performed by positive displacement using a 10-μl micropipette (Drummond Scientific Co., Broomall, PA). While the gradations on this instrument theoretically allow for distinction between 10-nl aliquots of solution, it is best to initially calibrate the actual volume injected and then to assume that this volume is constant per division for further injections. Electrophysiology is performed with a standard two-microelectrode voltage-clamp circuit with oocytes that have been incubated for 2–7 d in ND96 supplemented with antibiotics as previously described (Leonard et al., 1987).

Results and Discussion

Detection of Voltage-gated Ca^{2+} Channels in *Xenopus* Oocytes

Under voltage-clamp conditions at potentials near the normal resting potential of the oocyte (−50 to −70 mV), an inward current can be evoked by intracellular injection of Ca^{2+} (Barish, 1983). In addition, depolarization to potentials more positive than −20 mV elicits a transient outward current (Miledi, 1982). These two currents appear to result from activation of endogenous Ca^{2+}-dependent Cl^- channels. The difference in the direction of current flow in these two cases reflects the fact that the Nernst equilibrium potential for Cl^- in *Xenopus* oocytes is approximately −25 mV. In normal physiological solution, depolarization from a holding potential of −100 to 0 mV produces a transient outward current, usually <100 nA in amplitude and 0.5–1 s in duration. The current is inactivated by >75% at a holding potential of −40 mV (Fig. 1*A*). The absolute dependence of the detected Cl^- current on both external and intracellular Ca^{2+} and its voltage-dependent activation properties suggest that it occurs because Ca^{2+}, entering the cell through voltage-activated Ca^{2+} channels, in turn activates the Cl^- channel $I_{Cl(Ca)}$ (Barish, 1983; Miledi, 1982).

In order to characterize voltage-gated Ca^{2+} channels in oocytes injected with RNA from excitable tissues, it was first necessary to identify and characterize any endogenous Ca^{2+} currents in uninjected oocytes. Substitution of Ca^{2+} with high-Ba^{2+} saline results in a significant reduction in the amplitude of $I_{Cl(Ca)}$, allowing the initial detection of a net inward current upon depolarization from −80 to +10 mV. The replacement of Ca^{2+} with Ba^{2+} does not completely inhibit $I_{Cl(Ca)}$, however, as evidenced by the rapid (within 50 ms) appearance of the outward $I_{Cl(Ca)}$ (for example, see Fig. 3 with a brain RNA–injected oocyte). In order to eliminate the remaining $I_{Cl(Ca)}$, it was also necessary to replace external Cl^- with methanesulfonate. Under these conditions, the endogenous Ca^{2+} channels could be studied (with Ba^{2+} as the current carrier; Dascal et al., 1986; Leonard et al., 1987). I_{Ba} in control oocytes is transient, inactivating with an exponential time constant of ~0.12 s.

Like $I_{Cl(Ca)}$, the endogenous Ba^{2+} current is almost completely inactivated by holding potentials of −40 mV. I_{Ba} is not blocked by removal of Na^+ from the bath, nor is it inhibited by 1 μM tetrodotoxin (TTX). It is, however, completely blocked by 50 μM Cd^{2+}, which suggests that the current does indeed involve voltage-gated Ca^{2+} channels.

In oocytes that have been injected with mRNA isolated from excitable tissues, the amplitude and waveform of $I_{Cl(Ca)}$ are dramatically altered as compared with

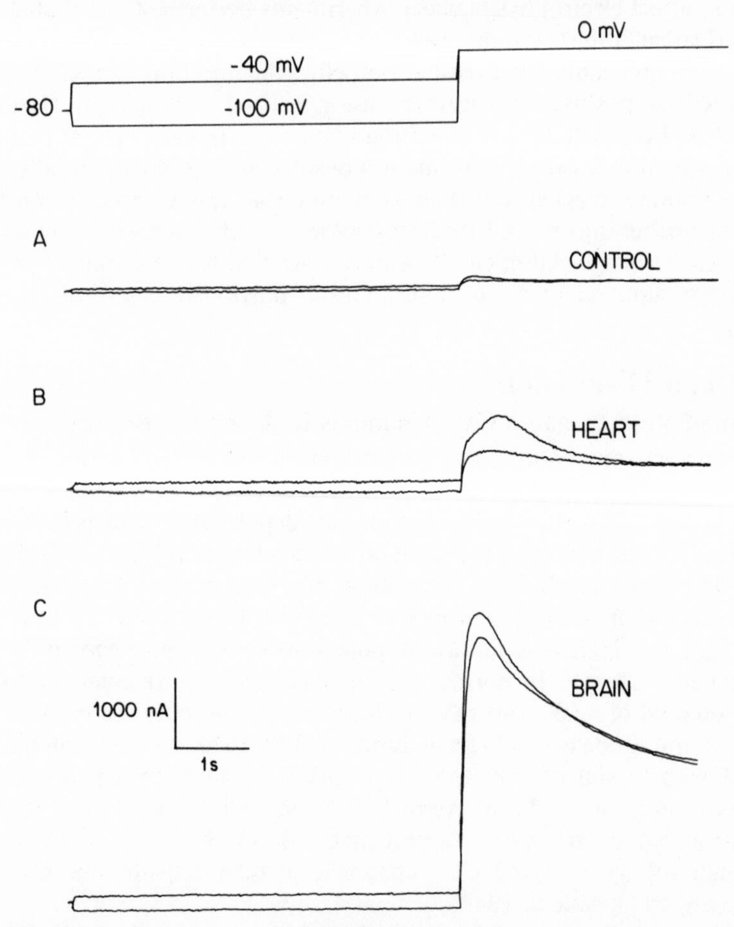

Figure 1. Comparison of $I_{Cl(Ca)}$ in noninjected oocytes (*A*) and oocytes injected with either rat heart RNA (*B*) or rat brain RNA (*C*). The pulse protocol is shown at the top of the figure. Oocytes were bathed in normal saline during recording.

those of control cells (Dascal et al., 1986; Leonard et al., 1987). Fig. 1 shows, for example, that oocytes injected with rat heart RNA often show an $I_{Cl(Ca)}$ with two partially overlapping peaks. For brain RNA in particular, the amplitude of $I_{Cl(Ca)}$ is typically 30–50 times greater and lasts several times longer than in control oocytes (Fig. 1 *C*). It is also noteworthy that in RNA-injected oocytes, $I_{Cl(Ca)}$ inactivates less than in control cells when depolarized from a holding potential of −40 mV.

Clearly, the translation of exogenous RNA from electrically excitable tissues by the oocyte provides for additional components that modify $I_{Cl(Ca)}$. The simplest explanations to account for these observations include the following. (*a*) RNA encodes for a distinct population of voltage-dependent Ca^{2+} channels, which, upon membrane depolarization, open and allow additional Ca^{2+} to enter the cell and activate the endogenous Cl^- channels. (*b*) RNA encodes a distinct population of Cl^- channels that are either voltage-activated upon membrane depolarization or are Ca^{2+}-activated, and have different properties than the endogenous Ca^{2+}-activated Cl^- channels. (*c*) RNA encodes proteins that modify the endogenous components of $I_{Cl(Ca)}$. (*d*) Any combination of the above may be possible.

An examination of I_{Ba} in oocytes injected with brain, heart, and skeletal muscle RNA shows that, at the very least, a newly synthesized population of voltage-gated Ca^{2+} channels can accoun* for the modified $I_{Cl(Ca)}$ (Dascal et al., 1986; Leonard et al., 1987). In oocytes injected with these RNAs, I_{Ba} is significantly enhanced, shows different activation and inactivation properties, and responds to the application of drugs and neurotransmitters differently than does the endogenous I_{Ba}. In addition, oocytes injected with RNA from each of these tissues display a unique set of Ba^{2+} currents. The next section describes the Ca^{2+} channels synthesized by *Xenopus* oocytes after the injection of rat heart and brain RNAs.

Heart RNA–directed Ca^{2+} Channels

Fig. 2 shows a typical I_{Ba} resulting from the injection of heart RNA isolated from neonatal rats (2–8 d). Depolarization from a holding potential of −80 mV to a test potential of −10 mV activates an inward current that is maintained for several seconds. The current-voltage relationship for the heart I_{Ba} shows that it activates at potentials more positive than −30 mV and peaks at approximately +15 mV. It is completely blocked by 20 μM Cd^{2+}.

Several 1,4-dihydropyridine compounds, for instance nifedipine and nitrendipine, are used clinically to treat a number of cardiovascular disorders, including angina, hypertension, and arrhythmias. These "Ca^{2+} channel antagonists" (Fleckenstein, 1977, 1983) bind specifically to voltage-gated Ca^{2+} channels and block the voltage-dependent flow of Ca^{2+} into the cell. Other dihydropyridines, for instance BAY K 8644, act as Ca^{2+} channel agonists and enhance the activation of Ca^{2+} channels (Schramm et al., 1983; Hess et al., 1984). Because the dihydropyridines are thought to bind specifically to voltage-gated Ca^{2+} channels in cardiac tissue, one might expect these compounds to affect the I_{Ba} observed in oocytes injected with heart RNA. Indeed, the noninactivating I_{Ba} observed in heart RNA–injected oocytes is specifically inhibited by nifedipine (1–10 μM; Dascal et al., 1986). In addition, the Ca^{2+} channel agonist BAY K 8644 selectively enhances the heart I_{Ba} by ~65% for a depolarization from −80 to −10 mV (Fig. 2; $n = 2$). Significantly, of the two types of Ca^{2+} channels observed in heart atrial and ventricular cells, it is the noninactivating "L-type" channel that is sensitive to dihydropyridines (Nowycky et al., 1985; Bean, 1985; Nilius et al., 1985). Thus, the I_{Ba} we observe in heart RNA–injected oocytes appears similar to the heart L-type channels in its activation and inactivation properties and in its sensitivity to dihydropyridines. The observed interactions of the dihydropyridines with the voltage-gated Ca^{2+} channels synthesized in *Xenopus* oocytes suggest that the oocytes have assembled functional Ca^{2+} polypeptides with many normal functional characteristics.

In previous studies with oocytes injected with heart RNA, we reported a transient I_{Ba} that decays in <0.5 s (Dascal et al., 1986). This current was insensitive to dihydropyridines and showed a slightly different current-voltage relationship from the noninactivating I_{Ba}. The transient I_{Ba} was not detected in all heart RNA–injected oocytes and was smaller than the noninactivating component when it was present. The transient I_{Ba} possessed some properties similar to those of the transient I_{Ba} observed in control oocytes and some properties similar to the transient T-type Ca^{2+} channel observed in ventricular tissue (Nilius et al., 1985). For these reasons, at this time we cannot positively identify the transient I_{Ba} observed in some heart RNA–injected oocytes as specific to the heart RNA preparation rather than to an unusual endogenous I_{Ba}.

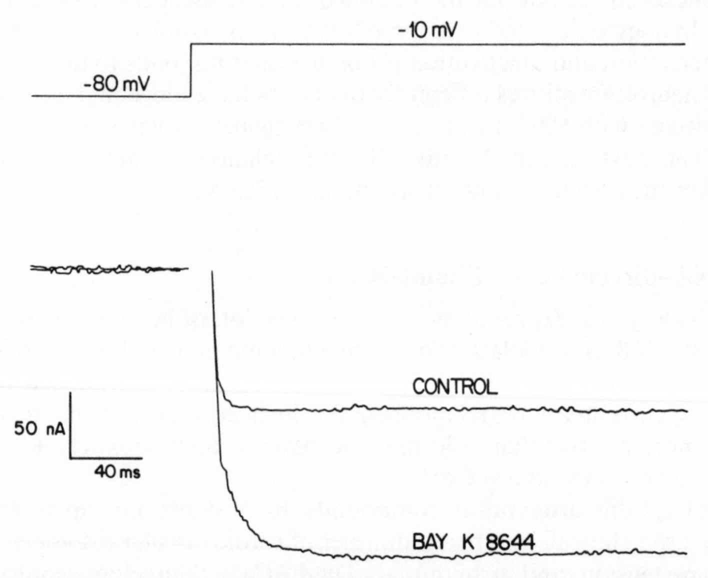

Figure 2. Enhancement of I_{Ba} by BAY K 8644 in an oocyte injected with rat heart RNA. I_{Ba} was elicited by a depolarization to −10 mV from a holding potential of −80 mV. The bathing solution contained Ba methanesulfonate saline (40 mM Ba, 50 mM Na, 2 mM K, 5 mM HEPES, pH 7.2, with methanesulfonate) and 1 μM TTX.

In the heart, Ca^{2+} currents are modulated by the antagonistic effects of acetylcholine and β-adrenergic agonists (Vassort et al., 1969; Reuter, 1983; Tsien, 1983). Ca^{2+} currents are enhanced by β-adrenergic agonists through the elevation of cAMP levels and the subsequent activation of a cAMP-dependent protein kinase. In fact, the slow, L-type Ca^{2+} channel in cardiac tissue is specifically modulated by cAMP-dependent phosphorylation (Bean, 1985). In oocytes injected with heart RNA (Dascal et al., 1986), enhancement of the noninactivating I_{Ba} was observed with three independent activators of the β-adrenergic pathway. (*a*) The application of the β-adrenergic agonist isoproterenol (10 μM) enhanced I_{Ba} to ~135% of the control level. The effect was reversible and was inhibited by 10 μM propranolol. (*b*) The application of the adenylate cyclase activator forskolin (40 μM) increased I_{Ba} to ~150% of the control level. (*c*) The injection of cAMP into the oocyte (2–4 pmol/oocyte) enhanced I_{Ba} to ~130% of the control level.

One effect of acetylcholine on heart is to decrease Ca^{2+} currents, possibly by inhibition of adenylate cyclase (Murad et al., 1962) and possibly by activating cAMP phosphodiesterase (Hartzell and Fischmeister, 1986). The application of 10 μM acetylcholine to heart RNA-injected oocytes reduced I_{Ba} to ~65% of the control level, which is consistent with its effect on intact cardiac tissue.

The modulation of the heart I_{Ba} in RNA-injected oocytes is quite striking: all the necessary components for a number of multistep regulatory pathways are functional and are able to correctly modulate the newly synthesized Ca^{2+} channels. It should be noted that *Xenopus* oocytes possess endogenous muscarinic, adrenergic (Kusano et al., 1977; Dascal and Landau, 1980; Dascal et al., 1984), β-adrenergic, and purinergic (Lotan et al., 1982) responses, and that at least two second-messenger pathways (including the cAMP–A kinase pathway) operate to mediate these responses (reviewed by Dascal, 1986). Thus, we do not know whether the heart RNA encodes the necessary modulatory components or whether all of the observed modulations of I_{Ba} are directed by the oocyte's inherent machinery. This question might be approached by size fractionation of the heart RNA into discrete classes of mRNAs before injection.

Brain RNA–directed Ca^{2+} Channels

Voltage-dependent Ca^{2+} channels are found in the cell bodies, dendrites, and synaptic termini of neurons (reviewed by Miller, 1986). At the presynaptic nerve terminal, they perform the important function of introducing depolarization-dependent bursts of Ca^{2+} into the cytoplasm, which in turn initiate the release of neurotransmitter into the synaptic cleft. Because it is presently not possible to measure directly the Ca^{2+} currents at the presynaptic termini of mammalian central nervous system (CNS) neurons, much of the present knowledge of presynaptic currents is based on the study of pinched-off nerve endings called synaptosomes. While the analysis of synaptosomes has provided much information regarding presynaptic Ca^{2+} channels, it would be beneficial to have an independent method with which to study these currents. The analysis of Ca^{2+} channels synthesized from CNS-derived RNA after injection into *Xenopus* oocytes may offer a practical alternative.

Oocytes injected with rat brain RNA (14–16 d postnatal) show both an increased $I_{Cl(Ca)}$ (up to 4 μA from holding potentials of −100 mV; e.g., Fig. 1*C*) and increased I_{Ba} (up to 1.2 μA) compared with uninjected oocytes. Both currents also show less inactivation at a holding potential of −40 mV as compared with the endogenous I_{Ba} (Leonard et al., 1987). Fig. 3*A* shows a record of a brain RNA–induced I_{Ba} in the presence of 40 mM $BaCl_2$ saline and 1 μM TTX. In this solution, a large, rapidly activating inward current that peaks at +10 mV is observed. Within 20–30 ms of the maximum inward peak, an outward current appears. Fig. 3*B* shows that the majority of the outward current can be blocked by the Cl channel blocker 9-anthracene carboxylic acid (9-AC), which suggests that $I_{Cl(Ca)}$ is responsible for a large portion of the outward current. In 9-AC with the addition of 100 μM Cd^{2+} to block the inward I_{Ba}, a small amount of outward current remains at more positive potentials (Fig. 3*C*). This is probably a brain RNA–encoded K^+ current, which complicates the detection of I_{Ba}. In order to circumvent this problem, we routinely incubate the oocytes in saline in which K^+ has been replaced by Cs^+ for 1–2 d before voltage-clamping. This Cs^+ treatment blocks ~70% of the outward current remaining after block with 9-AC and Cd^{2+}, as shown in Fig. 3. The brain RNA I_{Ba} measured in Ba methanesulfonate is similar to that shown in Fig. 3*B* for

$BaCl_2$ and 9-AC. The current peaks at approximately +15 mV when depolarized from a holding potential of −80 mV and is very sensitive to low concentrations of Cd^{2+} (50% block at 6 μM). The brain RNA I_{Ba} is insensitive to the effects of dihydropyridine antagonists and agonists (Leonard et al., 1987).

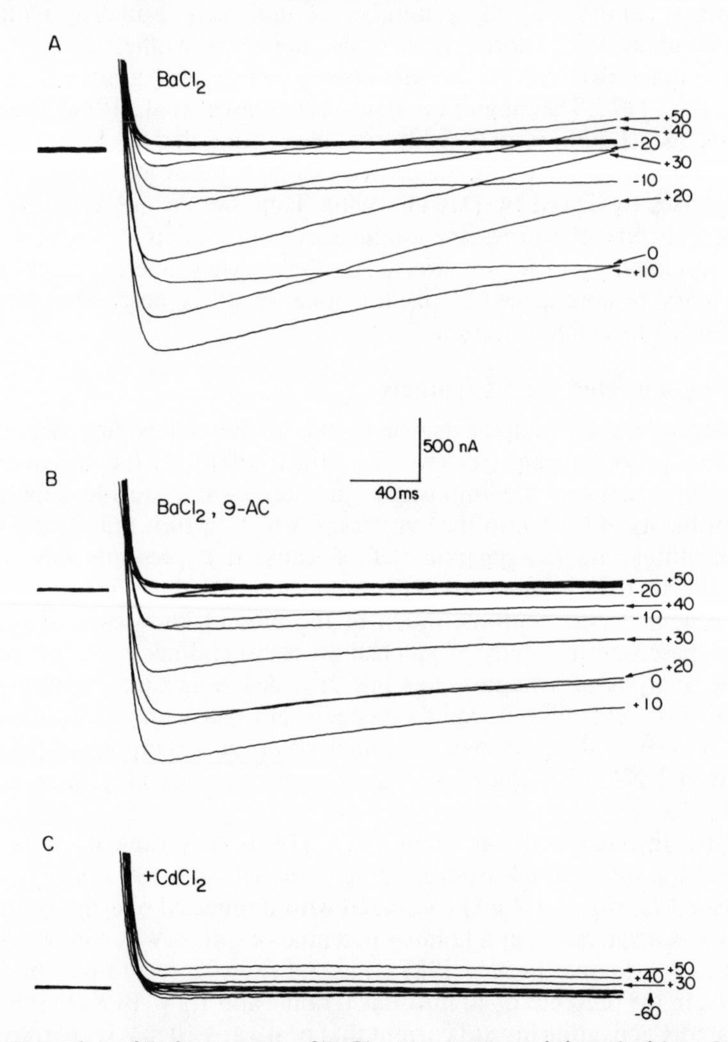

Figure 3. Detection of I_{Ba} in presence of $BaCl_2$. An oocyte was injected with rat brain RNA and voltage-clamped in the presence of $BaCl_2$ saline with 1 μM TTX (*A*). With the addition of 1 mM 9-AC to block Cl^- channels, most of the outward current disappeared (*B*). The further addition of 100 μM Cd^{2+} eliminated the inward I_{Ba}, leaving an outward I_K that activated at positive potentials (*C*). The oocyte was incubated in saline in which the K^+ was replaced by Cs^+ for 24 h before recording.

As in mammalian heart, smooth muscle, and skeletal muscle, high-affinity dihydropyridine-binding sites exist in mammalian brain (Miller and Freedman, 1984). A recent review by Miller (1986) argues that some of these sites represent dihydropyridine-sensitive Ca^{2+} channels. To attempt to determine whether the difference in drug sensitivity between the I_{Ba} induced by heart and brain RNA

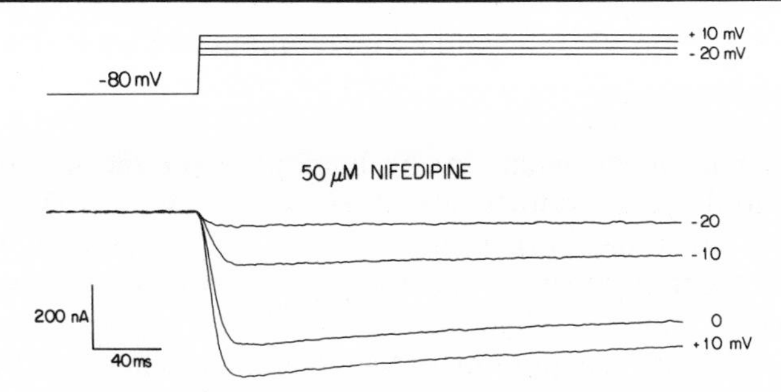

Figure 4. Effect of nifedipine on I_{Ba} in an oocyte injected with RNA isolated from rat brain hippocampus. The current observed in the presence of 50 μM nifedipine is identical to the current elicited in the absence of this antagonist. The pulse protocol is shown at top of the figure. I_{Ba} was detected in Ba methanesulfonate saline with 1 μM TTX.

simply reflects a much lower abundance of dihydropyridine-sensitive Ca^{2+} mRNA in the brain, we also tested RNA isolated from a region of the brain enriched in dihydropyridine-binding sites. Autoradiographic mapping of [^{3}H]dihydropyridine-binding sites in rat brain shows that one of the regions with the highest density of binding sites is the hippocampus (Cortes et al., 1984). If these dihydropyridine-binding sites represent functional Ca^{2+} channels, we might expect that RNA isolated from the hippocampus and injected into oocytes would show a nifedipine-sensitive I_{Ba}. As demonstrated in Fig. 4, this is not the case. The only detectable I_{Ba} in oocytes injected with RNA isolated from rat brain hippocampus is insensitive to nifedipine. Figs. 5 and 6 show that the I_{Ba} detected in oocytes injected with hippocampal RNA resembles the I_{Ba} detected with whole brain RNA with regard to current-voltage characteristics: the threshold for activation is around −40 mV and the peak is between +10 and +20 mV (Fig. 5). Similarly, the inactivation kinetics are similar: one-third of the peak brain I_{Ba} at +10 mV inactivates with a single-exponential time constant of 650 ms, while the remaining fraction inactivates very little over a 3-s depolarization. Also, the steady state inactivation profile of I_{Ba}

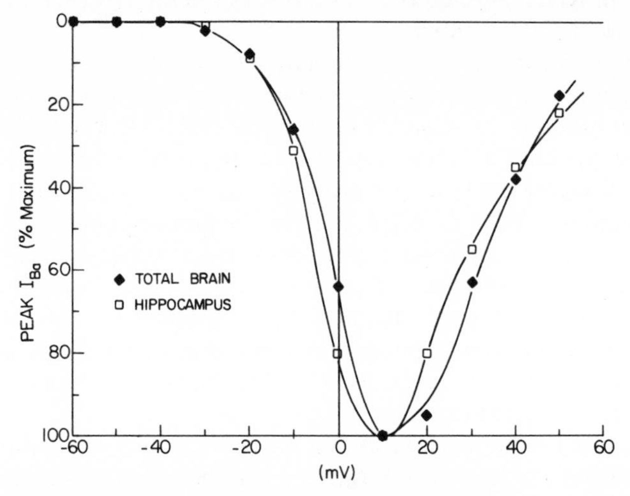

Figure 5. Comparison of the current-voltage relationship of I_{Ba} in oocytes injected with RNA isolated from either intact rat brain or the hippocampus.

in oocytes injected with RNA from the hippocampus and whole brain are similar (Fig. 6). Approximately one-third of I_{Ba} does not inactivate, even with prepulses to +10 mV.

There is some disagreement in the literature over the sensitivity of brain synaptosomes to dihydropyridines (Nachshen and Blaustein, 1979; Daniell et al., 1983; Rampe et al., 1984; Turner and Goldin, 1985; Suszkiw et al., 1986). Our results showing the insensitivity of the brain I_{Ba} to dihydropyridines are similar to those reporting the failure of dihydropyridines to block the depolarization-dependent uptake of Ca^{2+} into synaptosomes (Nachshen and Blaustein, 1979; Daniell et al., 1983; Rampe et al., 1984; Suszkiw et al., 1986). The time courses for inactivation of the I_{Ba} in brain RNA–injected oocytes and that reported for Ca^{2+} uptake into synaptosomes are similar, which suggests the possibility that the I_{Ba} that we

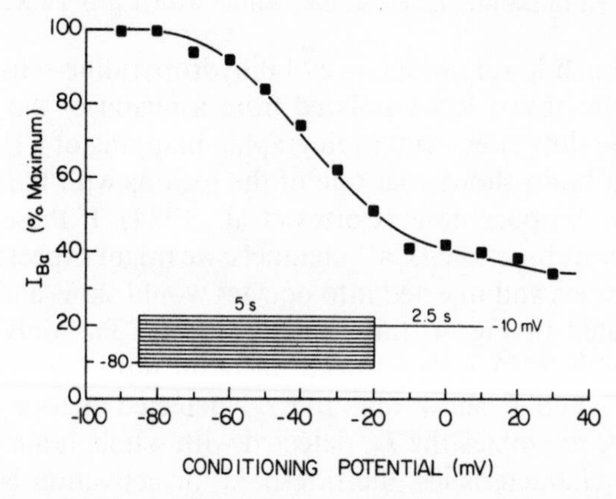

Figure 6. Inactivation of I_{Ba} in oocytes injected with hippocampus RNA. Note that I_{Ba} does not completely inactivate. The result is identical to that found with RNA from intact rat brain.

observe in brain RNA–injected oocytes is due to the mRNA encoding presynaptic voltage-dependent Ca^{2+} channels.

It should be noted that several alternative explanations exist for these results. The predominant heart and brain Ca^{2+} channels may have different primary amino acid sequences and may be encoded by different mRNAs. The brain mRNA may encode several types of Ca^{2+} channels, which in brain are dihydropyridine resistant and sensitive. Possibly the latter are properly expressed in oocytes but are too low in relative abundance to be detected, even in the hippocampus RNA preparation. Alternatively, the brain dihydropyridine-sensitive Ca^{2+} channels may not undergo the correct post-translational modifications to acquire dihydropyridine sensitivity, although the heart RNA Ca^{2+} channels do so. Finally, we cannot rule out the possibility that another polypeptide, either encoded in the RNA preparation or endogenous to the oocyte, interacts with the brain Ca^{2+} channel to alter its sensitivity to dihydropyridines.

In addition to kinetic differences and the lack of effect of nifedipine, the brain I_{Ba} is distinct from the heart RNA I_{Ba} by a number of other criteria. Unlike the heart RNA I_{Ba}, the brain RNA I_{Ba} is insensitive to forskolin, which activates adenylate cyclase, and is not blocked by the application of acetylcholine. Table I

Table I
Ba^{2+} Currents in *Xenopus* Oocytes

	Endogenous	Induced by Rat mRNA	
		Heart (2–8 d)	Brain (14–16 d)
Time course	Transient (0.12 s)	Maintained	Partial inactivation (0.65 s)
50% inactivation	−60 mV	—	−20 mV
Peak current (*mV*)	+15±5	+15±5	+15±5
Cd^{2+} block	<50 μM	~20 μM	6 μM
Nifedipine	No effect*	Block	No effect
Bay K 8644	?	Increase	No effect
A kinase	No effect	Increase	No effect
Acetylcholine	No effect	Decrease	No effect
C kinase	?	?	Increase

* $I_{Cl(Ca)}$ measurements.

summarizes these and other characteristics of the Ba^{2+} currents we have studied in *Xenopus* oocytes.

Conclusions

The main conclusion from our results is that one can distinguish among diverse types of Ca^{2+} channels directed by exogenous mRNA in *Xenopus* oocytes. In theory, it should be possible to extract mRNA from any number of tissues or tissue culture cell lines, to inject the mRNA into *Xenopus* oocytes, and to characterize the Ca^{2+} channels synthesized. A more thorough analysis of the Ca^{2+} channels than that reported here, including the study of channel modulation, should be possible using the "big patch" technique (Leonard et al., 1986) and single channel recording methods. The fact that, in both heart and brain RNA–injected oocytes, one type of I_{Ba} dominates could reflect mRNA abundances in these tissues or represent a bias on the part of the oocyte's translational and post-translational machinery for certain mRNAs and polypeptides.

Acknowledgments

We thank Dr. Douglas Krafte and Dr. Nathan Dascal for their comments on this manuscript. We also thank Moira Fearey for excellent technical assistance.

This research was supported by fellowships from the American Heart Association, Greater Los Angeles Affiliate, and the Natural Science and Engineering Research Council of Canada, and by grants GM-10991, GM-29836, and HL-35782 from the National Institutes of Health. J.N. thanks Laboratoire Servier (Neuilly-sur-Seine) for a travel grant.

References

Armstrong, C. M., and D. R. Matteson. 1985. Two distinct populations of calcium channels in a clonal line of pituitary cells. *Science.* 227:65–67.

Auffray, C., and F. Rougeon. 1980. Purification of mouse immunoglobulin heavy-chain messenger RNAs from total myeloma tumor RNA. *European Journal of Biochemistry.* 107:303–324.

Barish, M. E. 1983. A transient calcium-dependent chloride current in the immature *Xenopus* oocyte. *Journal of Physiology.* 342:309–325.

Barnard, E. A., R. Miledi, and K. Sumikawa. 1982. Translation of exogenous messenger RNA coding for nicotinic acetylcholine receptors produces functional receptors in *Xenopus* oocytes. *Proceedings of the Royal Society of London, Series B.* 215:241–246.

Bean, B. P. 1985. Two kinds of calcium channels in canine atrial cells. Differences in kinetics, selectivity, and pharmacology. *Journal of General Physiology.* 86:1–30.

Brown, A. L. 1970. The African Clawed Toad *Xenopus laevis*: a Guide for Laboratory Practical Work. Butterworths, London. 140 pp.

Chirgwin, J. M., A. E. Przybyla, R. J. MacDonald, and W. J. Rutter. 1979. Isolation of biologically active ribonucleic acid from sources enriched in ribonuclease. *Biochemistry.* 18:5294–5299.

Cortes, R., P. Supavilai, M. Karobath, and J. M. Palacios. 1984. Calcium antagonist binding sites in the rat brain: quantitative autoradiographic mapping using the 1,4 dihydropyridines [^{3}H]N 200-110 and [^{3}H]Y 108-068. *Journal of Neural Transmission.* 60:169–197.

Daniell, L. C., E. M. Barr, and S. W. Leslie. 1983. $^{45}Ca^{2+}$ uptake into rat whole brain synaptosomes unaltered by dihydropyridine calcium antagonists. *Journal of Neurochemistry.* 41:1455–1459.

Dascal, N. 1986. Use of the *Xenopus* oocyte system to study ion channels. *CRC Critical Reviews in Biochemistry.* In press.

Dascal, N., and E. M. Landau. 1980. Types of muscarinic response in *Xenopus* oocytes. *Life Sciences.* 27:1423–1428.

Dascal, N., E. M. Landau, and Y. Lass. 1984. *Xenopus* oocyte resting potential, muscarinic responses and the role of calcium and guanosine 3′,5′-cyclic monophosphate. *Journal of Physiology.* 352:551–574.

Dascal, N., T. P. Snutch, H. Lubbert, N. Davidson, and H. A. Lester. 1986. Expression and modulation of voltage gated-calcium channels after RNA injection in *Xenopus* oocytes. *Science.* 231:1147–1150.

Dumont, J. N. 1972. Oogenesis in *Xenopus laevis* (Daudin). I. Stages of oocyte development in laboratory maintained animals. *Journal of Morphology.* 136:153–180.

Fleckenstein, A. 1977. Specific pharmacology of calcium in myocardium, cardiac pacemakers, and vascular smooth muscle. *Annual Review of Pharmacology and Toxicology.* 17:149–166.

Fleckenstein, A. 1983. History of calcium antagonists. *Circulation Research.* 53(Suppl. 1):3–16.

Fox, A. P., and S. Krasne. 1981. Two calcium currents in egg cells. *Biophysical Journal.* 33:145*a*. (Abstr.)

Gundersen, C. B., R. Miledi, and I. Parker. 1983. Voltage-operated channels induced by foreign messenger RNA in *Xenopus* oocytes. *Proceedings of the Royal Society of London, Series B.* 220:131–140.

Gurdon, J. B., C. D. Lane, H. R. Woodland, and G. Marbaix. 1971. Use of frog eggs and oocytes for the study of messenger RNA and its translation in living cells. *Nature.* 233:177–182.

Hartzell, H. C., and R. Fischmeister. 1986. Opposite effects of cyclic GMP and cyclic AMP on Ca^{2+} current in single heart cells. *Nature.* 323:273–275.

Hess, P., J. B. Lansman, and R. W. Tsien. 1984. Different modes of Ca channel gating behaviour favoured by dihydropyridine Ca agonists and antagonists. *Nature.* 311:538–544.

Kusano, K., R. Miledi, and J. Stinnakre. 1977. Acetylcholine receptors in the oocyte membrane. *Nature.* 270:739–741.

Leonard, J. P., J. Nargeot, T. P. Snutch, N. Davidson, and H. A. Lester. 1987. Ca channels induced in *Xenopus* oocytes by rat brain mRNA. *Journal of Neuroscience.* 7:875–881.

Leonard, J. P., T. Snutch, H. Lübbert, N. Davidson, and H. A. Lester. 1986. Macroscopic Na currents with gigaohm seals on mRNA-injected *Xenopus* oocytes. *Biophysical Journal.* 49:386*a*. (Abstr.)

Lotan, I., N. Dascal, S. Cohen, and Y. Lass. 1982. Adenosine-induced slow inward currents in the *Xenopus* oocyte. *Nature.* 298:572–574.

Miledi, R. 1982. A calcium-dependent transient outward current in *Xenopus laevis* oocytes. *Proceedings of the Royal Society of London, Series B.* 215:491–497.

Miledi, R., I. Parker, and K. Sumikawa. 1982. Synthesis of chick brain GABA receptors by frog oocytes. *Proceedings of the Royal Society of London, Series B.* 216:509–515.

Miller, R. J. 1986. Calcium channels in neurones. *Receptor Biochemistry and Methodology.* In press.

Miller, R. J., and S. B. Freedman. 1984. Are dihydropyridine binding sites voltage sensitive calcium channels? *Life Sciences.* 34:1205–1221.

Mishina, M., M. T. Tobimatsu, K. Imoto, K.-I. Tanaka, Y. Fujita, K. Fukuda, M. Kurasaki, H. Takahashi, Y. Morimoto, T. Hirose, S. Inayama, T. Takahashi, M. Kuno, and S. Numa. 1985. Localization of functional regions of acetylcholine receptor α-subunit by site-directed mutagenesis. *Nature.* 313:364–369.

Murad, F., Y.-M. Chi, T. W. Rall, and E. W. Sutherland. 1962. Adenyl cyclase. III. The effects of catecholamines and choline esters on the formation of adenosine 3′,5′-phosphate by preparations from cardiac muscle and liver. *Journal of Biological Chemistry.* 237:1233–1238.

Nachshen, D. A., and M. P. Blaustein. 1979. The effects of some organic "calcium antagonists" on calcium influx in presynaptic nerve terminals. *Molecular Pharmacology.* 16:579–586.

Nieuwkoop, P. D., and J. Faber. 1967. Normal Table of *Xenopus laevis* (Daudin). Elsevier/North-Holland, Amsterdam. 252 pp.

Nilius, B., P. Hess, J. B. Lansman, and R. W. Tsien. 1985. A novel type of cardiac calcium channel in ventricular cells. *Nature.* 316:443–446.

Noda, M., T. Ikeda, T. Kayano, H., Suzuki, H. Takeshima, M. Kurasaki, H. Takahashi, and S. Numa. 1986. Existence of distinct sodium channel messenger RNAs in rat brain. *Nature.* 320:188–192.

Nowycky, M. C., A. P. Fox, and R. W. Tsien. 1985. Long-opening mode of gating of neuronal calcium channels and its promotion by the dihydropyridine calcium agonist Bay K 8644. *Proceedings of the National Academy of Sciences.* 82:2178–2182.

Rampe, D., R. A. Janis, and D. J. Triggle. 1984. Bay K8644, a 1,4-dihydropyridine Ca^{2+} channel activator: dissociation of binding and functional effects in brain synaptosomes. *Journal of Neurochemistry.* 43:1688–1691.

Reuter, H. 1983. Calcium channel modulation by neurotransmitters, enzymes, and drugs. *Nature.* 301:569–574.

Schramm, M., G. Thomas, R. Towart, and G. Franckowiak. 1983. Novel dihydropyridines with positive inotropic action through inactivation of Ca^{2+} channels. *Nature.* 303:535–537.

Suszkiw, J. B., M. E. O'Learey, M. M. Murawsky, and T. Wang. 1986. Presynaptic calcium channels in rat cortical synaptosomes: fast-kinetics of phasic calcium influx, channel inactivation, and relationship to nitrendipine receptors. *Journal of Neuroscience.* 6:1349–1357.

Tsien, R. W. 1983. Calcium channels in excitable cell membranes. *Annual Review of Physiology.* 45:341–358.

Turner, T. J., and S. M. Goldin. 1985. Effects of dihydropyridine agonists and antagonists in Ca^{2+} uptake and neurotransmitter release by rat brain synaptosomes. *Society of Neuroscience Abstracts.* 11:579.

Vassort, G., O. Rougier, D. Garnier, M. P. Sauviat, E. Coraboeuf, and Y. M. Gargouil. 1969. Effects of adrenalin on membrane inward currents during the cardiac action potential. *Pflügers Archiv.* 309:70–81.

White, M. M., K. Mixter-Mayne, H. A. Lester, and N. Davidson. 1985. Mouse-*Torpedo* hybrid acetylcholine receptors: functional homology does not equal sequence homology. *Proceedings of the National Academy of Sciences.* 82:4852–4856.

Chapter 11

The Role of Cyclic AMP–dependent Phosphorylation in the Maintenance and Modulation of Voltage-activated Calcium Channels

John Chad, Daniel Kalman, and David Armstrong

Department of Biology, University of California, Los Angeles, California

In Memoriam: Roger Eckert, 1934–1986

Roger Eckert died of cancer on June 16, 1986. A few words cannot give a true account of the man. Although each of us knew him in a different way, we all have fond memories of Roger. Suffice it to say that he was a dedicated scientist, a proud father, and a warm and loyal friend to many. He is sadly missed by his family, friends, and colleagues.

The experiments reported here were carried out in Roger's laboratory at UCLA with his support and guidance. They were based on a decade of research by Roger and his students on Ca-mediated inactivation of Ca currents in *Paramecium* and molluscan neurons. Roger was also involved in the work presented here, but his untimely death prevented him from participating in writing this article. Therefore, the responsibility for any errors or omissions rests with the authors.

Introduction

Ca ions regulate vital processes in a wide variety of cells, and voltage-gated, Ca-selective ion channels in the plasma membrane provide a means of rapidly injecting Ca into the cell. In this chapter, we will review the evidence from our experiments on molluscan neurons and a mammalian cell line showing (*a*) that one class of voltage-activated Ca channels inactivates when Ca ion accumulation inside the cell results in dephosphorylation of the channels by a Ca-dependent phosphatase, and (*b*) that dephosphorylated channels remain inactivated until they are rephosphorylated by the cyclic AMP–dependent protein kinase. This dependence of Ca channel function on phosphorylation provides cells with a mechanism for modulating Ca fluxes across the cell membrane in response to external signals and to changes in the intracellular concentration of Ca or the metabolic state of the cell.

The inactivation of Ca currents in molluscan neurons is a Ca-dependent process; it depends on the entry and accumulation of Ca ions inside the cell (Tillotson, 1979; Eckert and Tillotson, 1981). Thus, depolarizing sufficiently to reduce Ca influx, substituting Ba for Ca, and buffering internal Ca with EGTA all reduce inactivation of the Ca current in those cells. This type of Ca current inactivation was analyzed in *Paramecium* by Brehm and Eckert (1978) and has been described subsequently in a wide variety of cells (reviewed by Eckert and Chad, 1984), including cardiac muscle (Mentrard et al., 1984; Lee et al., 1985; Nilius and Roder, 1985). The initial experiments implicating cAMP-dependent phosphorylation in the regulation of Ca currents were carried out on cardiac muscle cells over 10 years ago (Tsien et al., 1972), and subsequent studies have confirmed that activation of the cAMP-dependent protein kinase in the heart increases peak Ca currents and slows their inactivation by increasing the probability of channel opening on depolarization (Osterrieder et al., 1982; Reuter, 1983; Bean et al., 1984).

Further studies of Ca channel regulation have been hindered by the rapid loss of Ca current that occurs when the cytoplasmic side of the membrane is exposed to standard physiological saline solutions during whole-cell dialysis (Hagiwara and Nakajima, 1966; Byerly and Hagiwara, 1982) or after excision of an isolated patch of membrane (Fenwick et al., 1982; Cavalie et al., 1983; Nilius et al., 1985). Like the reversible inactivation of the Ca current in intact cells, this irreversible loss of activity, or "washout," in dialyzed cells is slowed by Ca buffers (Byerly and

Hagiwara, 1982; Fenwick et al., 1982) and substances that promote cAMP-dependent phosphorylation (Doroshenko et al., 1982, 1984; Forscher and Oxford, 1985; Byerly and Yazejian, 1986; Chad and Eckert, 1986).

Phosphorylation Maintains Ca Channel Activity

Dialyzed Molluscan Neurons

We have investigated the processes underlying washout in voltage-clamped *Helix aspersa* neurons dialyzed intracellularly with artificial saline solutions (Chad and Eckert, 1984, 1985*a*, 1986). In these experiments, Ca was used as the charge carrier and Ca buffers were omitted from the intracellular solutions to permit investigation of Ca-mediated processes such as inactivation. Voltage-activated Ca currents were isolated by ion substitutions and pharmacological blocking agents. Fig. 1*A* plots the normalized peak current in 50 mM Ca, during 100-ms voltage steps to +10 mV from a holding potential of −40 mV, vs. time during dialysis. When the dialysis solution contained only Cs-asparate, HEPES, and ATP-Mg, the peak Ca current declined rapidly (Fig. 1*A*, curve *a*). We have also observed that this time-dependent loss of Ca current, or washout, can be slowed by adding EGTA or the catalytic subunit (CS) of cAMP-dependent protein kinase to the dialysis solution; however, neither of those additions eliminated washout entirely, nor did they restore current that had already washed out (Chad and Eckert, 1986).

The results of such experiments were very different when the dialysis solution was supplemented with 0.1 mM leupeptin, an inhibitor of Ca-dependent proteases (Murachi, 1983). In the experiment illustrated in Fig. 1*A*, curve *b*, there was initially some instability in the peak current as the last traces of contaminating currents were blocked, but after 8 min a steady washout of Ca current was observed. Thus, leupeptin by itself has no effect on washout. CS was then added to the dialysis solution at the time indicated by the arrow. In the presence of leupeptin, the peak Ca current rapidly returned to its original value and remained stable for >1 h. Washout was fully reversible only when a protease inhibitor such as leupeptin was present throughout the perfusion. The restoration of Ca current by CS also depended on the presence of a hydrolyzable form of ATP with Mg^{2+} and was not observed when CS had been heat-inactivated (Chad and Eckert, 1986).

The Ca currents maintained by phosphorylation in the presence of leupeptin were activated maximally at +20 mV in 50 mM Ca, and were not altered by more negative holding potentials. Fig. 1*B* shows plots of the peak current-voltage relationship for one cell in three different divalent solutions of equal molarity (50 mM). In each case, the curve has a simple bell shape, but the order of maximum current is Ba > Sr > Ca. The substitution of Ba for Ca in the extracellular solution also caused a shift of the curve to more negative potentials. A similar but smaller shift was seen with Sr. Presumably, the shifts reflect differences in the efficacy with which different divalent ions screen surface charges. Thus, Ca channels that have a greater effective conductance to Ba than to Ca and are not affected by changes in holding potential seem to underlie the Ca currents maintained by phosphorylation.

Leak-corrected Ca currents recorded at different times during a dialysis experiment are shown in Fig. 2. The control trace (*a*), taken right after perfusion had been established with the minimal solution plus leupeptin, shows an inward current that peaked and then inactivated during the pulse; no outward tail current is visible. The current was then allowed to wash out partially (40%, not shown) before the addition of CS to the dialysis solution. The current restored by CS (trace *b*) had activation kinetics similar to the control but had slower inactivation kinetics. When

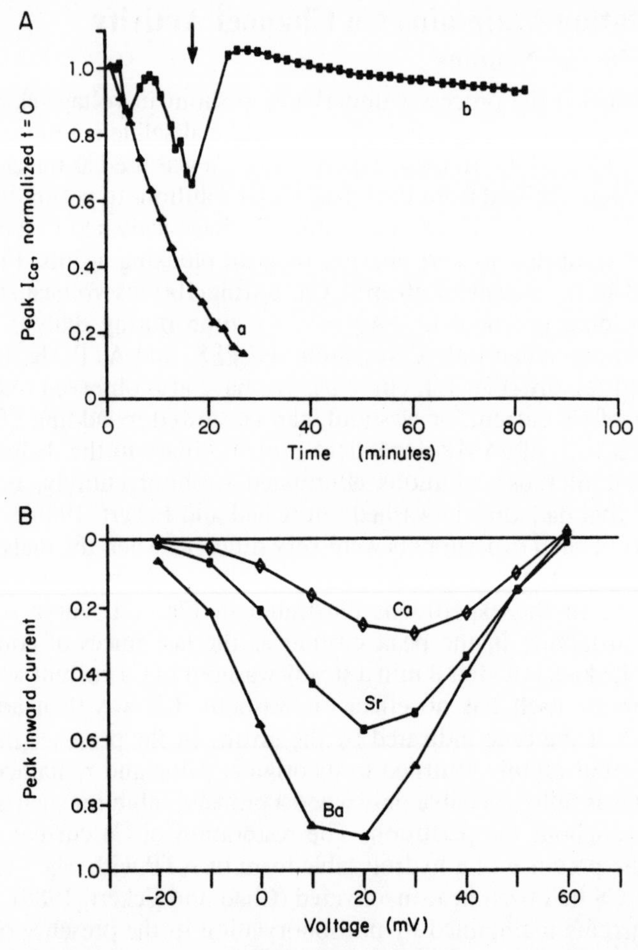

Figure 1. (*A*) The Ca current washes out during dialysis of a *Helix* neuron in the absence of kinase activity. Normalized peak amplitudes of Ca currents (leak-corrected) recorded during 100-ms steps to +10 mV are plotted vs. time during dialysis for two different cells. The extracellular concentration of Ca was 50 mM and the holding potential was −40 mV. (*a*) Dialysis of the cell with a Cs-aspartate solution that contained TEA (10 mM), HEPES (100 mM, pH 7.8), ATP-Mg (7 mM), dibutyryl cAMP (5 mM), and dithiothreitol (5 mg/ml) did not sustain the Ca current. (*b*) Another experiment in which 100 μM leupeptin was added to the dialysis solution. Initially, washout proceeded at the same rate as the control in trace *a*; however, the addition of the CS (20 μg/ml) of the cAMP-dependent protein kinase at the arrow produced full recovery of the Ca current, which persisted for the remainder of the experiment. (*B*) The amplitude of the inward current depends on the divalent ion carrying the charge. Peak current (×10 nA; leak-corrected) was plotted vs. step potential for a series of depolarizing pulses from a holding potential of −40 mV for the three divalent species. The external divalent concentration was 50 mM and the order of maximum current was Ba > Sr > Ca.

CS was again removed from the dialysis solution, the Ca current washed out until only a small fraction of the original current remained (trace *c*). The remaining current was inward throughout the pulse, which indicates an absence of contaminating outward currents.

Thus, it appears that dialysis disrupts the maintenance of Ca current in at least two ways, one reversible and another irreversible. Reversible washout can be prevented by the phosphorylating activity of cAMP-dependent protein kinase. We have interpreted this result to imply that the Ca channel, or a protein closely associated with it in the membrane, must be phosphorylated to respond fully to membrane depolarization. The irreversible component can be prevented by the protease inhibitor leupeptin. Since CS stabilized the current in the absence of protease inhibitors, but did not fully reverse the washout that had occurred before

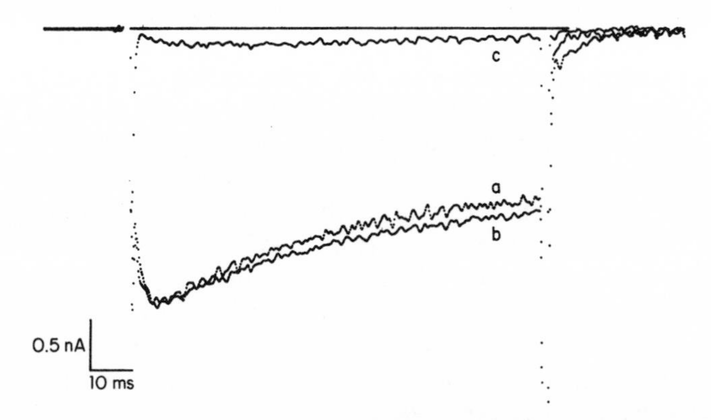

Figure 2. Ca currents recorded at different times during a dialysis experiment similar to that shown in Fig. 1*A*. Inward currents were elicited by 100-ms pulses to +10 mV from a holding potential of −40 mV and were leak-corrected. (*a*) After establishment of dialysis, contaminating currents were blocked and an initial Ca current was recorded. Note that all tail currents are inward. The Ca current washed out by 40%, although the dialysis solution contained both ATP-Mg and leupeptin (not shown). (*b*) The addition of CS (20 μg/ml) restored the peak current to its original value with no change in activation kinetics, although inactivation was slower. (*c*) When CS was removed from the dialysis solution, washout resumed at its original rate until the current was reduced to a small fraction of its original value. This small current remained inward throughout the pulse, showing an absence of voltage-activated, contaminating outward currents.

it was added, we also conclude that dephosphorylated Ca channels are more susceptible to proteolysis than phosphorylated channels (cf. Holzer and Heinrich, 1980).

Cell-free Membrane Patches from a Mammalian Cell Line

Further support for the role of phosphorylation in the maintenance of Ca channel activity has been obtained from analogous experiments on individual Ca channels in cell-free patches of membrane (Armstrong and Eckert, 1985, 1987) from an electrically excitable cell line (GH_3) derived from a rat pituitary tumor (Tashjian, 1979). To avoid the irreversible loss of activity attributed to proteolysis, these experiments were all conducted in the presence of 0.1 mM leupeptin.

The recent application of the patch-clamp technique to the measurement of currents flowing through individual ion channels (Hamill et al., 1981) has revealed at least three classes of voltage-activated Ca channels in mammalian cells (Carbone and Lux, 1984; Nowycky et al., 1985; Nilius et al., 1985). These classes are distinguished on the basis of their relative conductance to Ca and Ba, the voltage dependence of their activation, their sensitivity to antagonists, and their mechanism of inactivation (Matteson and Armstrong, 1986; McClesky et al., 1986). All electrically excitable mammalian cells tested thus far appear to have at least two of the classes in common. One class undergoes voltage-dependent activation and inactivation in a manner analogous to Na channels. These low-conductance (~10 pS in 90 mM Ba^{2+}), low-threshold (−40 mV), dihydropyridine-insensitive Ca channels inactivate rapidly at the relatively depolarized holding potential (−40 mV) used in our studies, but do not require ATP or exogenous kinases to remain active in cell-free patches of membrane (Armstrong and Eckert, 1985; Nilius et al., 1985).

The other widely distributed class of Ca channels (~23 pS in 90 mM Ba^{2+}) also exhibits voltage-dependent activation (threshold $\simeq -20$ mV) but differs in its mechanism of inactivation and its sensitivity to dihydropyridines. This class shows little voltage-dependent inactivation in Ba, but stops responding to depolarization altogether when the cytoplasmic side of the membrane is exposed to standard physiological saline solutions (Fenwick et al., 1982; Cavalie et al., 1983; Nilius et al., 1985). Our studies focus on this latter class of Ca channels, which closely resemble the phosphorylation-dependent Ca channels in *Helix* in their voltage-dependent properties. Therefore, the conditions that enabled the maintenance of Ca current in *Helix* neurons (Chad and Eckert, 1986) were modified for use with GH_3 cells (Armstrong and Eckert, 1987).

Fig. 3 illustrates the phosphorylation dependence of the unitary Ba currents through Ca channels in a patch of membrane from a GH_3 cell. Within 1 min of excising the patch of membrane with its cytoplasmic side facing a CsCl solution in the bath (inside out), the channel stopped responding to depolarization. This loss of activity occurred even though the concentration of Ca ions on the cytoplasmic side was buffered below 10 nM with 5 mM EGTA. The loss of activity was complete; no channel openings could be elicited at any voltage. Adding ATP to the solution perfusing the bath restored a small amount of activity in this patch, but when CS was added with ATP, activity eventually resumed at its original level.

In other patches, 2 mM ATP by itself prevented the loss of activity without added exogenous CS (Armstrong, 1986). An example is shown in Fig. 4*A*. However, unlike ATP plus CS, ATP by itself was not always sufficient to sustain activity. The effect of ATP did appear to be mediated by phosphorylation, as the nonhydrolyzable analogue of ATP, AMP-PCP, did not sustain channel activity (Fig. 4*B*). Because CS in brain tissue is often membrane-bound (Nairn et al., 1985), we investigated the possibility that ATP sustained activity by acting as a substrate for an endogenous kinase isolated with the patch. A specific protein inhibitor (PKI) of CS, with a dissociation constant near 1 nM, has been purified from skeletal muscle (Walsh et al., 1971). When PKI was added with ATP, Ca channel activity in most patches from GH_3 cells disappeared as though no ATP had been added (Fig. 4*C*). This result suggests that an endogenous cAMP-dependent kinase is often attached to the membrane in sufficient proximity to Ca channels to regulate their activity. That may help explain why extended recordings of Ca channel activity have been obtained in other preparations in the presence of ATP and Ca buffers but no exogenous CS (Carbone and Lux, 1984; Forscher and Oxford, 1985; Byerly and Yazejian, 1986).

An endogenous kinase cannot explain the successful reconstitution of Ca channel activity in the absence of ATP (Affolter and Coronado, 1985; Ehrlich et

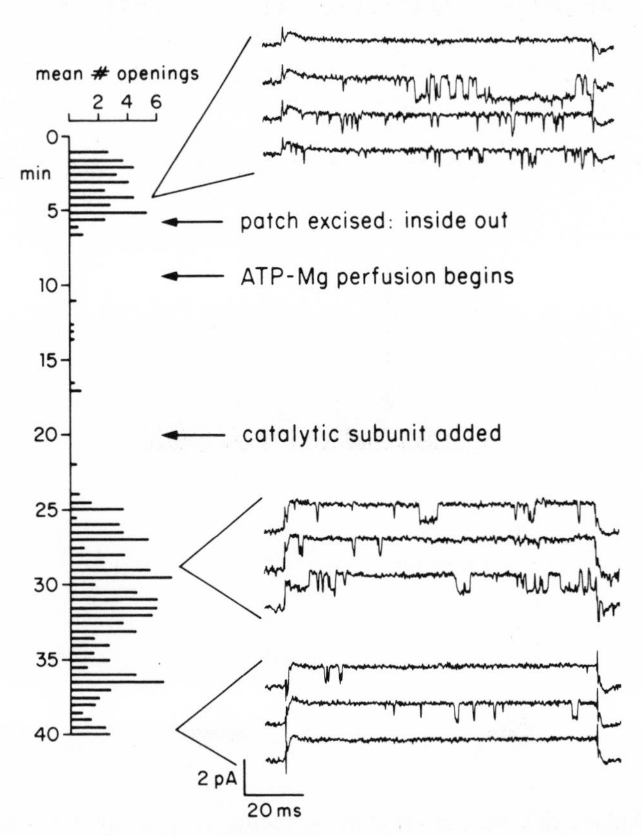

Figure 3. The CS of the cAMP-dependent protein kinase restores voltage-activated Ca channel activity in a cell-free patch of membrane from a GH_3 cell. The graph on the left plots the average number of channel openings during voltage steps from −40 to 0 mV vs. time after the formation of a gigohm seal with a patch pipette containing 90 mM Ba, 20 mM TEA, and 2 μM TTX. On the right, representative traces from the time indicated are displayed. Ca channel activity ceased within 1 min of excising the patch away from the cell and exposing the cytoplasmic side of the membrane to a minimal CsCl solution containing 40 mM HEPES, pH 7.2, 5 mM EGTA, pCa 8, and 0.1 mM leupeptin. Adding 2 mM ATP-Mg to the solution perfusing the bath restored very little activity. At the time indicated by the third arrow, perfusion of the bath was stopped and ~1 μg of purified CS was added. The experiment was halted after 40 min when the gigohm seal was lost. (Reproduced with permission from Armstrong and Eckert, 1987.)

al., 1986; Rosenberg et al., 1986). In those studies, the Ca channels in disrupted muscle membrane preparations were isolated by their affinity for dihydropyridines, clinically important drugs that modulate the class of Ca channels whose activity appears to depend on phosphorylation in our experiments (Hess et al., 1984). In particular, one dihydropyridine, BAY K 8644 (Schramm et al., 1983), produces a dramatic increase in the open time of these Ca channels. In the absence of BAY K 8644, the probability of successful reconstitution is very low (Affolter and Coronado, 1985). This could reflect the proportion of Ca channels that remain phosphorylated during the reconstitution procedure and/or some effect of BAY K 8644 on the susceptibility of the channels to dephosphorylation.

Therefore, we decided to test the ability of BAY K 8644 to prevent the loss of activity in patches exposed to the minimal CsCl solution buffered with HEPES and EGTA (Armstrong et al., 1987). As illustrated in Fig. 5, 1 μM BAY K 8644 delayed the loss of activity in cell-free patches exposed to the minimal saline solution, but only when it was added before the patches were excised (Armstrong and Eckert, 1987). In the absence of ATP, BAY K 8644 (0.1–10 μM) did not

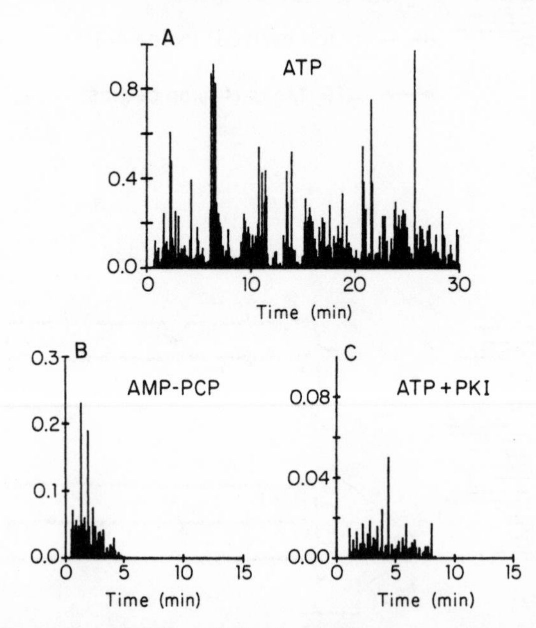

Figure 4. ATP alone can maintain activity by serving as a substrate for an endogenous kinase. In this figure, the fraction of time Ca channels spend in the open state during voltage steps to 0 mV is plotted vs. the time elapsed after the formation of a cell-free patch from a GH_3 cell. The CsCl solution bathing the cytoplasmic side of the membrane in these outside-out patches contained 2 mM Mg^{2+}, 40 mM HEPES, pH 7.2, 5 mM EGTA, pCa 8, 0.1 mM leupeptin, and the reagents indicated.

restore activity to Ca channels that had already stopped responding to depolarization (Fig. 5*B*). Since our experiments suggest that the loss of activity results from dephosphorylation of the channels, we postulate that Ca channels modulated by BAY K 8644 are less susceptible to dephosphorylation. This could explain how BAY K 8644 facilitates the reconstitution of these phosphorylation-dependent channels in the absence of ATP, and may underlie the effect of BAY K 8644 on gating.

Ca-dependent Inactivation

The experiments reported here all support the hypothesis that Ca channels in dialyzed molluscan neurons and cell-free membrane patches from a mammalian cell line do not respond to membrane depolarization under conditions that prevent cAMP-dependent phosphorylation. Under normal conditions, the inactivation of the phosphorylation-dependent Ca current in molluscan neurons results from Ca ion entry and accumulation inside the cell (Tillotson, 1979; Eckert and Tillotson, 1981). Consequently, Eckert and Chad (1984) hypothesized that inactivation might be produced by an endogenous, Ca-dependent phosphatase.

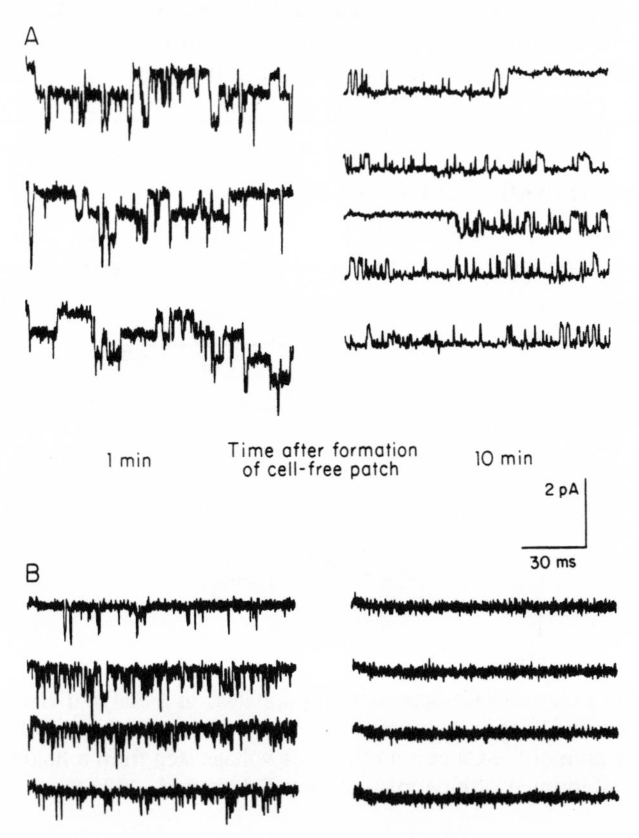

Figure 5. (*A*) Previous exposure to 1 μM BAY K 8644 delays the loss of voltage-activated Ca channel activity in cell-free membrane patches from GH_3 cells exposed to a minimal CsCl solution containing 40 mM HEPES, pH 7.2, 5 mM EGTA, pCa 8, and 0.1 mM leupeptin. (*B*) Adding 1 μM BAY K 8644 to another outside-out patch after activity had ceased did not restore activity in the absence of ATP. The records display representative activity during voltage steps from −40 to −20 mV in *A* (left), to 0 mV (right), to −10 mV in *B* (left), and to +10 mV (right).

Calcineurin Enhances Inactivation in Molluscan Neurons

Calcineurin is a Ca- and calmodulin-dependent phosphatase (IIb), originally purified from mammalian brain, which is stimulated directly by micromolar concentrations of Ca and indirectly through calmodulin (Klee et al., 1979; Stewart et al., 1982). Thus, calcineurin could act as a link between elevated intracellular Ca activity and the dephosphorylation that appears to lead to inactivation of the Ca channel. We have tested that possibility in voltage-clamped *Helix* neurons dialyzed with CsCl, HEPES, leupeptin, ATP-Mg, and CS to stabilize the current, but with no Ca buffers to reduce Ca ion accumulation (Chad and Eckert, 1985*b*, 1986). An example is shown in Fig. 6. After obtaining a stable current, the perfusate was changed to one containing the same amount of CS but with added calcineurin (40 μg/ml protein) and calmodulin (10 μM). The augmentation of Ca buffering normally slows inactivation; however, calcineurin (Fig. 6) markedly increased the rate of inactivation. In other experiments, addition of EGTA to the internal solution or substitution of Ba for external Ca resulted in a much slower rate of inactivation. Under those conditions, however, exogenous calcineurin produced no increase in the rate of inactivation. Thus, calcineurin-enhanced inactivation of the Ca current in dialyzed cells has the same Ca dependence as the inactivation of the Ca current in intact cells (Chad and Eckert, 1986).

Ca-dependent Inactivation in GH_3 Cells

Like the Ca current in *Helix* and *Aplysia* neurons, the Ca current through phosphorylation-dependent Ca channels in GH_3 cells, voltage-clamped with patch

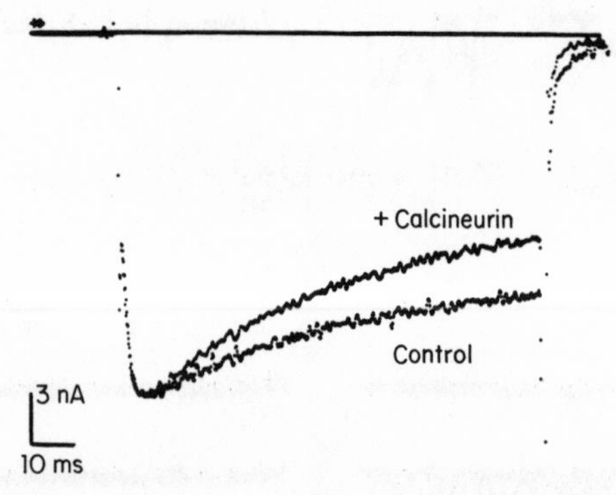

Figure 6. Effect of exogenous Ca-dependent phosphatase in a dialyzed *Helix* neuron. The addition of calcineurin (40 μg/ml) plus calmodulin (10 μM) to the dialysis solution increased the rate of inactivation of the Ca current during a voltage step from a holding potential of −40 to +10 mV. (Reprinted with permission from Eckert et al., 1986.)

pipettes in the whole-cell configuration, also exhibits Ca-dependent inactivation (Kalman et al., 1987). Although 5 mM EGTA in the pipette solution did not suppress inactivation completely (cf. Matteson and Armstrong, 1984), substituting Ba for Ca as the charge carrier did eliminate the inactivation observed during steps from −40 to 0 mV (Fig. 7*A*). Thus, inactivation is specific to Ca as the charge carrier, and 5 mM BAPTA (Tsien, 1980), a more efficient buffer of rapid Ca transients (Neher and Marty, 1985), suppressed inactivation completely (Fig. 7*B*). In contrast to *Helix* and *Aplysia* neurons, holding GH_3 cells at more negative potentials revealed the presence of another class of Ca channels that exhibit voltage-dependent inactivation. However, in most GH_3 cells, those channels were largely inactivated at a holding potential of −40 mV, and depolarizing steps from −40 mV revealed little remaining current after the phosphorylation-dependent channels were allowed to wash out or after they were blocked with 1 μM nimodipine (Fig. 7*B*). GH_3 cells also contain calcineurin (Farber et al., 1985), and we have postulated that the basal activity of this membrane-bound enzyme may be responsible for the

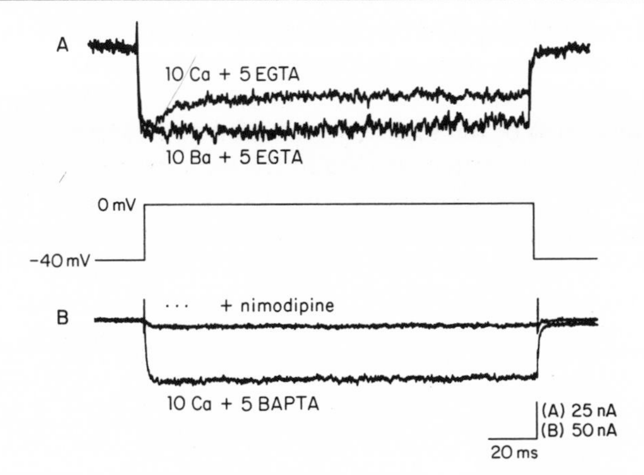

Figure 7. Voltage-activated Ca currents recorded in the whole-cell configuration from GH_3 cells during voltage steps from −40 to 0 mV. The patch pipette contained a CsCl solution buffered to pH 7.2 with 40 mM HEPES and to pCa 8 with either 5 mM EGTA or BAPTA. (*A*) Although EGTA does not suppress inactivation of the current completely, substituting Ba for Ca as the charge carrier demonstrates that inactivation is Ca specific. (*B*) 5 mM BAPTA completely suppresses inactivation of the dihydropyridine-sensitive Ca current.

loss of Ca channel activity we observed in cell-free patches even in the presence of Ca buffers (Armstrong and Eckert, 1987).

Endogenous Enzymes Also Modulate Inactivation

To investigate whether endogenous enzyme systems participate in inactivation, we have injected agents that modulate the cAMP-dependent kinase into metabolically intact *Aplysia* neurons (Kalman and Eckert, 1985; Eckert et al., 1986). In these experiments, *Aplysia* neurons were voltage-clamped with a conventional two-electrode system and Ca currents were isolated by ion substitution and pharmacological blockers (cf. Chad et al., 1984). Intracellular injection of the purified CS of the kinase increased the peak Ca current and, in some cases, slowed the rate of inactivation (Fig. 8), even though larger Ca currents normally inactivate more rapidly because of the increased influx of Ca ions into the cell (Chad et al., 1984).

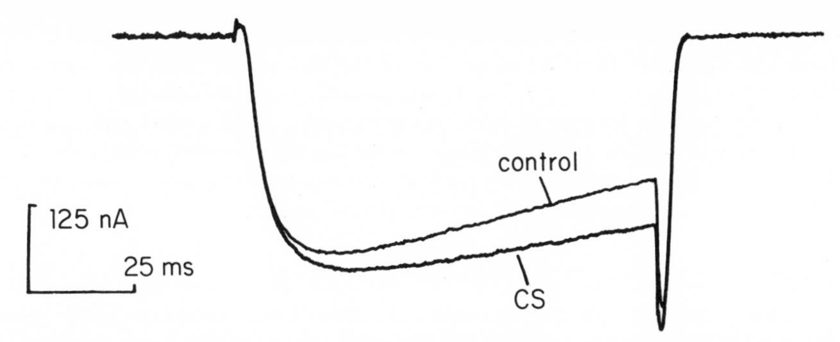

Figure 8. Effect of pressure injection of the exogenous CS of cAMP-dependent protein kinase into an *Aplysia* neuron. Ca currents elicited by voltage steps to +10 mV from a holding potential of −40 mV. After injection (CS), the peak current increased and the rate of inactivation slowed.

The effects of CS were mimicked by injections of 8-bromo-cAMP (8-Br-cAMP), a nonhydrolyzable derivative of cAMP. Further evidence implicating an endogenous kinase in the effect of cAMP on inactivation was obtained by injecting the regulatory subunit (RS) of the kinase, which inhibits cAMP-dependent phosphorylation by binding to the catalytic subunit. In the experiment illustrated in Fig. 9*A*, the first injection resulted in a faster inactivation rate during the pulse but no change in the peak current. A subsequent injection decreased both the peak current

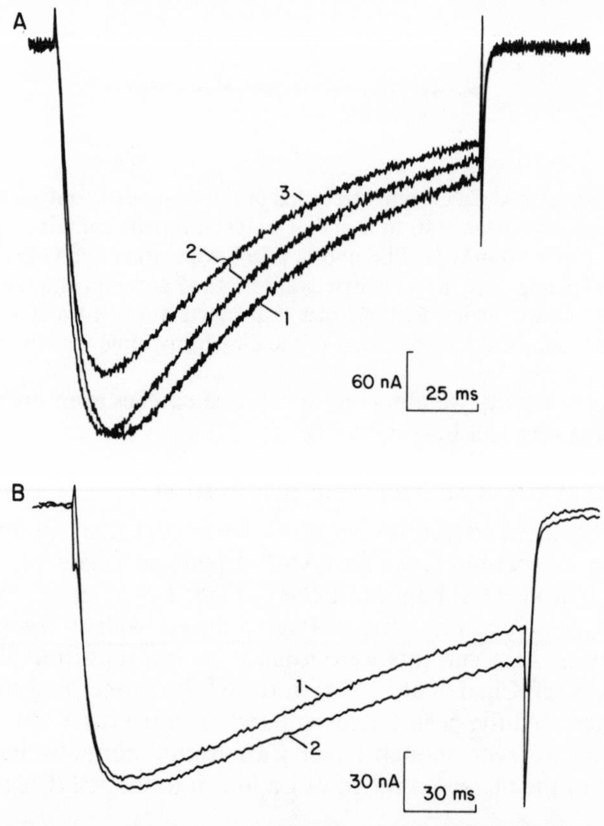

Figure 9. Effect of pressure injection of the RS of cAMP-dependent protein kinase (0.2 mg/ml) followed by iontophoretic injection of 8-Br-cAMP into *Aplysia* neurons. Ca currents were elicited by steps to +10 mV. (*A*) The largest current (1) was recorded before injection of RS. After injection (2), the rate of inactivation increased. With further injection of RS (3), the peak current decreased and the rate of decline of the current slowed. Cell L2. (*B*) The effects of prior RS injection (1) were partially reversed by iontophoretic injection of 8-Br-cAMP (2). Cell L4. (Reprinted with permission from Eckert et al., 1986.)

and the rate of inactivation. Since the rate of inactivation also depends on how much Ca enters the cell, we presume that the slower rate of inactivation results from the reduction in peak current. Finally, the effects of RS were reversed by injecting 8-Br-cAMP (Fig. 9*B*), which relieves CS from inhibition by binding to RS and promoting its dissociation. The Walsh inhibitor protein (PKI: Walsh et al., 1971) produced effects analogous to those produced by RS (Eckert et al., 1986).

Although cAMP and CS always reversed the effects of previous injections of RS, they only produced effects on their own in about half the cells tested. Therefore, it appears that in many molluscan neurons the phosphorylating activity of endogenous CS is sufficient to keep the majority of Ca channels in an activatable state. Consequently, the addition of more cAMP or exogenous CS has little effect. When the basal level of phosphorylation by the cAMP-dependent kinase was first reduced by lowering the amount of free CS with exogenous RS, then the subsequent addition of cAMP always enhanced the Ca current and slowed inactivation. As the effects of cAMP analogues will depend on the metabolic state of the cell, this variation in the resting level of Ca channel phosphorylation may account for published observations that cAMP analogues do not enhance Ca current in certain metabolically intact molluscan neurons (Gerschenfeld et al., 1986), but do enhance Ca currents in other molluscan neurons (Lotshaw et al., 1986) and mammalian cells (Reuter, 1983; Bean et al., 1984; Luini et al., 1985).

Discussion

Our tentative conclusions from these experiments are summarized schematically in Fig. 10. Only phosphorylated channels open when the membrane is depolarized; dephosphorylated channels are inactivated. The Ca influx through open channels promotes inactivation by stimulating a Ca-dependent phosphatase. To remove inactivation, dephosphorylated channels must be rephosphorylated by the cAMP-dependent kinase, and that requires ATP-Mg. Phosphorylated channels do not inactivate until they are dephosphorylated, and Ba ions are unable to stimulate the phosphatase.

Phosphorylation

The experimental results from dialyzed neurons and cell-free membrane patches indicate that the dihydropyridine-sensitive Ca channels that have a higher conductance in Ba and a higher threshold of activation, and do not exhibit voltage-dependent inactivation, also have a much lower probability of opening when the membrane is depolarized in the absence of phosphorylation. If that probability is not zero, it appears to be too low to measure for individual channels in their native membrane (Armstrong and Eckert, 1987). Since the kinase and ATP-Mg are sufficient to restore activity to cell-free patches that have been perfused for several minutes with minimal saline solutions (Fig. 3), the phosphorylated site that regulates the activity of the channel must be closely associated with the proteins forming the Ca-selective pore in the membrane.

Biochemical studies of a putative Ca channel protein isolated from skeletal muscle by its affinity for dihydropyridines reveals a multisubunit complex, one subunit of which is extremely sensitive to proteolysis (Curtis and Catterall, 1984). One or more subunits are also substrates for cAMP-dependent protein kinase (Curtis and Catterall, 1985; Hosey et al., 1986). Furthermore, the site on the protein that is phosphorylated by the cAMP-dependent protein kinase can be dephosphorylated by calcineurin in a Ca/calmodulin-dependent manner (Hosey et al., 1986). Thus, biochemical evidence also suggests that the phosphorylated site postulated to control channel activity is part of the channel itself. Finally, biochemical evidence also indicates that both the CS of cAMP-dependent kinase and calcineurin are membrane-bound enzymes, often in close association with one another (Hathaway et al., 1981; Aitken et al., 1982). Therefore, it is not implausible to imagine that Ca channels in the membrane could be regulated by these enzymes on a physiologically relevant time scale. Other types of voltage-activated Ca channels may require phosphorylation by different kinases to respond fully to depolarization (DeRiemer et al., 1985; Paupardin-Tritsch et al., 1986).

Another type of voltage-activated Ca channel in mammalian cells does not appear to require phosphorylation for sustained activity in cell-free patches (Armstrong and Eckert, 1985; Nilius et al., 1985), but they inactivate rapidly at the

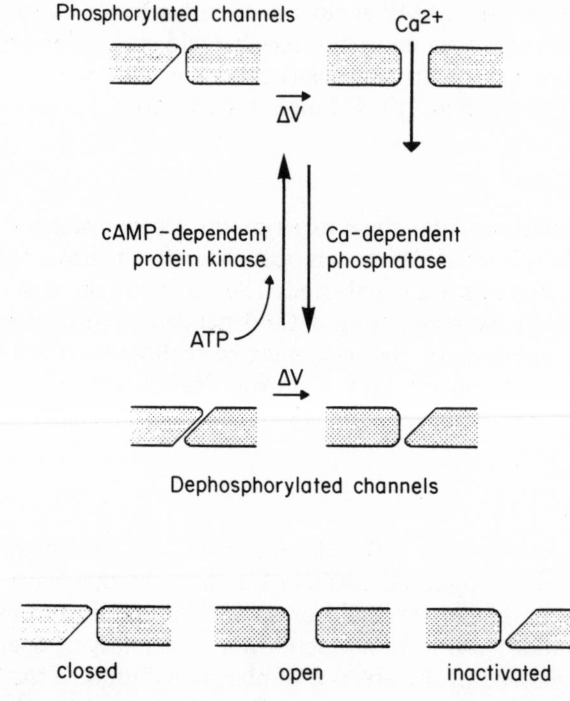

Figure 10. Enzymatic hypothesis of Ca-dependent inactivation and removal of inactivation. (*Top*) A Ca channel opens with high probability in response to depolarization if the regulatory site is phosphorylated. The subsequent rise in internal Ca activates a Ca-dependent phosphatase, which increases the probability of the enzyme dephosphorylating the channel. (*Bottom*) The dephosphorylated channel has a lower probability of opening in response to depolarization and is therefore inactivated. As internal Ca decreases, the channel is rephosphorylated by a cAMP-dependent protein kinase and can open again in response to depolarization.

relatively depolarized values of holding potential (−40 mV) used in our studies (Kalman et al., 1987). The biological roles of the different types of Ca channel described in mammalian cells (Carbone and Lux, 1984; Nowycky et al., 1985; Nilius et al., 1985) have been discussed (Matteson and Armstrong, 1986; McClesky et al., 1986), and their participation in presynaptic release of neurotransmitters is presently under investigation (Dunlap et al., 1986; Chad and Yeats, 1986; Hirning et al., 1986; Perney et al., 1986).

Dephosphorylation

The effect of elevated intracellular Ca activity on both channel maintenance and inactivation suggests that the dephosphorylation that appears to underlie inactivation of the channel may be due to a Ca-dependent phosphatase. Calcineurin, a Ca-dependent phosphatase (Klee et al., 1979), has been localized immunohistochemically in neurons of *Aplysia* (Saitoh and Schwartz, 1983) and GH_3 cells (Farber et al., 1985). When *Helix* neurons were dialyzed with exogenous calcineurin and calmodulin, inactivation of the Ca current was enhanced in a divalent-specific manner. Thus, we have hypothesized that dephosphorylation of a site on, or closely associated with, the Ca channel by an endogenous Ca-dependent phosphatase produces inactivation (Chad and Eckert, 1986). Biochemical studies (Hosey et al., 1986) demonstrating dephosphorylation of a putative Ca channel protein by calcineurin support that conclusion. In our experiments, recovery from inactivation was accelerated more by factors that increased buffering of intracellular Ca than by factors that stimulated phosphorylation. Thus, we suspect that the rate of recovery from inactivation is governed predominantly by the decrease of phosphatase activity as the Ca concentration inside the cell returns to its resting level.

Proteolysis

The prevention of irreversible washout in *Helix* neurons by the protease inhibitor leupeptin (Chad and Eckert, 1986) and the slowing of washout by Ca chelators (Byerly and Yazejian, 1986) suggest that a Ca-dependent protease could be responsible for the irreversible loss of channel activity. One family of endogenous Ca-dependent proteases, the calpains (Murachi et al., 1983), is inhibited by leupeptin and EGTA. Calpains are normally regulated by another endogenous protein, calpstatin, that is not membrane bound (Aoyagi et al., 1969; Sasaki et al., 1984). Thus, intracellular dialysis may remove calpstatin and, hence, disinhibit the calpains. The sensitivity of this postulated proteolysis of Ca channels to intracellular Ca levels could play an important role in the feedback regulation of the number and turnover of Ca channels in normal cells.

In conclusion, we have implicated a cAMP-dependent kinase, a Ca-dependent phosphatase, and a Ca-dependent protease in the maintenance and inactivation of an important class of voltage-activated Ca channels. These enzymes provide the cell with many potential mechanisms for modulating the influx of Ca from the environment.

Summary

The predominant class of voltage-activated Ca channels in molluscan neurons (*Helix*) and a mammalian cell line (GH_3) do not respond to membrane depolarization under conditions that prevent cAMP-dependent phosphorylation when the cytoplasm is replaced with standard physiological saline solutions. Under normal conditions, inactivation of these channels results from Ca ion entry and accumulation inside the cell. In dialyzed neurons, inactivation is enhanced in the presence of an exogenous Ca-dependent phosphatase, and inactivated channels appear to be susceptible to Ca-dependent proteolysis. Rephosphorylation of the channels by a cAMP-dependent kinase removes inactivation and protects the channels from proteolysis. Thus, phosphorylation/dephosphorylation reactions appear to be an important means of modulating Ca channel activity and may underlie both Ca-dependent inactivation and the metabolic maintenance of the channels.

Acknowledgments

We are grateful for the gifts of calcineurin and calmodulin from C. Klee, catalytic subunit from S. Halegoua, I. Levitan, and A. Nairn, and protein kinase inhibitor protein from K. Diltz. We thank our colleagues in the Eckert laboratory for invaluable discussions during the course of this work.

This work was supported by a Javits Neuroscience Investigator Award to R. Eckert from the National Institute of Neurological and Communicative Disorders and Stroke, and by U.S. Public Health Service grants NS-08364 and GM-07185.

References

Affolter, H., and R. Coronado. 1985. Agonists BAY-K-8644 and CGP-28392 open calcium channels reconstituted from skeletal muscle transverse tubules. *Biophysical Journal.* 48:341–347.

Aitken, A., P. Cohen, S. Santikarn, D. H. Williams, A. G. Calder, A. Smith, and C. B. Klee. 1982. Identification of the NH_2-terminal blocking group of calcineurin B as myristic acid. *FEBS Letters.* 150:314–318.

Aoyagi, T., S. Miyatu, M. Nanbo, F. Koyima, M. Matsuyaki, M. Ishizuka, T. Takeuchi, and H. Umezawa. 1969. Biological activities of leupeptins. *Journal of Antibiotics.* 22:558–561.

Armstrong, D. 1986. An endogenous cyclic AMP-dependent protein kinase modulates the activity of voltage-dependent calcium channels. *Journal of General Physiology.* 88:11*a.* (Abstr.)

Armstrong, D., and R. Eckert. 1985. Phosphorylating agents prevent wash-out of unitary calcium currents in excised membrane patches. *Journal of General Physiology.* 86:25*a*–26*a.* (Abstr.)

Armstrong, D., and R. Eckert. 1987. Voltage-activated calcium channels that must be phosphorylated to respond to membrane depolarization. *Proceedings of the National Academy of Sciences.* 84:2518–2522.

Armstrong, D., C. Erxleben, and D. Kalman. 1987. Calcium channels modulated by BAY-K-8644 appear less susceptible to dephosphorylation. *Biophysical Journal.* 51:233*a.* (Abstr.)

Bean, P. B., M. C. Nowycky, and R. W. Tsien. 1984. Beta-adrenergic modulation of calcium channels in frog ventricular heart cells. *Nature.* 307:371–375.

Brehm, P., and R. Eckert. 1978. Calcium entry leads to inactivation of calcium current in *Paramecium. Science.* 202:1203–1206.

Byerly, L., and S. Hagiwara. 1982. Calcium currents in internally perfused nerve cell bodies of *Lymnaea stagnalis. Journal of Physiology.* 322:503–528.

Byerly, L., and B. Yazejian. 1986. Intracellular factors for the maintenance of calcium currents in perfused neurons of the snail *Lymnaea stagnalis. Journal of Physiology.* 370:631–650.

Carbone, E., and H. D. Lux. 1984. A low voltage-activated fully inactivating Ca channel in vertebrate sensory neurons. *Nature.* 310:501–502.

Cavalie, A., R. Ochi, D. Pelzer, and W. Trautwein. 1983. Elementary currents through Ca^{2+} channels in guinea-pig myocytes. *Pflügers Archiv.* 398:284–297.

Chad, J. E., and R. Eckert. 1984. Stimulation of cAMP-dependent protein phosphorylation retards both inactivation and "wash-out" of Ca current in dialyzed *Helix* neurons. *Society for Neuroscience Abstracts.* 10:866.

Chad, J. E., and R. Eckert. 1985*a*. Leupeptin, an inhibitor of Ca-dependent proteases, retards the kinase-irreversible Ca-dependent loss of calcium current in perfused snail neurons. *Biophysical Journal.* 47:266*a*. (Abstr.)

Chad, J. E., and R. Eckert. 1985*b*. Calcineurin, a calcium-dependent phosphatase, enhances Ca-mediated inactivation of Ca current in perfused snail neurons. *Biophysical Journal.* 47:266*a*. (Abstr.)

Chad, J. E., and R. Eckert. 1986. An enzymatic mechanism for calcium current inactivation in dialyzed *Helix* neurons. *Journal of Physiology.* 378:31–51.

Chad, J. E., R. Eckert, and D. Ewald. 1984. Kinetics of Ca-dependent inactivation of calcium current in neurones of *Aplysia californica. Journal of Physiology.* 347:279–300.

Chad, J. E., and J. Yeats. 1986. Calcium fluxes activated during spiking induced by bradykinin. *Journal of General Physiology.* 88:15*a*. (Abstr.)

Curtis, B. M., and W. A. Catterall. 1984. Purification of the calcium antagonist receptor of the voltage-sensitive calcium channel from skeletal muscle transverse tubules. *Biochemistry.* 23:2113–2118.

Curtis, B. M., and W. A. Catterall. 1985. Phosphorylation of the calcium antagonist receptor of the voltage sensitive calcium channel by cAMP-dependent protein kinase. *Proceedings of the National Academy of Sciences.* 82:2528–2532.

DeRiemer, S. A., J. A. Strong, K. A. Albert, P. Greengard, and L. K. Kaczmarek. 1985. Enhancement of calcium current in *Aplysia* neurons by protein kinase C. *Nature.* 313:313–316.

Doroshenko, P. A., P. G. Kostyuk, and A. I. Martynyuk. 1982. Intracellular metabolism of adenosine 3′-5′-cyclic monophosphate and calcium inward current in perfused neurones of *Helix pomatia. Neuroscience.* 7:2125–2134.

Doroshenko, P. A., P. G. Kostyuk, A. E. Martynyuk, M. D. Kursky, and Z. D. Vorobetz. 1984. Intracellular protein kinase and calcium inward currents in perfused neurones of the snail *Helix pomatia. Neuroscience.* 11:263–267.

Dunlap, K., R. M. Kream, and G. G. Holy. 1986. Alpha-2 adrenergic and GABA-B receptors mediate inhibition of neuropeptide secretion from dorsal root ganglion cells. *Neuroscience Abstracts.* 12:1195.

Eckert, R., and J. E. Chad. 1984. Inactivation of calcium channels. *Progress in Biophysics and Molecular Biology.* 44:215–267.

Eckert, R., J. E. Chad, and D. Kalman. 1986. Enzymatic regulation of the calcium current in dialyzed and intact molluscan neurons. *Journal de Physiologie.* 81:318–324.

Eckert, R., and D. Tillotson. 1981. Calcium-mediated inactivation of the calcium conductance in caesium-loaded giant neurones of *Aplysia californica. Journal of Physiology.* 314:265–280.

Ehrlich, B. E., C. R. Schen, M. L. Garcia, and G. J. Kaczorowski. 1986. Incorporation of calcium channels from cardiac sarcolemmal membrane vesicles into planar lipid bilayers. *Proceedings of the National Academy of Sciences.* 83:193–197.

Farber, L., F. Iannetta, T. Kirby, and D. J. Wolff. 1985. Calmodulin dependent phosphatase of PC-12, C-6-glioma, and GH_3 pituitary adenoma cell lines. *Neuroscience Abstracts.* 11:855.

Fenwick, E. M., A. Marty, and E. Neher. 1982. Sodium and calcium channels in bovine chromaffin cells. *Journal of Physiology.* 331:599–635.

Forscher, P., and G. S. Oxford. 1985. Modulation of calcium channels by norepinephrine in internally dialyzed avian sensory neurons. *Journal of General Physiology.* 85:743–763.

Gerschenfeld, H. M., C. Hammond, and D. Paupardin-Tritsch. 1986. Modulation of the calcium current of molluscan neurons by neurotransmitters. *Journal of Experimental Biology.* 124:73–91.

Hagiwara, S., and S. Nakajima. 1966. Effects of the intracellular Ca ion concentration upon the excitability of the muscle fiber membrane of a barnacle. *Journal of General Physiology.* 49:807–818.

Hamill, O. P., A. Marty, E. Neher, B. Sakmann, and F. Sigworth. 1981. Improved patch-clamp techniques for high resolution recordings from cells and from cell free membrane patches. *Pflügers Archiv.* 391:85–100.

Hathaway, D. R., R. S. Adelstein, and C. B. Klee. 1981. Interaction of calmodulin with myosin light chain kinase and cyclic AMP-dependent protein kinase in bovine brain. *Journal of Biological Chemistry.* 256:8183–8189.

Hess, P., J. B. Lansman, and R. W. Tsien. 1984. Different modes of calcium channel gating behaviour favoured by dihydropyridine Ca agonists and antagonists. *Nature.* 311:538–544.

Hirning, L. D., A. P. Fox, E. W. McClesky, R. J. Miller, B. M. Olivera, S. A. Thayer, and R. W. Tsien. 1986. Dominant role of N-type calcium channels in K-evoked release of norepinephrine from rat sympathetic neurons. *Neuroscience Abstracts.* 12:28.

Holzer, H., and P. C. Heinrich. 1980. Control of proteolysis. *Annual Review of Biochemistry.* 49:63–91.

Hosey, M. M., M. Borsetto, and M. Lazdunski. 1986. Phosphorylation and dephosphorylation of dihydropyridine sensitive voltage-dependent calcium channel in skeletal muscle membranes by cAMP- and Ca-dependent processes. *Proceedings of the National Academy of Sciences.* 83:3733–3737.

Kalman, D., and R. Eckert. 1985. Injection of the catalytic and regulatory subunits of protein kinase into *Aplysia* neurons alters calcium current inactivation. *Journal of General Physiology.* 86:26*a*–27*a*. (Abstr.)

Kalman, D., C. Erxleben, and D. Armstrong. 1987. Inactivation of the dihydropyridine-sensitive calcium current in GH_3 cells is a calcium-dependent process. *Biophysical Journal.* 51:432*a*. (Abstr.)

Klee, C. B., T. H. Crouch, and M. H. Krinks. 1979. Calcineurin: a calcium- and calmodulin-binding protein of the nervous system. *Proceedings of the National Academy of Sciences.* 76:6270–6273.

Lee, K. S., E. Marban, and R. W. Tsien. 1985. Inactivation of calcium channels in mammalian heart cells: joint dependence on membrane potential and intracellular calcium. *Journal of Physiology.* 364:395–411.

Lotshaw, D. P., E. S. Levitan, and I. B. Levitan. 1986. Fine tuning of neuronal electrical activity: modulation of several ion channels by intracellular messengers in a single identified nerve cell. *Journal of Experimental Biology.* 124:307–322.

Luini, A., D. Lewis, S. Guild, D. Corda, and J. Axelrod. 1985. Hormone secretagogues increase cytosolic calcium by increasing cyclic AMP in corticotropin-secreting cells. *Proceedings of the National Academy of Sciences.* 82:8034–8038.

Matteson, D. R., and C. M. Armstrong. 1984. Na and Ca channels in a transformed line of anterior pituitary cells. *Journal of General Physiology.* 83:371–394.

Matteson, D. R., and C. M. Armstrong. 1986. Properties of two types of calcium channels in clonal pituitary cells. *Journal of General Physiology.* 87:161–182.

McCleskey, E. W., A. P. Fox, D. Feldman, and R. W. Tsien. 1986. Different types of calcium channels. *Journal of Experimental Biology.* 124:177–190.

Mentrard, D., G. Vassort, and R. Fischmeister. 1984. Calcium-mediated inactivation of the calcium conductance in cesium-loaded frog heart cells. *Journal of General Physiology.* 83:105–131.

Murachi, T. 1983. Intracellular Ca^{2+} protease and its inhibitor protein: calpain and calpstatin. *In* Calcium and Cell Function. W. Y. Cheung, editor. Academic Press, Inc., New York. IV:377–410.

Nairn, A. C., H. C. Hennings, and P. Greengard. 1985. Protein kinases in the brain. *Annual Review of Biochemistry.* 54:931–976.

Neher, E., and A. Marty. 1985. BAPTA, unlike EGTA, efficiently suppresses Ca-transients in chromaffin cells. *Biophysical Journal.* 47:278*a*. (Abstr.)

Nilius, B., P. Hess, J. B. Lansman, and R. W. Tsien. 1985. A novel type of cardiac calcium channel in ventricular cells. *Nature.* 316:443–446.

Nilius, B., and A. Roder. 1985. Direct evidence of Ca-sensitive inactivation of slow inward channels in frog atrial myocardium. *Biomedica Biochimica Acta.* 44:1151–1161.

Nowycky, M. C., A. P. Fox, and R. W. Tsien. 1985. Three types of neuronal calcium channels with different calcium agonist sensitivity. *Nature.* 316:440–443.

Osterrieder, W., G. Brum, J. Hescheler, W. Trautwein, V. Flockerzi, and R. Hofmann. 1982. Injection of subunits of cyclic AMP-dependent protein kinase into cardiac myocytes modulates Ca^{2+} current. *Nature.* 298:576–578.

Paupardin-Tritsch, D., C. Hammond, H. M. Gerschenfeld, A. C. Nairn, and P. Greengard. 1986. Cyclic GMP-dependent protein kinase enhances calcium current and potentiates the serotonin-induced Ca current increase in snail neurones. *Nature.* 323:812–814.

Perney, T. M., D. A. Ewald, and R. J. Miller. 1986. Regulation of substance-P release from cultured rat sensory neurones. *Neuroscience Abstracts.* 12:1195.

Reuter, H. 1983. Calcium channel modulation by neurotransmitters, enzymes and drugs. *Nature.* 301:569–574.

Rosenberg, R. L., P. Hess, J. P. Reeves, H. Smilowitz, and R. W. Tsien. 1986. Calcium channels in planar lipid bilayers: insights into mechanisms of ion permeation and gating. *Science.* 231:1564–1566.

Saitoh, T., and J. H. Schwartz. 1983. Serotonin alters the subcellular distribution of a Ca/calmodulin-binding protein in neurons of *Aplysia. Proceedings of the National Academy of Sciences.* 80:6708–6712.

Sasaki, T., T. Kikuchi, N. Yomoto, N. Yoshimura, and T. Murachi. 1984. Comparative specificity and kinetic studies on porcine calpain I and calpain II with naturally occurring peptides and synthetic fluorogenic substrates. *Journal of Biological Chemistry.* 259:12489–12494.

Schramm, M., G. Thomas, R. Towart, and G. Frankowiak. 1983. Novel dihydropyridines with positive inotropic action through activation of Ca channels. *Nature.* 311:538–544.

Stewart, A. A., T. S. Ingbretsen, A. Manalan, C. B. Klee, and P. Cohen. 1982. Discovery of a Ca- and calmodulin-dependent protein phosphatase: probable identity with calcineurin (CaM-BP_{80}). *FEBS Letters.* 137:80–84.

Tashjian, A. H. 1979. Clonal strains of hormone-producing cells. *Methods in Enzymology.* 58:527–535.

Tillotson, D. 1979. Inactivation of Ca conductance dependent on entry of Ca ions in molluscan neurons. *Proceedings of the National Academy of Sciences.* 77:1497–1500.

Tsien, R. W., W. Giles, and P. Greengard. 1972. Cyclic AMP mediates the effects of adrenaline on cardiac Purkinje fibres. *Nature New Biology.* 240:181–183.

Tsien, R. Y. 1980. New calcium indicators and buffers with high selectivity against magnesium and protons: design, synthesis, and properties of prototype structures. *Biochemistry.* 19:2396–2404.

Walsh, D. A., C. D. Ashby, C. Gonzalez, D. Calkins, E. Fisher, and E. G. Krebs. 1971. Purification and characterization of a protein kinase inhibitor of adenosine 3′,5′-monophosphate-dependent protein kinases. *Journal of Biological Chemistry.* 246:1977–1985.

Chapter 12

Multiple Roles for Calcium and Calcium-dependent Enzymes in the Activation of Peptidergic Neurons of *Aplysia*

J. A. Strong, L. Fink, A. Fox, and L. K. Kaczmarek

Departments of Pharmacology and Physiology, Yale University, New Haven, Connecticut

Introduction

Intracellular Ca^{2+} regulates a variety of cellular processes in nerve cells. In addition to its rapid effects on secretion and excitability, a rise in intracellular Ca^{2+} activates a variety of enzymes that may produce long-term changes in excitability and influence parameters such as neurotransmitter synthesis and the growth of neurites. Such Ca^{2+}-dependent alterations in excitability may result in long-term alterations in behaviors that are controlled by specific neurons.

Molluscan neurons have provided several important model systems for the investigation of processes that are regulated by intracellular Ca^{2+}. Pioneering studies of the voltage-activated Ca^{2+} current, Ca^{2+}-activated K^+ current, Na^+-Ca^{2+} exchange, and secretion have been carried out in molluscan cells. In this chapter, we discuss some of the Ca^{2+}-dependent mechanisms that lead to long-term changes in the excitability of a simple system of invertebrate neurons. In particular, we describe the role of the second messengers inositol trisphosphate and diacylglycerol in the control of K^+ and Ca^{2+} currents and the existence of three species of Ca^{2+} channels, two of which may be regulated by intracellular enzymes.

The Bag Cell Neurons

The bag cell neurons of the mollusk *Aplysia* provide several advantages for the investigation of prolonged changes in neuronal excitability. These cells are the "command" neurons for a prolonged sequence of reproductive behaviors culminating in egg-laying. They are found in two anatomically distinct clusters in the abdominal ganglion. Each cluster contains 200–400 large neurons (40–120 μm diam), which are electrically and morphologically homogeneous (Kupfermann and Kandel, 1970; Chiu and Strumwasser, 1981). Because of these structural features, the cells can be studied in isolation from all other neurons, using both electrophysiological and biochemical techniques. In addition, the cells can be enzymatically dispersed and grown in primary culture, where they maintain many of their electrical characteristics (Kaczmarek et al., 1979; Kaczmarek and Strumwasser, 1981). Voltage-clamp and patch-clamp techniques can be used to study such single neurons in culture.

The bag cell neurons in the abdominal ganglion normally have very negative resting potentials and display no spontaneous electrical activity. However, a brief extracellular stimulus applied to the adjacent pleuroabdominal connective nerve will cause the cells to depolarize and fire for 30 min (Fig. 1 *A*). Such afterdischarges can also be triggered by the application of identified peptides isolated from the reproductive tract (Heller et al., 1980). During the afterdischarge, the cells release several peptides, which act on other neurons and on the reproductive tract to bring about egg-laying behavior (Arch et al., 1976; Chiu et al., 1979; Rothman et al., 1983; Scheller et al., 1983).

The afterdischarge is not just a simple consequence of depolarization, but represents a major transformation of the cells' electrical properties. The action potentials during an afterdischarge are greatly enhanced in height and width compared with action potentials elicited by current injection in resting cells (Fig. 1 *B*). In addition, the cells acquire the ability to fire repetitively (for a review, see Kaczmarek et al., 1986).

The stimulation of an afterdischarge is associated with the activation of at least two major enzyme systems. One of these, the adenylate cyclase system, has a well-characterized role in initiating and maintaining the afterdischarge. Cyclic AMP levels rise during an afterdischarge, and afterdischarges can be initiated and prolonged by application of exogenous cAMP analogues or by forskolin (Kaczmarek et al., 1978). The application of cAMP analogues or the direct microinjection of the catalytic subunit of cAMP-dependent protein kinase leads to similar changes in the electrical properties of isolated bag cell neurons in cell culture (Kaczmarek et al., 1980; Kaczmarek and Strumwasser, 1981). At least some of the electrical changes seen during an afterdischarge are likely to be due to the cAMP-induced reduction of three voltage-dependent K^+ currents, an effect that has been extensively characterized in voltage-clamp studies on isolated cells (Strong, 1984; Strong

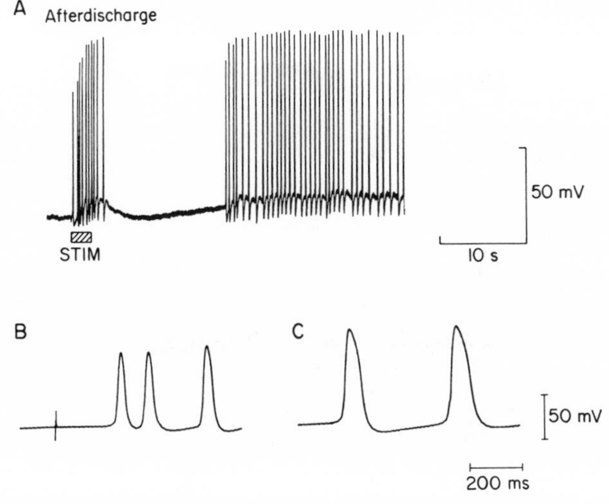

Figure 1. (*A*) Onset of afterdischarge in the bag cell neurons triggered by brief electrical stimulation of the pleuroabdominal connective nerve (STIM). (*B*) Action potentials evoked during the stimulus train that triggered a discharge. (*C*) Enhanced action potentials 10 min after the onset of afterdischarge.

and Kaczmarek, 1986). In addition, elevations of cAMP lead to changes in synthesis of the egg-laying hormone (ELH) precursor as well as the subsequent proteolytic processing of this molecule (Bruehl and Berry, 1985; Azhderian and Kaczmarek, 1986).

The second major enzyme system that plays a role in the transformation of the electrical properties of the bag cell neurons is that linked to the turnover of phosphoinositides. A primary action of many hormones and neurotransmitters is to increase the rate of hydrolysis of membrane phosphoinositides. The hydrolysis of phosphatidylinositol bisphosphate leads to the formation of two potential second messengers, inositol trisphosphate (IP_3) and diacylglycerol (Berridge, 1984). Evidence obtained using a variety of cell types has indicated that an elevation of IP_3 causes the release of Ca^{2+} from intracellular stores. The formation of diacylglycerol, on the other hand, is believed to be the major intracellular pathway for the activation of the Ca^{2+}/phospholipid-dependent protein kinase (protein kinase C) (Nishizuka, 1984). This chapter discusses the way in which each of these second messengers modulates the excitability of the bag cell neurons.

Inositol Trisphosphate

Stimuli that trigger afterdischarges in the bag cell neurons cause increased hydrolysis of phosphoinositides in these cells, as measured by the increased incorporation of tritiated inositol into membrane lipids and inositol polyphosphates. This result is seen after two different procedures known to activate the cells: electrical stimulation of the afferent connective nerve and application of reproductive tract extracts (Fink, L., and L. K. Kaczmarek, unpublished results).

The electrophysiological effects of IP_3 injection into cultured bag cell neurons are consistent with the hypothesis that IP_3 causes the release of Ca^{2+} from intracel-

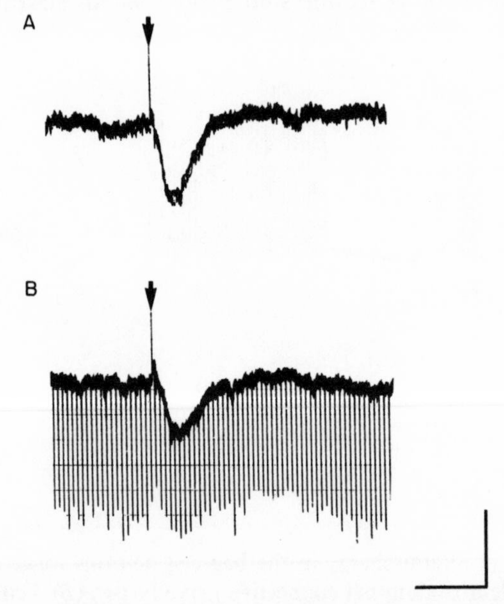

Figure 2. (*A*) Hyperpolarizing voltage response of an isolated bag cell neuron to intracellular pressure injection of IP_3 (arrow). (*B*) The effect of IP_3 injection on the response of the cell to a series of constant hyperpolarizing current pulses. The injection produces an apparent increase in membrane conductance that is not mimicked by control injections or by passive hyperpolarization of the membrane. (Calibration bars: 1 min, 15 mV.)

lular stores, as has been observed in several non-neuronal cell types. Direct microinjection of IP_3 into the soma of an isolated bag cell neuron causes a transient hyperpolarization of the membrane that occurs with a conductance increase (Fig. 2) (Fink et al., 1985). This response to IP_3 is abolished by 50 mM tetraethylammonium. Single channel recordings from these cells (Fig. 3) have demonstrated that the hyperpolarization is associated with an IP_3-induced opening of channels that are likely to be Ca^{2+}-dependent K^+ channels. These observations can be explained most simply by postulating that IP_3 induces the release of intracellular Ca^{2+}, which then activates a Ca^{2+}-dependent K^+ current. In fact, in preliminary experiments using bag cell neurons preloaded with the Ca^{2+} indicator fura-2 (Fink et al., 1985), IP_3 injection was followed by an increase in the Ca^{2+} signal. Moreover, the direct injection of Ca^{2+} elicits a hyperpolarization very similar to that seen after IP_3 injection. It is interesting that the magnitude, time course, and pharmacological sensitivity of the IP_3-induced hyperpolarization closely match those of the hyperpolarization that immediately follows synaptic stimulation of an afterdischarge in intact clusters of bag cell neurons (Fig. 1 *A*). This raises the possibility that the production of IP_3 is one of the earliest events in the transition from the resting to the active state.

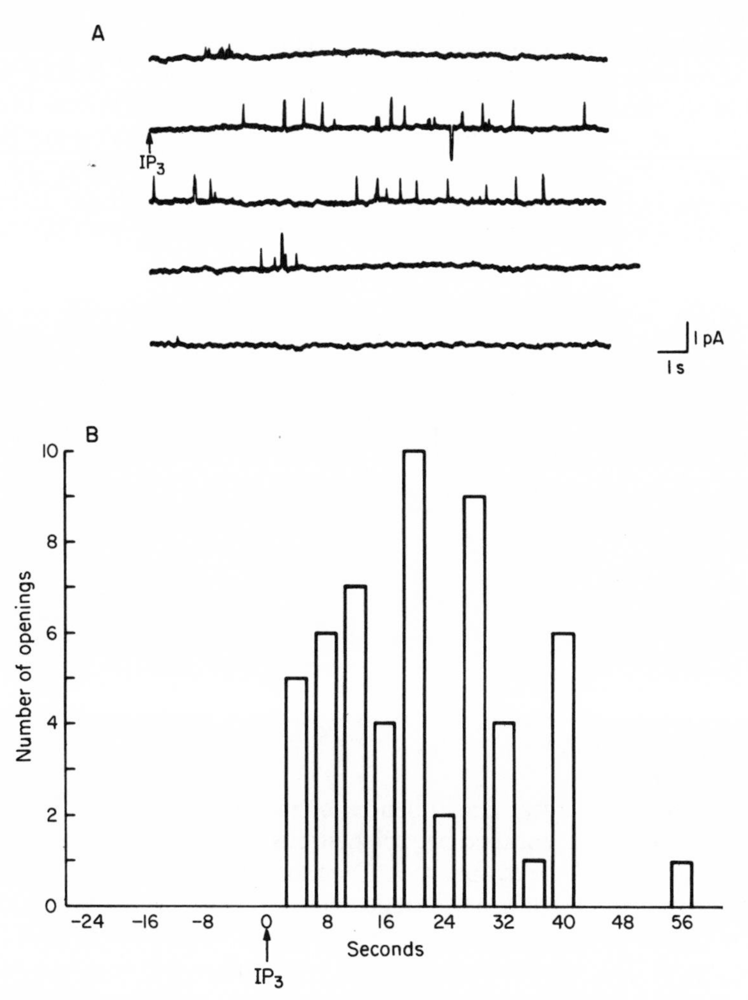

Figure 3. Effect of IP_3 injection on single channel activity recorded in cell-attached patches on isolated bag cell neurons. (*A*) In a patch maintained at the resting potential of the cell (−40 mV), IP_3 injection elicits the repeated opening of channels, presumed to be Ca^{2+}-dependent K^+ channels. (*B*) Pooled data from three experiments illustrating the time course of opening of single channels carrying outward current after IP_3 injection.

Ca^{2+}-dependent Protein Kinases

The elevation of intracellular Ca^{2+} may have multiple consequences, one of which is the activation of enzymes that require Ca^{2+} as a cofactor. A major Ca^{2+}-dependent enzyme in the *Aplysia* nervous system is a protein kinase that also requires calmodulin for its activation (Novak-Hofer and Levitan, 1983). By the criteria of peptide mapping, pharmacology, and substrate specificity, this enzyme shows marked homology to the mammalian enzyme Ca/calmodulin kinase II (DeRiemer et al., 1984). The injection of the purified kinase into single bag cell neurons has not been found to cause any reproducible changes in the cells' electrical characteristics. However, the enzyme may play an important role in secretion, as has been proposed for the squid giant synapse (Llinas et al., 1985).

A second major Ca^{2+}-dependent enzyme is protein kinase C, whose activation requires diacylglycerol and phosphatidylserine as well as Ca^{2+} (Kikkawa et al., 1982). Protein kinase C has been characterized in the nervous system of *Aplysia* and has been shown to have properties quite similar to the enzyme found in mammalian brain. In particular, the activation of the enzyme by the second messenger diacylglycerol may be mimicked by low concentrations (10 nM) of phorbol esters or by synthetic diacylglycerols (DeRiemer et al., 1985*a*).

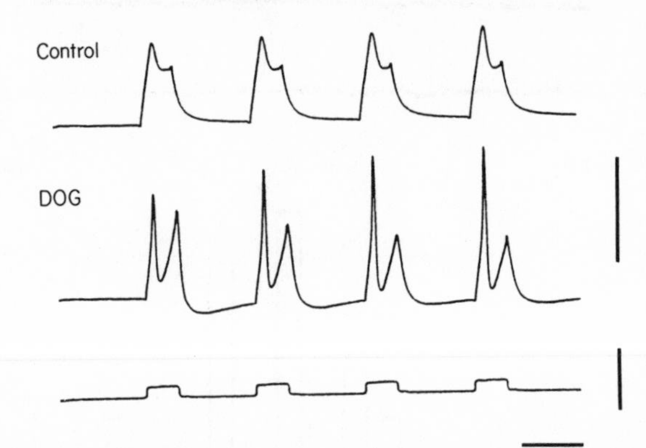

Figure 4. The response of an isolated bag cell neuron to DOG, an activator of protein kinase C. Action potentials were evoked by a series of constant current pulses (lower trace). (Calibration bars: 50 mV, 0.1 nA, 350 ms.)

The electrophysiological consequences of protein kinase C activation have been characterized in the cultured bag cell neurons. The application of the phorbol ester TPA (12-*O*-tetradecanoyl-phorbol 13-acetate) (DeRiemer et al., 1985*b*), or 1,2-dioctanoylglycerol (DOG), a synthetic diacylglycerol that is a potent activator of protein kinase C, causes a marked enhancement of the height of the action potential (Fig. 4). This enhanced spike height was mimicked by injection of protein kinase C purified from bovine brain (DeRiemer et al., 1985*b*). (In contrast, the elevation of cAMP causes a marked enhancement of the action potential width in these cells [Kaczmarek et al., 1985].)

The ionic basis of enhancement of spike height by protein kinase C activation has been investigated using the whole-cell voltage-clamp technique. Pretreatment with the phorbol ester TPA causes an increase in the magnitude of the voltage-dependent Ca^{2+} current (Fig. 5), with no obvious changes in its voltage dependence or kinetic properties. None of the three major voltage-dependent K^+ currents found in these cells is affected by pretreatment with phorbol ester (DeRiemer et al., 1985*b*). In contrast, elevations of cAMP reduce all three of these K^+ currents but have no effect on the Ca^{2+} current.

The enhancement of Ca^{2+} current after protein kinase C activation has been characterized at the single channel level. Single channel recordings have been made on isolated bag cell neurons using the cell-attached pipette configuration (Hamill et al., 1981). To optimize the recording of currents through single Ca^{2+} channels, the pipettes contained 185 mM Ba^{2+} as the charge carrier. In control, untreated cells, or in cells treated with inactive phorbol analogues, inward current events such as those shown on the left-hand side of Fig. 6 are observed. These correspond to a voltage-activated channel with a single channel conductance of 12 pS. However, after pretreatment of the cells with activators of protein kinase C, another voltage-activated channel is expressed, one with a single channel conductance of 24 pS (Fig. 6). This larger channel is never seen in control cells or in cells treated with phorbol compounds that are not activators of protein kinase C (Table I). The

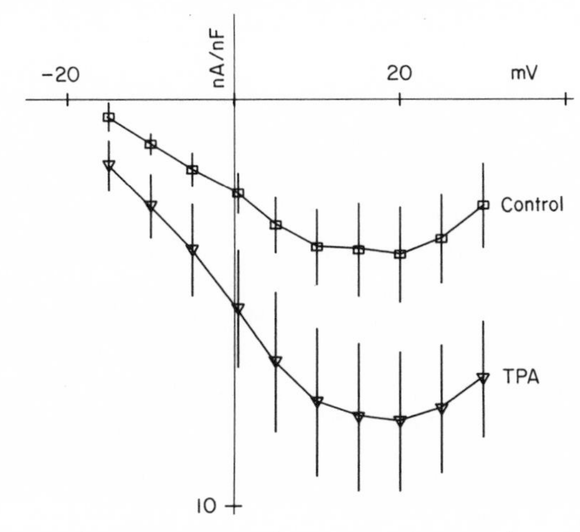

Figure 5. Enhancement of inward current in voltage-clamped, internally dialyzed bag cell neurons by pretreatment with the protein kinase C activator TPA. Inward current density as a function of voltage is plotted for control and TPA-treated cells isolated from the same bag cell cluster (DeRiemer et al., 1985*b*).

properties of the two channels can qualitatively account for the observations made with whole-cell recordings: both channels activate on depolarization of a patch to near 0 mV, both show little inactivation during 100-ms depolarizations, and the kinetics of activation for both channels are relatively rapid (Strong et al., 1987). The most straightforward interpretation of these results is that the enhancement of the Ca^{2+} current by protein kinase C is due to the activation of the large channel, a covert channel whose activity is undetectable before the stimulation of protein kinase C.

An interesting feature of the enhancement of the Ca^{2+} current by activation of protein kinase C is that the process is disrupted by both intracellular dialysis (whole-cell recording) and cell-attached recording. The enhancement of the Ca^{2+} current can be readily observed when a cell penetrated by microelectrodes is exposed to activators of protein kinase C. However, these agents have little or no effect if they are added after whole-cell patch recording has been initiated (DeRiemer et al., 1985*b*), although, as described above, pretreatment of a bag cell neuron with phorbol esters reliably increases the amplitude of the Ca^{2+} current

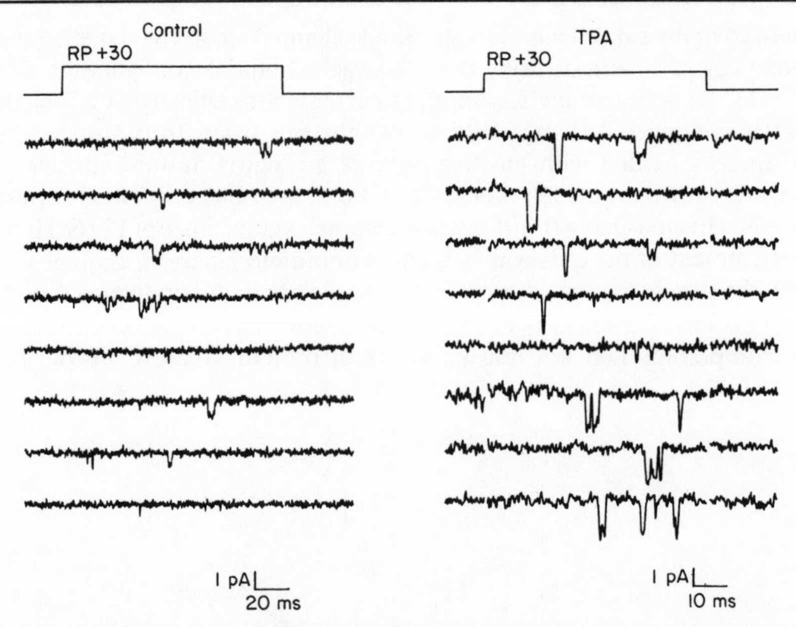

Figure 6. Activators of protein kinase C induce the appearance of a species of Ca^{2+} channel that is not observed in control cells. Unitary currents were recorded in cell-attached patches using Ba^{2+}-containing pipettes. The left-hand panel shows recordings of unitary channel openings during depolarizations of a patch on a control cell. The right-hand panel shows recordings from a cell exposed to TPA. In the latter, two sizes of openings are observed, one of which has no counterpart in control cells.

that is measured using this technique. In addition, although 36% of the patches in cells pretreated with a phorbol ester contain the large, 24-pS channel, we have never observed the appearance of the large channel when phorbol esters were added to the bath or to the inside of the pipette after the cell-attached patch had already been formed (Strong et al., 1987).

It is not yet clear why the protein kinase C enzyme system or the large channel is especially sensitive to both of these patch-clamp recording techniques. The sensitivity appears to be specific to the regulation of Ca^{2+} current by this enzyme, as intracellular dialysis does not disrupt the modulation of K^+ currents by cAMP

Table I
Summary of Observations of Single Ca Channels in the Absence and Presence of Activators of Protein Kinase C

Condition	Number of patches	No channels	Small channels	Large channels
		%	%	%
Control	33	64	36	0
TPA pretreated (10–20 nM)	28	43	36	36
DOG pretreated (1.5 μg/ml)	3	0	33	66

Inward channels were observed in cell-attached patches using 185 mM Ba^{2+} as the charge carrier, and were classified as small (<16 pS) or large (>20 pS) according to their slope conductance. Cells were pretreated with TPA or DOG for 10–120 min. Some patches in TPA-treated cells contained both large and small channels (note that the sum is >100%).

in these cells (Strong, 1984; Strong and Kaczmarek, 1986). One interesting possibility is that the techniques interfere with the translocation of protein kinase C from the cytoplasm to the plasma membrane. Such a translocation occurs with activation of the enzyme in other systems (Kraft and Anderson, 1983; Wolf et al., 1985), and preliminary evidence indicates that this process also occurs in the bag cell neurons. A second possibility is that the covert channels are inserted into the plasma membrane from an intracellular pool by an exocytotic event. Such a process could well be disrupted by the patch-clamp techniques.

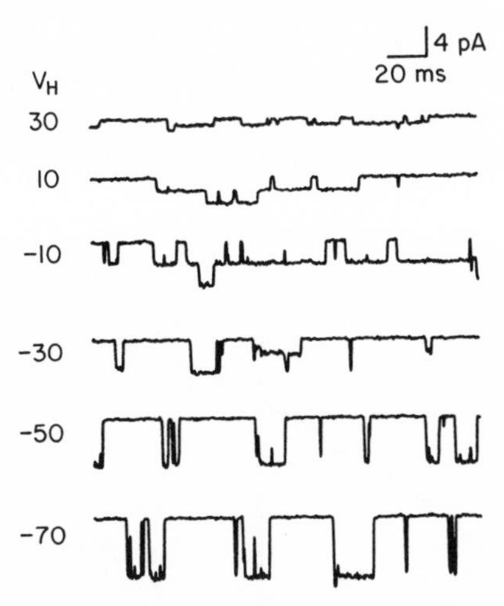

Figure 7. Recordings of unitary currents in a cell-free, inside-out patch from a bag cell neuron, showing the openings of a large-conductance, relatively voltage-independent inward channel. The recording was made using 185 mM Ba^{2+} (along with K^+ blockers) in the pipette and 535 mM K-aspartate, 20 mM EGTA at the cytoplasmic face of the membrane patch.

A Second Covert Ca^{2+} Channel

Although the two channels described above are the only two types of Ca^{2+} channel that can normally be observed in cell-attached patches on the somata of bag cell neurons, we discovered, in the course of these experiments, that a third type of Ca^{2+} channel is also present in the plasma membrane of these cells. Examples of the activity of this channel, recorded in an inside-out, cell-free patch, are shown in Fig. 7. Two of the striking characteristics of this channel are its very large unitary conductance (85 pS, Fig. 8) and its relative voltage independence; channel openings have been observed at all potentials examined (−150 to 150 mV). These characteristics indicate that it is probably the same channel that has been characterized by Chesnoy-Marchais (1985) in other *Aplysia* neurons using outside-out recording techniques. We have found that openings of this channel during cell-attached recording are observed extremely rarely. However, as soon as a patch is removed from the cell (e.g., to make an inside-out recording in K-aspartate), this large voltage-independent channel appears. Thus, it is likely that the channel is always present in the membrane, but is maintained in its closed state by cytoplasmic or cytoskeletal factors. In search of a physiological regulator for this channel, we have applied ATP, Ca^{2+}, and GTP to the cytoplasmic face of the channel, but have seen no effects of these agents. As reported by Chesnoy-Marchais (1985), the channel is present in large numbers (there are usually several in each patch). Thus, if significant numbers of this second type of covert Ca^{2+} channel were to open on an intact cell, they would induce a very large and voltage-independent influx of Ca^{2+}. The physiological conditions under which such influx is likely to occur are currently being investigated.

Conclusions

The sequence of behaviors that occur in vivo after a discharge in the bag cell neurons results from the action of several neuroactive peptides that are secreted from these cells (Kupfermann, 1970; Dudek et al., 1979; Strumwasser et al., 1980). The release of these peptides is likely to be tightly regulated by the levels of

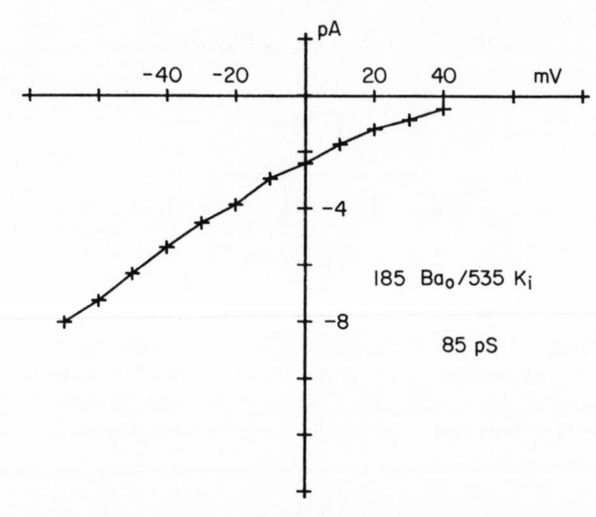

Figure 8. Single channel current-voltage relation for the channel shown in Fig. 7.

intracellular Ca^{2+} during the discharge (Stuart et al., 1980; Kaczmarek and Kauer, 1983). One mechanism that may induce rapid changes in Ca^{2+} levels, and which may play a role at the very onset of a discharge, is the release of Ca^{2+} from intracellular stores by the second messenger IP_3.

The bag cell neurons use several independent pathways to regulate the influx of Ca^{2+} through the plasma membrane. The very large voltage-independent Ca^{2+} channel represents one pathway for the elevation of intracellular Ca^{2+} whose regulation is not yet understood. The action potentials of these cells have a prominent Ca^{2+} component and at least two second-messenger systems, linked to the activation of protein kinases, which regulate the height and width of these action potentials. The width of action potentials may, in large part, be determined by K^+ currents whose amplitude and kinetics are regulated by cAMP-dependent mechanisms. On the other hand, as described in this article, the activation of protein kinase C enhances the voltage-dependent Ca^{2+} current and hence the influx of Ca^{2+} during the action potential. Since the latter is a Ca^{2+}-activated enzyme, this system may provide a form of positive feedback that promotes the "all-or-none" character of the afterdischarge. Moreover, the cAMP and protein kinase C pathways are likely to act in parallel during the afterdischarge to enhance excitability and Ca^{2+} influx.

References

Arch, S., P. Early, and T. Smock. 1976. Biochemical isolation and physiological identification of the egg-laying hormone in *Aplysia californica. Journal of General Physiology.* 68:197–210.

Azhderian, E., and L. K. Kaczmarek. 1986. Onset of discharge and elevations of cyclic AMP stimulate prohormone processing in bag cell neurons. *Journal of Cell Biology.* 103:191*a.* (Abstr.)

Berridge, M. J. 1984. Inositol triphosphate and diacylglycerol as second messengers. *Biochemical Journal.* 220:345–360.

Bruehl, C. L., and R. W. Berry. 1985. Regulation of synthesis of the neurosecretory egg-laying hormone of *Aplysia*: antagonistic roles of calcium and cyclic adenosine 3′:5′-monophosphate. *Journal of Neuroscience.* 5:1233–1238.

Chesnoy-Marchais, D. 1985. Kinetic properties and selectivity of calcium-permeable single channels in *Aplysia* neurones. *Journal of Physiology.* 367:457–488.

Chiu, A. Y., M. W. Hunkapiller, E. Heller, D. K. Stuart, L. E. Hood, and F. Strumwasser. 1979. Purification and primary structure of neuroactive egg-laying hormone of *Aplysia californica. Proceedings of the National Academy of Sciences.* 76:6656–6660.

Chiu, A. Y., and F. Strumwasser. 1981. An immunohistochemical study of the neuropeptidergic bag cells of *Aplysia. Journal of Neuroscience.* 1:812–826.

DeRiemer, S. A., P. Greengard, and L. K. Kaczmarek. 1985*a.* Calcium/phosphatidylserine/diacylglycerol-dependent protein phosphorylation in the *Aplysia* nervous system. *Journal of Neuroscience.* 5:2672–2676.

DeRiemer, S. A., J. A. Strong, K. A. Albert, P. Greengard, and L. K. Kaczmarek. 1985*b.* Phorbol ester and protein kinase C enhance calcium current in *Aplysia* neurones. *Nature.* 313:313–316.

DeRiemer, S. A., L. K. Kaczmarek, Y. Lai, and T. L. McGuinness, and P. Greengard. 1984. Calcium/calmodulin-dependent protein phosphorylation in the nervous system of *Aplysia. Journal of Neuroscience.* 4:1618–1625.

Dudek, F. E., J. S. Cobbs, and H. M. Pinsker. 1979. Bag cell electrical activity underlying spontaneous egg laying in freely behaving *Aplysia brasiliana. Journal of Neurophysiology.* 42:804–817.

Fink, L., J. A. Connor, J. A. Strong, and L. K. Kaczmarek. 1985. Inositol trisphosphate injections and imaging of the calcium indicator fura-2 in molluscan neurons. *Society for Neuroscience Abstracts.* 11:854.

Hamill, O. P., A. Marty, E. Neher, B. Sakmann, and F. J. Sigworth. 1981. Improved patch-clamp techniques for high-resolution current recording from cells and cell-free membrane patches. *Pflügers Archiv.* 391:85–100.

Heller, E., L. K. Kaczmarek, M. W. Hunkapiller, L. E. Hood, and F. Strumwasser. 1980. Purification and primary structure of two neuroactive peptides that cause bag cell afterdischarge and egg-laying in *Aplysia. Proceedings of the National Academy of Sciences.* 77:2328–2332.

Kaczmarek, L. K., M. Finbow, J. Revel, and F. Strumwasser. 1979. The morphology and coupling of *Aplysia* bag cells within the abdominal ganglion and in cell culture. *Journal of Neurobiology.* 10:535–550.

Kaczmarek, L. K., K. Jennings, and F. Strumwasser. 1978. Neurotransmitter modulation, phosphodiesterase inhibitor effects, and cAMP correlates of afterdischarge in peptidergic neurites. *Proceedings of the National Academy of Sciences.* 75:5200–5204.

Kaczmarek, L. K., K. R. Jennings, F. Strumwasser, A. C. Nairn, U. Walter, F. D. Wilson, and P. Greengard. 1980. Microinjection of catalytic subunit of cyclic AMP-dependent protein kinase enhances calcium action potentials of bag cell neurons in cell culture. *Proceedings of the National Academy of Sciences.* 77:7487–7491.

Kaczmarek, L. K., and J. A. Kauer. 1983. The role of calcium entry in the prolonged refractory period that follows afterdischarge in peptidergic neurons of *Aplysia. Journal of Neuroscience.* 3:2230–2239.

Kaczmarek, L. K., J. A. Strong, and S. A. DeRiemer. 1985. Biochemical mechanisms that modulate potassium and calcium currents in peptidergic neurons. *In* Neurosecretion and the Biology of Neuropeptides. H. Kobayashi, H. A. Bern, and A. Urano, editors. Springer-Verlag, Berlin. 80–87.

Kaczmarek, L. K., J. A. Strong, and J. A. Kauer. 1986. The role of protein kinases in the control of prolonged changes in neuronal excitability. *Progress in Brain Research.* 69:77–90.

Kaczmarek, L. K., and F. Strumwasser. 1981. The expression of long lasting afterdischarge by isolated *Aplysia* bag cell neurons. *Journal of Neuroscience.* 1:626–634.

Kikkawa, U., Y. Takai, R. Minakuchi, S. Inohara and Y. Nishizuka. 1982. Ca-activated, phospholipid-dependent protein kinase from rat brain. *Journal of Biological Chemistry.* 257:13341–13348.

Kraft, A. S., and W. B. Anderson. 1983. Phorbol esters increase the amount of Ca^{2+}-phospholipid-dependent protein kinase associated with plasma membrane. *Nature.* 301:621–623.

Kupferman, I. 1980. Stimulation of egg laying by extracts of neuroendocrine cells (bag cells) of abdominal ganglion of *Aplysia. Journal of Neurophysiology.* 33:877–881.

Kupferman, I., and E. R. Kandel. 1970. Electrophysiological properties and functional interconnections of two symmetrical neurosecretory clusters (bag cells) in abdominal ganglion of *Aplysia. Journal of Neurophysiology.* 33:865–876.

Llinas, R., T. L. McGuinness, C. S. Leonard, M. Sugimori, and P. Greengard. 1985. Intraterminal injection of synapsin I or calcium/calmodulin-dependent protein kinase II alters neurotransmitter release at the squid giant synapse. *Proceedings of the National Academy of Sciences.* 82:3035–3039.

Nishizuka, Y. 1984. The role of protein kinase C in cell surface signal transduction and tumour promotion. *Nature.* 308:693–697.

Novak-Hofer, I., and I. B. Levitan. 1983. Ca^{++}/calmodulin-regulated protein phosphorylation in the *Aplysia* nervous system. *Journal of Neuroscience.* 3:473–481.

Rothman, B. S., E. Mayeri, R. O. Brown, P. Yuan, and J. Shively. 1983. Primary structure and neuronal effects of α-bag cell peptide, a second neurotransmitter candidate encoded by a single gene in bag cell neurons of *Aplysia. Proceedings of the National Academy of Sciences.* 80:5733–5757.

Scheller, R. H., J. F. Jackson, L. B. McAllister, B. S. Rothman, E. Mayeri, and R. Axel.

1983. A single gene encodes multiple neuropeptides mediating a stereotyped behavior. *Cell.* 32:7–22.

Strong, J. A. 1984. Modulation of potassium current kinetics in bag cell neurons of *Aplysia* by an activator of adenylate cyclase. *Journal of Neuroscience.* 4:2772–2783.

Strong, J. A., A. P. Fox, R. W. Tsien, and L. K. Kaczmarek. 1987. Stimulation of protein kinase C recruits covert calcium channels in *Aplysia* bag cell neurones. *Nature.* 325:714–717.

Strong, J. A., and L. K. Kaczmarek. 1986. Multiple components of delayed potassium current in peptidergic neurons of *Aplysia*: modulation by an activator of adenylate cyclase. *Journal of Neuroscience.* 6:814–822.

Strumwasser, F., L. K. Kaczmarek, A. Y. Chiu, E. Heller, K. R. Jennings, and D. P. Viele. 1980. Peptides controlling behavior in *Aplysia. In* Peptides: Integrators of Cell and Tissue Functions. F. E. Bloom, editor. Raven Press, New York. 197–218.

Stuart, D. K., A. Y. Chiu, and F. Strumwasser. 1980. Neurosecretion of egg-laying hormone and other peptides from electrically active bag cell neurons of *Aplysia. Journal of Neurophysiology.* 43:488–498.

Wolf, M., H. Levine, W. S. May, P. Cuatrecasas, and N. Sahyoun. 1985. A model for intracellular translocation of protein kinase C involving synergism between Ca^{2+} and phorbol esters. *Nature.* 317:546–549.

Chapter 13

Intracellular Calcium in Cardiac Myocytes: Calcium Transients Measured Using Fluorescence Imaging

M. B. Cannell, J. R. Berlin, and W. J. Lederer

Department of Physiology, University of Maryland School of Medicine, Baltimore, Maryland

Introduction

Depolarization of the heart leads to a transient increase in the intracellular calcium ion concentration, $[Ca^{2+}]_i$, which activates muscle contraction (Allen and Blinks, 1978). Much has been learned about the $[Ca^{2+}]_i$ transient in cardiac ventricular muscle with the use of Ca^{2+} indicators such as the bioluminescent protein aequorin (Allen and Blinks, 1978; Fabiato, 1985). Nevertheless, the description of the mechanisms that control changes in $[Ca^{2+}]_i$ is incomplete (e.g., Eisner and Lederer, 1985). For example, the voltage dependence of the magnitude of the $[Ca^{2+}]_i$ transient is unknown. This question has been extremely difficult to address for many reasons, including the difficulties in obtaining large and interpretable signals from previously used Ca^{2+} indicators and the difficulty in adequately voltage-clamping multicellular ventricular muscle preparations. Recently, a new family of Ca^{2+} indicators, which includes fura-2, has been introduced that overcomes many of these limitations. The high quantum efficiency of fura-2 allows for $[Ca^{2+}]_i$ determinations in single cells (Grynkiewicz et al., 1985). Using patch-clamp technology to voltage-clamp single cardiac cells (Hamill et al., 1981), it is now possible for us to determine the effects of membrane potential on the $[Ca^{2+}]_i$ transient. This technique should therefore allow us to study in a more thorough manner the mechanisms that control $[Ca^{2+}]_i$.

We have examined the Ca^{2+} transient in voltage-clamped single rat ventricular myocytes using the fluorescent Ca^{2+} indicator fura-2. The indicator was introduced into each cell by allowing the K^+ salt of fura-2–free acid to diffuse into the cell from the patch electrode used to voltage-clamp the myocyte. The spatial distribution of $[Ca^{2+}]_i$ was examined with the use of a digital imaging fluorescence microscope (Tanasugarn et al., 1984; Williams et al., 1985; Bright and Taylor, 1986). We have found that Ca^{2+} is uniformly distributed in the cytoplasm at rest under physiologically normal conditions. With depolarization, $[Ca^{2+}]_i$ rises rapidly but uniformly throughout the cell. Furthermore, the peak level of $[Ca^{2+}]_i$ reached during the transient is dependent on the size of the depolarization. In addition, under conditions of "Ca^{2+} overload" (Kass et al., 1978; Lakatta and Lappé, 1981), propagated waves of elevated $[Ca^{2+}]_i$ are observed within the cell.

Methods

Preparation of Cells

Isolated single rat ventricular muscle cells were obtained using a conventional enzymatic isolation procedure (Powell et al., 1980; Hume and Giles, 1981). Rat hearts were rapidly removed from 4–6-wk-old rats that had been anesthetized with intraperitoneal pentobarbital (50 mg/kg). The hearts were perfused using the Langendorff method. After a brief perfusion with Ca^{2+}-free solution, 1 mg/ml of collagenase (type II, Worthington Biochemical Corp., Freehold, NJ) and sufficient $CaCl_2$ to increase $CaCl_2$ to 50 μM were added to the perfusion solution. Perfusion was continued for an additional 20 min. The heart was then minced and gently triturated in an enzyme-free solution containing 50 μM $CaCl_2$. Free-floating cells were separated from the larger remaining pieces of tissue and Ca^{2+} was elevated to 1 mM over 30 min. The cells were stored in 1 mM solution at 37°C awaiting experimental use.

Solutions

The dissociation solutions contained (in mM): 118 NaCl, 25 $NaHCO_3$, 4.8 KCl, 1.2 $MgSO_4$, 1 KH_2PO_4, 10 glucose (pH 7.4 at 37°C with 95% O_2/5% CO_2). $CaCl_2$ was added as indicated. The storage solutions contained (in mM): 145 NaCl, 4 KCl, 1 $MgCl_2$, 10 glucose, 10 HEPES (pH 7.4 at 37°C), 1 $CaCl_2$. The pipette-filling solutions contained (in mM): 120 K-glutamate, 20 CsCl, 10 K_2ATP, 1 $MgCl_2$, 10 glucose, 3 PIPES (pH 7.2 at 37°C), 0.07 K_5^+–fura-2.

Voltage Clamp

We recorded membrane currents using a single-microelectrode voltage clamp in the whole-cell clamp mode (Hamill et al., 1981). Before any series resistance compensation, the electrode resistance with the usual pipette-filling solution was 1.5–3.0 MΩ.

Digital Imaging Fluorescence Microscope

A Nikon Inc. (Garden City, NY) Diaphot microscope with an epifluorescence illumination system containing a 100-W Hg light source was modified to have quartz optics and selectable 10-nm-wide interference filters centered at 340 or 380 nm in the illumination pathway. Fluorescence at 510 nm was measured by obtaining fluorescent images of cells with a DAGE intensified silicon-intensified target (ISIT) video camera (model 66, DAGE-MTI, Inc., Michigan City, IN) or by detecting the fluorescence signal from a 10-μm-diam spot on a cell with a Thorn-EMI photomultiplier tube (EMI 9698). A more complete description of the system is given in Cannell, M. B., J. R. Berlin, and W. J. Lederer (manuscript in preparation).

Recording and Analysis of Data

The video data with timing and calibration information were recorded on a video cassette recorder (PV8500, Panasonic Industrial Co., Secaucus, NJ). Video signals were digitized using PC Vision hardware (Imaging Technology, Woburn, MA) on a PC-AT computer (PC's Limited, Austin, TX) and image-processing software written by us. Membrane currents and potentials, as well as light signals from the photomultiplier tube data, were recorded on FM tape and later digitized with a Data Translation (Marlboro, MA) 2805 analog-to-digital converter and Vacuum MkII (c) software (Micro-Based Computer Systems, Baltimore, MD).

Ca^{2+} Measurement Procedure

After a gigohm seal was formed between the microelectrode and the cell, background autofluorescence was measured when the cell was illuminated with 340-nm and 380-nm light. The membrane under the electrode was then ruptured to form a low-resistance electrical pathway into the cell and to allow fura-2 to diffuse into the cell. A steady level of fluorescence was generally observed within 10 min. At this point, autofluorescence was a minor component of the total light signal. To obtain ratio images of fura-2 fluorescence, a single video frame obtained during illumination with 340-nm light was divided pixel by pixel with a video frame obtained during 380-nm illumination after subtraction of the appropriate background images.

Fluorescence Ratio and Calibration of Ca^{2+} Signal

The measurement of Ca^{2+} levels with fura-2 uses the ratiometric technique (Heiple and Taylor, 1982). This technique involves measuring fluorescence at two illumination wavelengths, $L1$ and $L2$. The fura-2 fluorescence at a single wavelength is described in Eqs. 1 and 2. For a 1:1 complexation of Ca^{2+} and fura-2 (Eq. 1), the total fluorescence at a single wavelength depends on the sum of the fluorescence from the Ca^{2+}-bound form of the indicator and the Ca^{2+}-free form of the indicator (Eq. 2). The fluorescence from the Ca^{2+}-bound form of the indicator illuminated with light at wavelength $L1$ (F_{L1}) is given by the product of M_{L1}^{Ca}, the molar fluorescence coefficient for the Ca^{2+}-bound form of the indicator at $L1$, [fura-Ca], the concentration of the Ca^{2+}-bound form of the indicator, and T, the effective thickness of the sample containing the Ca^{2+}-bound form of the indicator. Similarly, the fluorescence from the Ca^{2+}-free indicator is given by the product of M_{L1}, the molar fluorescence coefficient for the Ca^{2+}-free form of the indicator at $L1$, [fura], the concentration of the Ca^{2+}-free form of the indicator, and T, the effective thickness of the sample containing the Ca^{2+}-free form of the indicator.

$$Ca^{2+} + \text{fura} \rightleftharpoons (\text{fura-Ca}); \qquad (1)$$

$$F_{L1} = (M_{L1}^{Ca} \times [\text{fura-Ca}] \times T) + (M_{L1} \times [\text{fura}] \times T). \qquad (2)$$

Dividing the fluorescence measured during illumination with light of wavelength 1 (F_{L1}) by fluorescence with illumination at wavelength 2 (F_{L2}) and substituting Eq. 3 for the equilibrium dissociation constant,

$$K_d = ([Ca^{2+}] \times [\text{fura}])/[\text{fura-Ca}], \qquad (3)$$

will yield Eq. 4. The fluorescence intensity ratio is

$$\frac{F_{L1}}{F_{L2}} = \frac{\{[Ca^{2+}] \times (M_{L1}^{Ca}/K_d) + M_{L1}\}}{\{[Ca^{2+}] \times (M_{L2}^{Ca}/K_d) + M_{L2}\}}. \qquad (4)$$

Eq. 4 expresses the fluorescence ratio as a function of Ca^{2+} and a number of constants. Neither the concentration of the indicator nor the effective thickness of the sample play any role in this relationship. Using the same equipment and a series of EGTA-buffered test solutions instead of cells, one can measure the relationship between the fluorescence ratio and Ca^{2+}. This calibration curve implicitly defines the constants that relate $[Ca^{2+}]_i$ to the fluorescence ratio and permit one to calibrate the fluorescence ratio in terms of Ca^{2+} from the experimentally constructed curve. It is important to remember, however, that if the calibration solution used to establish the calibration curve does not adequately represent the intracellular environment, then the calibration of Ca^{2+} will be in error.

Results and Discussion

Intracellular Ca^{2+} was measured in voltage-clamped myocytes (Cannell et al., 1986; Berlin et al., 1987). Fig. 1 shows records of the fluorescence of a cell in which the K^+ salt of fura-2 was allowed to diffuse into the cell from the pipette. The cell was held at a resting potential of −64 mV. The fluorescence intensity of the cell at rest is shown during illumination with 340-nm light (Fig. 1 *B*) and 380-nm light (Fig. 1 *C*). At either illumination wavelength, the fluorescence intensity is not uniform throughout the cell. This nonuniformity may reflect the spatial variation of cell thickness, the effective concentration of fura-2, or the intracellular Ca^{2+} concentration (as discussed in the Methods). The two nuclei seen in the transmitted light micrographs can also be seen as areas of increased fluorescence intensity when the cells are illuminated with 340-nm and 380-nm light. The ratio image, obtained by dividing (pixel by pixel) the 340-nm background-subtracted image by a similar image obtained with 380-nm illumination, is quite uniform (Fig. 1 *D*). Neither the nuclei nor the variations of cell thickness are associated with systematic variations

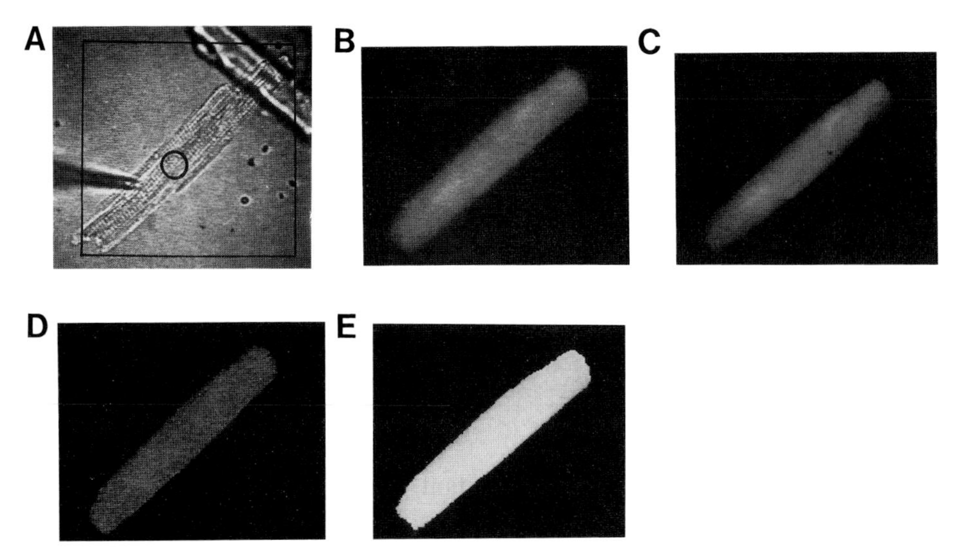

Figure 1. Fura-2 fluorescence images of a single cardiac cell. (*A*) Transmitted light micrograph of a single rat ventricular myocyte whose membrane potential is being clamped via the microelectrode at the lower left of the picture. Fura-2 diffused into the cell from the pipette as described in the Methods. The large box around the cell shows the region of the field that was illuminated by ultraviolet light. The small circle on the cell shows the area of the cell from which the photomultiplier tube measured fura-2 fluorescence. (*B* and *C*) Fluorescence micrographs of the cell held at −64 mV during illumination with 340-nm and 380-nm light, respectively. Note that the out-of-focus cell (which does not contain fura-2) at the top right of *A* is not visible. The two regions of brighter fluorescence approximately halfway between the center of the cell and its ends coincide with the two nuclei of the cell. (*D*) Image of fluorescence ratios obtained by dividing the image recorded during illumination with 340-nm light (*B*) by that recorded during illumination at 380 nm (*C*) after appropriate background subtraction. (*E*) Image of fluorescence ratios obtained in the same manner but at the end of a 100-ms depolarizing pulse to −4 mV. Note the uniformity of both ratio images. The mean ratio of the resting cell in *D* is 0.21 ± 0.03 (SD), while the mean ratio of the cell at the end of the depolarization is 0.43 ± 0.08 (SD). This shows that membrane depolarization produced a significant increase in $[Ca^{2+}]_i$.

in the magnitude of the ratio image. As the fluorescence ratio depends only on $[Ca^{2+}]_i$, the uniformity of the ratio image suggests that, at rest, $[Ca^{2+}]_i$ is uniform throughout the cytoplasm and that $[Ca^{2+}]_i$ is the same in the nuclear and cytosolic spaces. This result is in contrast to the nuclear and subsarcolemmal increase in intracellular Ca^{2+} seen in smooth muscle (Williams et al., 1985). The mean intensity of the ratio image shown in Fig. 1 *D* is 0.21 ± 0.03 (SD). The distribution around the mean intensity is approximately Gaussian, as would be expected for a fluorescence intensity variation caused by photon noise. The mean value of the intensity ratio obtained for Fig. 1 *D* has considerable precision, since each pixel of the ratio image constitutes an independent sample measurement and >4,500 pixels are contained within the cell boundary. The ratio images shown in Fig. 1 were obtained from one pair of video frames with no signal averaging. The degree of precision could be improved by signal averaging.

Fig. 2 shows the Ca^{2+} transient that results from depolarization of a cell from −50 to 0 mV for 100 ms. At each point, the mean ratio of fluorescence of the cell was determined by dividing a single video frame recorded during illumination with 340-nm light by a single frame at 380-nm illumination, as described above. An in vitro calibration curve was then used to estimate the level of $[Ca^{2+}]_i$ (see Methods).

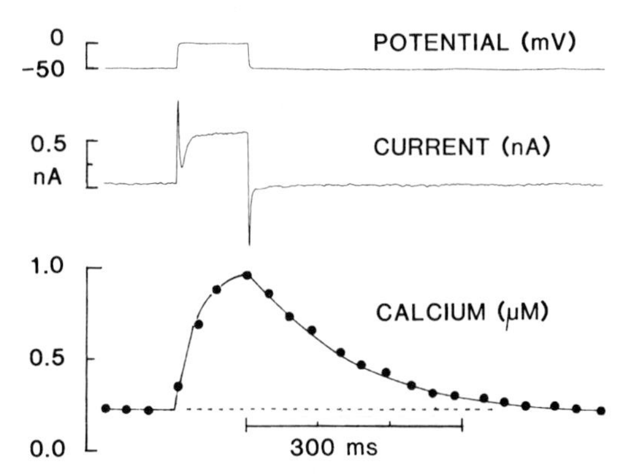

Figure 2. Time-dependent changes of $[Ca^{2+}]_i$ elicited by a voltage-clamp depolarization. The rat cardiac ventricular cell was loaded with fura-2 from the voltage-clamp pipette containing 0.2 mM fura-2. The cell was depolarized from a holding potential of −50 to 0 mV for 100 ms at 1.0 Hz. Background-subtracted images recorded during illumination with 340-nm light were divided (pixel by pixel) by images recorded during 380-nm illumination to obtain images of fluorescence ratios. The mean fluorescence ratio calculated from each image was converted to $[Ca^{2+}]_i$ with a calibration curve as described in the Methods. (Reprinted with permission from Cannell et al., 1986.)

The time course of the transient was determined by repeating these steps for a series of consecutive video frames recorded during illumination with 340-nm and 380-nm light. Note that $[Ca^{2+}]_i$ rises rapidly after depolarization until the end of the pulse. After membrane repolarization, $[Ca^{2+}]_i$ declines back to its resting level with a half-time of ~110 ms. In addition, Fig. 1 *E* shows that at the end of a 100-ms depolarization, $[Ca^{2+}]_i$ is uniform throughout the cell, even though the level of $[Ca^{2+}]_i$ is much higher than in the resting cell.

The above results indicate that $[Ca^{2+}]_i$ within the cell is spatially uniform at rest and at the peak of the Ca^{2+} transient. That $[Ca^{2+}]_i$ appears uniform at the peak of the Ca^{2+} transient may be somewhat surprising. During the normal Ca^{2+} transient, Ca^{2+} is released from the terminal cisternae of the sarcoplasmic reticulum, and it enters from the extracellular space via Ca^{2+} channels in the sarcolemma and t-tubular system. Such a net movement of Ca^{2+} from regions of high $[Ca^{2+}]$ to regions of low $[Ca^{2+}]$ clearly requires that the Ca^{2+} distribution within the cell be nonuniform during the period of this movement. It is equally clear, however, that for a normal transient, such inhomogeneities must exist at the sarcomere level (i.e., over distances of <2 μm). The spatial resolution of our imaging system for fluorescent sources is between 1 and 2 μm and consequently does not permit us to easily resolve such events, even if they persist for long periods (see below). In addition, our video imaging system cannot resolve rapidly changing events (i.e., milliseconds), so that we would not be able to resolve these predicted spatial inhomogeneities (see below). Indeed, we have not been able to detect systematic inhomogeneities in the spatial distribution of $[Ca^{2+}]_i$ during a normal Ca^{2+} transient, the ratio images obtained during the transient being about as uniform in the distribution of ratio values as the images of a cell at rest (data not shown).

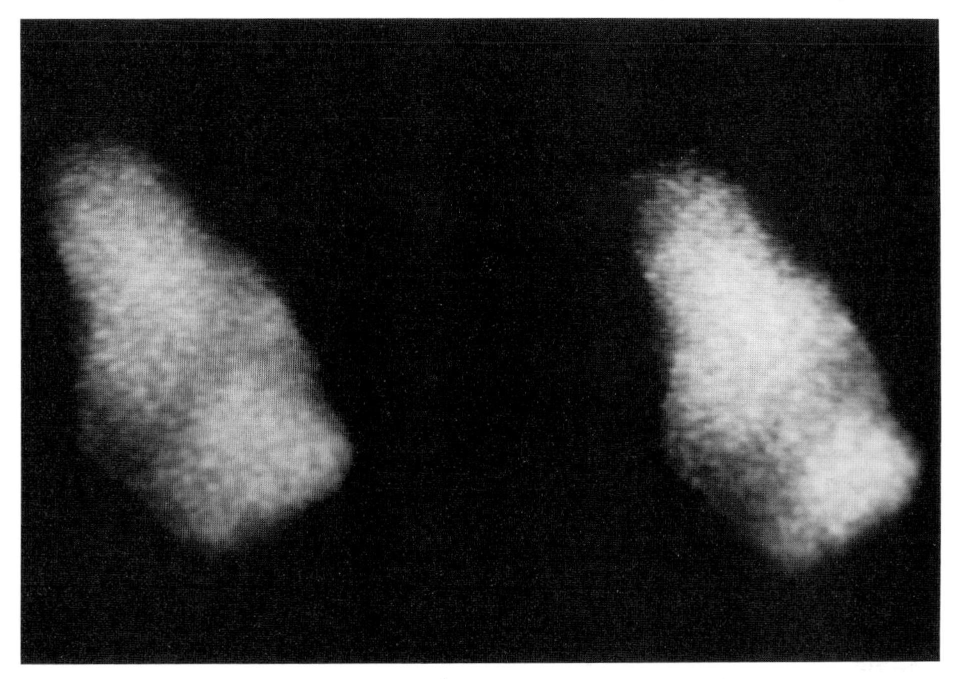

Figure 3. Propagated wave of elevated $[Ca^{2+}]_i$. The ventricular myocyte was voltage-clamped and loaded with fura-2 as described in the Methods. The cell was exposed to a low-K^+ (0.2 mM) solution to inhibit the Na^+ pump and produce Ca^{2+} overload. Fluorescence images of the cell were obtained during illumination with 380-nm light, but the image on the left was recorded 100 ms before the image on the right. The dark band (indicating elevated $[Ca^{2+}]_i$) is moving from the upper left end of the cell to the lower right end of the cell. This wave of elevated Ca^{2+} was traveling at ~100 μm/s. (Reprinted with permission from Lederer et al., 1987.)

In contrast to the case of the normal Ca^{2+} transient, there is a condition associated with marked spatial nonuniformity of $[Ca^{2+}]_i$. Fig. 3 shows an example of this phenomenon, which occurs when spontaneous waves of contraction are observed in a voltage-clamped rat myocyte under conditions of "Ca^{2+} overload" (produced, for example, by exposure to low extracellular K^+ to inhibit the Na^+ pump [Eisner and Lederer, 1979]). A wave of increased intracellular Ca^{2+} propagates spontaneously from the upper left to the lower right along the long axis of the cell. Because these images were obtained using 380-nm illumination, the dark band of decreased fluorescence indicates the regions of the cell where intracellular Ca^{2+} is relatively elevated. From the propagation velocity of the wave and the width of the elevated region of Ca^{2+}, we estimate that the duration of the wave phenomenon at a point within the cell is ~150 ms. This is comparable to the duration of the normal Ca^{2+} transient. We observed similar results in rat cells that were not voltage-clamped and that were loaded with fura-2 using the acetoxymethyl ester form of fura-2 (fura-2 AM) (Berlin et al., 1985; Wier et al., 1987).

It has been suggested that these fluctuations in intracellular Ca^{2+} arise from the spontaneous release of Ca^{2+} from the sarcoplasmic reticulum (Fabiato and Fabiato, 1975; Kass et al., 1978; Orchard et al., 1983; Wier et al., 1983; Allen et al., 1984). These fluctuations are believed to be responsible for a number of phenomena observed during Ca^{2+} overload, including Ca^{2+}-activated arrhythmogenic electrical currents, such as I_{TI} (Lederer and Tsien, 1976; Kass et al., 1978), increased membrane current fluctuations and tension fluctuations (Kass and Tsien, 1982; Cannell and Lederer, 1986), and altered contractile activity (Lakatta and Lappé, 1981; Capogrossi and Lakatta, 1985).

There are several important concerns associated with imaging time-dependent changes in $[Ca^{2+}]_i$ in heart cells, including (*a*) the speed of response of the video camera, (*b*) the temporal distortion produced by the sequential read-out of pixels during the video frame, and (*c*) the requirement for stereotyped responses locked to the video framing rate.

The temporal resolution of the imaging system is limited by the response time of the ISIT camera as well as by the video frame rate. During increases in fluorescence, there is a "lag" in the response because of the properties of the camera. Additionally, during a decrease in fluorescence, there is a "persistence" of fluorescence. Fig. 4 illustrates how the response time of a specific camera (model 66, DAGE-MTI) is affected by persistence and lag. The experiment was carried out by illuminating a fluorescent object with 380-nm light. A shutter controlled the illumination period as indicated in the figure. The ISIT used in this example has separate settings for the video amplifier gain and intensifier gain. At a fixed overall gain, the response time of the camera can be improved by decreasing the gain of the video amplifier while increasing the gain of the intensifier stage. These gain settings, which improve temporal resolution, are, however, associated with loss of spatial resolution.

A way to minimize many of these effects is to use gated illumination rather than continuous illumination. Gating the illumination to the vertical synchronization signal of the video camera means that a flash of light is applied at 30 Hz during the vertical retrace signal. Consequently, image persistence is the only one of these problems that remains (Cannell, M. B., J. R. Berlin, and W. J. Lederer, manuscript in preparation). All of the images shown in this article use continuous illumination.

The video framing rate is set at 30 Hz for most commercial systems available in the U.S. This means that each pixel is read out once every 33 ms. Consequently, the maximum temporal resolution of such systems is 30 Hz. (Faster video systems can be designed, but these systems require custom-built video and computer hardware and software. In addition, to maintain equal sensitivity at equal spatial resolution, the average fluorescence intensity must increase in direct proportion to the framing rate. Therefore, either the indicator concentration or the illumination intensity must increase proportionately. This introduces additional problems associated with Ca^{2+} buffering, indicator bleaching, and photon-induced damage to the cell. These topics are beyond the scope of this report.) Because each pixel is read out sequentially, there is ~30 ms between the time when the first pixel (upper left of the video frame) and the last pixel (lower right of the video frame) are determined. This will lead to an image composed of pixels within each video frame that are obtained at slightly different times. Thus, when Ca^{2+} is changing rapidly, some distortion of the $[Ca^{2+}]_i$ within the image may occur during the video frame because of the differing sample times. Since the cardiac cells are long and narrow, the cell can be placed horizontally in the video image to decrease the time between reading the first and last pixel containing the image of the cell. In addition, the persistence and lag of the video camera discussed above tend to make the camera

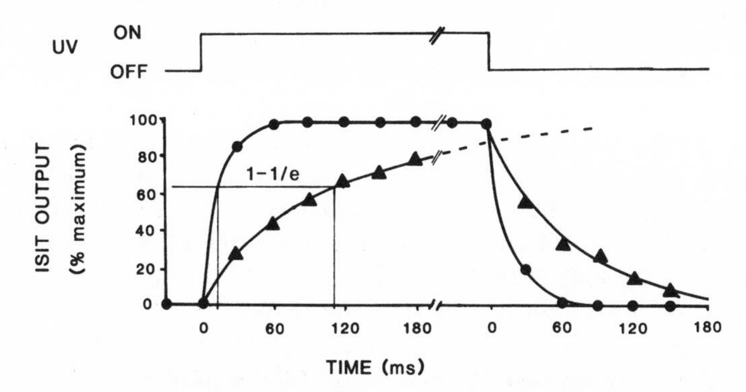

Figure 4. Response characteristics of the video camera. The output signal of the ISIT camera is plotted as a function of time. A fluorescent target was illuminated with ultraviolet light (380 nm) during the period indicated in the top panel of the figure. The ultraviolet illumination was controlled by an electromechanical shutter triggered synchronously with the vertical retrace of the camera. The response characteristics of the camera were examined at a fixed overall gain while the relative amount of gain between the video amplifier and the intensifier stages was varied. The average level of the camera output voltage during each video frame is plotted as a function of time. The filled circles show the camera response when the intensifier gain was set maximally and the video gain was set at a minimum. The filled triangles show the camera response when the intensifier gain was set at a minimum and the video gain was set maximally. The response time of the camera becomes slower as the intensifier gain is decreased and the video gain is increased.

act as a low-pass filter to the video information. During the Ca^{2+} transient, the degree of uniformity of $[Ca^{2+}]_i$ distribution was measured in records 100 ms after membrane depolarization. At this time, $[Ca^{2+}]_i$ is at or near its peak value and is changing relatively slowly, so that distortions introduced during the scan of the video frame are minimized. However, as pointed out above, we have not observed any systematic variations in $[Ca^{2+}]_i$ within the cell during the normal Ca^{2+} transient, so this type of distortion does not appear to be a major problem in these experiments.

In order to use fura-2 to image Ca^{2+} transients with our system, the Ca^{2+} transient must be a stereotyped event. This is required because of the poor response time of the video camera and because of the inability to rapidly switch the wavelength of the illuminating light between 340 nm and 380 nm. Consequently, the fura-2–loaded cell is first illuminated with 340-nm light while a voltage-clamp depolarization is carried out. The illuminating light is then changed to 380 nm and the same depolarization is repeated. The ratio is therefore obtained from matched pairs of video frames taken during illumination with 340-nm and 380-nm light. To correctly match the video frames recorded during illumination with different wavelengths of light, it is necessary that the Ca^{2+} transient be locked to the video framing rate. This is done by triggering the voltage-clamp circuit from the vertical synchronization pulse of the video frame. We have chosen to do this at rates between 0.1 and 1.0 Hz in these experiments. In heart cells, this adaptation of the ratiometric technique is possible because the Ca^{2+} transient behaves in a stereotyped way when voltage-clamp depolarizations are repeated at a fixed rate. The Ca^{2+} transient in Fig. 2 was obtained using the procedure outlined above.

This same limitation, however, has prevented a complete description of the spontaneous waves of Ca^{2+} observed under conditions of Ca^{2+} overload. In the absence of synchronization to the video framing rate, these waves of Ca^{2+} can only be viewed during illumination at a single wavelength (in Fig. 3, at 380 nm). There is no reason to believe that the effective concentration of fura-2 or cell thickness changes during the wave to a degree significantly different from that during the normal $[Ca^{2+}]_i$ transient. Thus, by comparing the fluorescence intensity change we observed during the normal transient with that observed during the wave of elevated Ca^{2+}, we find that the maximum level of Ca^{2+} reached during the wave in this cell is comparable to that seen during the Ca^{2+} transient.

To improve our ability to record rapidly changing fluorescence signals, we have also used a sensitive, high-speed detector (photomultiplier tube) to measure changes in fluorescence intensity. While the response time of the photomultiplier tube (PMT) is much faster, its spatial resolution is quite poor because it measures total fluorescence from a 10-μm-diam region of the cell. The region is chosen (see Fig. 1*A*) so that it is located near the center of the cell, avoiding the nuclei and edges of the cell. We saw no detectable movement artifact unless the area sampled by the PMT was close to the edge of the cell, a complication that is easy to avoid, given the ~20-μm width of the cells. Fig. 5 shows a typical Ca^{2+} transient recorded with the PMT. The rise in the Ca^{2+} transient is extremely fast (~100 s^{-1}) and is consistent with the idea that the rate of change of the fluorescence intensity signal may be limited by the reaction kinetics of the fura-Ca^{2+} reaction. Additionally, we observe that the peak level of $[Ca^{2+}]_i$ is achieved within 30–80 ms. This result suggests that while the measurement of Ca^{2+} using the ISIT camera will not accurately report the changes of Ca^{2+} during the time when Ca^{2+} is changing rapidly, at the time of the peak of the Ca^{2+} transient and at subsequent times, the ISIT camera and PMT will both give similar measurements of Ca^{2+}. This is in agreement with our other data.

How does the Ca^{2+} transient reported by fura-2 compare with other measurements of the Ca^{2+} transient? The peak level of intracellular Ca^{2+} revealed by the PMT detector, between 600 and 1,000 nM, is similar to the value obtained under similar conditions by Allen and Kurihara (1980, 1982) in aequorin-injected rat papillary muscle. Furthermore, the half-time of decay of Ca^{2+} determined from the fura-2 signal (~100 ms) is also similar to that calculated from the decay of the aequorin light signal reported by Allen and Kurihara (1982) under similar conditions. Thus, there is good evidence that the fura-2 signal is accurately reporting the time-dependent changes in cytosolic Ca^{2+}.

By using the method described in this article for measuring intracellular Ca^{2+}

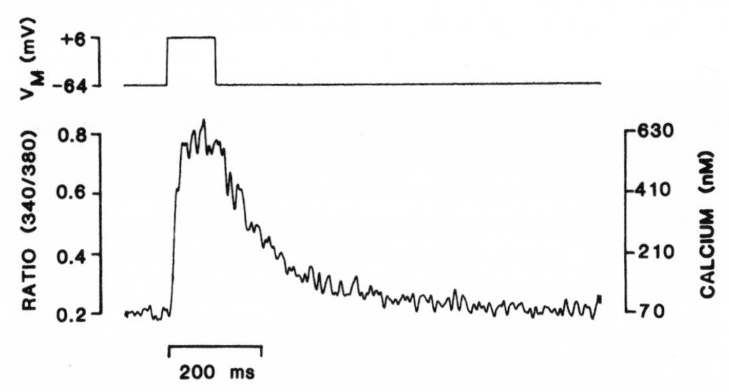

Figure 5. Ca^{2+} transient with high temporal resolution. Ca^{2+} transients in a ventricular myocyte were produced by 100-ms voltage-clamp depolarizations from −64 to 4 mV. A photomultiplier tube measured the fluorescence intensity (at 510 nm) from a 10-μm-diam region of the myocyte. The cell was illuminated sequentially with 340-nm and 380-nm light. The ratio of the fluorescence intensity (340/380) is plotted against time. Intracellular Ca^{2+} levels were estimated from the fluorescence intensity ratio and an in vitro calibration curve as described in the Methods. (Reprinted with permission from Lederer et al., 1987.)

while simultaneously voltage-clamping single cardiac cells, we are able to examine for the first time the voltage dependence of the Ca^{2+} transient in cardiac ventricular muscle. Fig. 6 shows the effect of graded depolarizations on the magnitude of the Ca^{2+} transient, monitored from the changes in fluorescence intensity while illuminating a fura-2–loaded cell with 380-nm light. The membrane was depolarized from −64 to −4 mV for 100 ms in 10-mV increments. With increasing depolari-

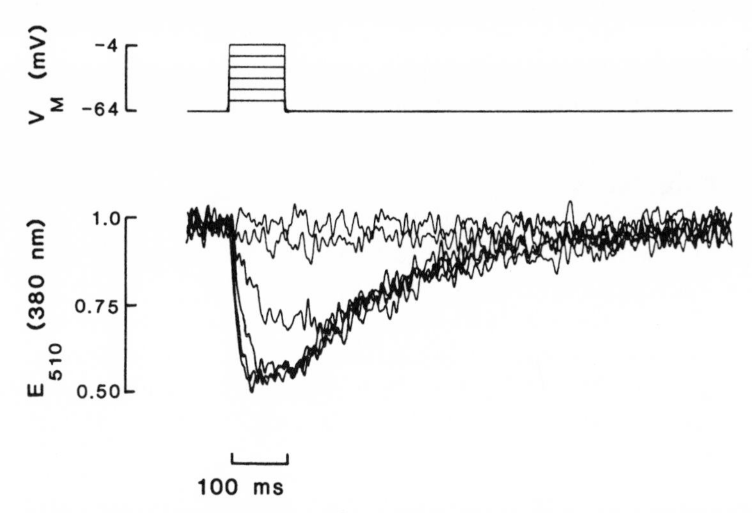

Figure 6. Voltage dependence of the Ca^{2+} transient. These data were recorded from the same cell as shown in Fig. 5. Fluorescence intensity was measured from a 10-μm-diam spot on the cell with a photomultiplier tube. The records shown were taken during illumination of the myocyte with 380-nm light only. During illumination with 340-nm light, changes in fluorescence intensity were in the opposite direction (i.e., an increase in intensity); however, the magnitude of change and the rate of the change in fluorescence showed the same voltage dependence. (Reprinted with permission from Berlin et al., 1987.)

zation, the initial rate of change of fluorescence and the peak change in fluorescence increased to a maximum that was voltage independent between −14 and +26 mV (the records from depolarizations to potentials more positive than −4 mV are not shown). The rate of recovery of fluorescence ($t_{0.5} \approx 120$ ms) was also independent of depolarization amplitude. The maximum rate of change in the fluorescence signal that accompanies the rising phase of the Ca^{2+} transient in Fig. 6 is ~100 s^{-1} and is probably dye-limited, as suggested for the results presented in Fig. 5. With 340-nm illumination, depolarization produced changes in fluorescence (i.e., an increase in fluorescence intensity) of similar time course but in the opposite direction to those obtained with 380-nm illumination. This confirms that the changes in fluorescence intensity shown in Fig. 6 reflect true changes in $[Ca^{2+}]_i$.

Summary

We have examined the distribution of Ca^{2+} in voltage-clamped cardiac myocytes under resting conditions and during the Ca^{2+} transient. We find that the resting Ca^{2+} level in a quiescent rat myocyte bathed in 1 mM extracellular Ca is relatively low (between 60 and 100 nM) and uniform. At the peak of the Ca^{2+} transient, Ca^{2+} can rise to a level as high as 600 nM to 1.0 μM. Furthermore, the magnitude of the Ca^{2+} transient is dependent on the size of the membrane depolarization. There is good agreement between measurements made using video imaging and those made using a photomultiplier tube for the value of intracellular Ca^{2+} at the peak of the Ca^{2+} transient and for the subsequent slow changes in intracellular Ca^{2+}. On repolarization, intracellular Ca^{2+} falls with a half-time of ~100 ms. The uniform distribution of Ca^{2+} reported in the Ca^{2+} images of myocytes at rest and at the peak of the Ca^{2+} transient under normal conditions is in contrast to what is observed during "Ca^{2+} overload" when subcellular regions of elevated Ca^{2+} are observed to propagate along the cell. Thus, the measurement of $[Ca^{2+}]_i$ in cardiac myocytes with fura-2 has already yielded important new information that was not available using other techniques to measure $[Ca^{2+}]_i$ in cardiac ventricular muscle.

Acknowledgments

We would like to thank Dr. R. J. Bloch for the use of his DAGE ISIT video camera and K. MacEachern for assistance in preparing the manuscript.

This work was supported by the National Institutes of Health (HL-25675) and by the American and Maryland Heart Associations. The work was carried out during the tenure of an Established Investigatorship of the American Heart Association and its Maryland Affiliate. J.R.B. is supported by a fellowship of the Maryland Heart Association. M.B.C. is the Young Investigator of the Maryland Heart Association.

References

Allen, D. G., and J. R. Blinks. 1978. Calcium transients in aequorin-injected frog cardiac muscle. *Nature.* 273:509–511.

Allen, D. G., D. A. Eisner, and C. H. Orchard. 1984. Characterization of oscillations of intracellular calcium concentration in ferret ventricular muscle. *Journal of Physiology.* 352:113–128.

Allen, D. G., and S. Kurihara. 1980. Calcium transients in mammalian ventricular muscle. *European Heart Journal.* 1A:5–15.

Allen, D. G., and S. Kurihara. 1982. The effects of muscle length on intracellular calcium transients in mammalian cardiac muscle. *Journal of Physiology.* 327:79–84.

Berlin, J. R., M. B. Cannell, W. F. Goldman, W. J. Lederer, E. Marban, and W. G. Wier. 1985. Subcellular calcium inhomogeneity indicated by fura-2 develops with calcium overload in single rat heart cells. *Journal of Physiology.* 371:200P.

Berlin, J. R., M. B. Cannell, and W. J. Lederer. 1987. Voltage-dependence of the Ca_i^{2+} transient in rat cardiac ventricular cells measured with fura-2. *Journal of Physiology.* 382:108P.

Bright, G. R., and D. L. Taylor. 1986. Imaging at low light level in fluorescence microscopy. *In* Applications of Fluorescence in the Biomedical Sciences. D. L. Taylor, A. S. Waggoner, R. F. Murphy, F. Lanni, and R. R. Birge, editors. Alan R. Liss, New York. 257–288.

Cannell, M. B., J. R. Berlin, and W. J. Lederer. 1986. Ca^{2+} transients elicited by voltage-clamp depolarizations in single rat ventricular muscle cells calculated from fura-2 fluorescence images. *Biophysical Journal.* 49:466*a.* (Abstr.)

Cannell, M. B., and W. J. Lederer. 1986. The arrhythmogenic current I_{TI} in the absence of electrogenic sodium-calcium exchange in sheep cardiac Purkinje fibres. *Journal of Physiology.* 374:201–220.

Capogrossi, M. C., and E. G. Lakatta. 1985. Frequency modulation and synchronization of spontaneous oscillations in cardiac cells. *American Journal of Physiology.* 248:H412–H418.

Eisner, D. A., and W. J. Lederer. 1979. Inotropic and arrhythmogenic effects of potassium depleted solutions on mammalian cardiac muscle. *Journal of Physiology.* 294:255–277.

Eisner, D. A., and W. J. Lederer. 1985. Na-Ca exchange: stoichiometry and electrogenicity. *American Journal of Physiology.* 248:C189–C202.

Fabiato, A. 1985. Simulated calcium current can both cause calcium loading in and trigger calcium release from the sarcoplasmic reticulum of a skinned canine cardiac Purkinje cell. *Journal of General Physiology.* 85:291–320.

Fabiato, A., and F. Fabiato. 1975. Contractions induced by a calcium-triggered release of calcium from the sarcoplasmic reticulum of single skinned cardiac cells. *Journal of Physiology.* 249:469–495.

Grynkiewicz, G., M. Poenie, and R. Y. Tsien. 1985. A new generation of Ca^{2+} indicators with greatly improved fluorescence properties. *Journal of Biological Chemistry.* 260:3440–3450.

Hamill, O. P., A. Marty, E. Neher, B. Sakmann, and F. J. Sigworth. 1981. Improved patch-clamp techniques for high-resolution recording from cells and cell-free membrane patches. *Pflügers Archiv.* 391:85–100.

Heiple, J. M., and D. L. Taylor. 1982. An optical technique for measurement of intracellular pH in single living cells. *In* Intracellular pH: Its Measurement, Regulation and Utilization in Cellular Functions. R. Nuccitelli and D. W. Deamer, editors. Alan R. Liss, New York. 21–54.

Hume, J. R., and W. Giles. 1981. Active and passive electrical properties of single bullfrog atrial cells. *Journal of General Physiology.* 78:19–42.

Kass, R. S., W. J. Lederer, R. W. Tsien, and R. Weingart. 1978. Role of calcium ions in transient inward currents and aftercontractions induced by strophanthidin in cardiac Purkinje fibres. *Journal of Physiology.* 281:187–208.

Kass, R. S., and R. W. Tsien. 1982. Fluctuations in membrane current driven by intracellular calcium in cardiac Purkinje fibers. *Biophysical Journal.* 38:259–269.

Lakatta, E. G., and E. M. Lappé. 1981. Diastolic scattered light fluctuations, resting force and twitch force in mammalian cardiac muscle. *Journal of Physiology.* 315:369–394.

Lederer, W. J., M. B. Cannell, and J. R. Berlin. 1987. Intracellular calcium measurements and calcium imaging in cardiac muscle using Fura-2: calcium fluctuations are linked to cardiac arrhythmias and negative inotropy. *In* Initiation and Conduction of the Cardiac Pacemaker Response. W. R. Giles, editor. Alan R. Liss, New York.

Lederer, W. J., and R. W. Tsien. 1976. Transient inward current underlying arrhythmogenic effects of cardiotonic steroids in Purkinje fibres. *Journal of Physiology.* 263:73–100.

Orchard, C. H., D. G. Allen, and D. A. Eisner. 1983. Oscillations of intracellular Ca^{2+} in mammalian cardiac muscle. *Nature.* 304:735–738.

Powell, T., D. A. Terrar, and V. W. Twist. 1980. Electrical properties of individual cells isolated from adult rat ventricular myocardium. *Journal of Physiology.* 302:131–153.

Tanasugarn, L., P. McNeil, G. Reynolds, and D. L. Taylor. 1984. Microspectrofluorometry 450 digital image processing: measurement of cytoplasmic pH. *Journal of Cell Biology.* 98:717–724.

Wier, W. G., M. B. Cannell, J. R. Berlin, E. Marban, and W. J. Lederer. 1987. Fura-2 fluorescence imaging reveals cellular and subcellular heterogeneity of $[Ca^{2+}]_i$ in single heart cells. *Science.* 235:325–328.

Wier, W. G., A. A. Kort, M. D. Stern, E. G. Lakatta, and E. Marban. 1983. Cellular calcium fluctuations in mammalian heart: direct evidence from noise analysis of aequorin signals in Purkinje fibers. *Proceedings of the National Academy of Sciences.* 80:7367–7371.

Williams, D. A., K. E. Fogarty, R. Y. Tsien, and F. S. Fay. 1985. Calcium gradients in single smooth muscle cells revealed by the digital imaging microscope using fura-2. *Nature.* 318:558–561.

Chapter 14

Cytoplasmic Free Ca^{2+} and the Intracellular pH of Lymphocytes

Sergio Grinstein, Julia D. Goetz-Smith, and Sara Cohen

Departments of Cell Biology and Immunology, Hospital for Sick Children, and Department of Biochemistry, University of Toronto, Toronto, Ontario, Canada

Introduction

The intracellular pH (pH_i) is actively and precisely regulated in mammalian cells. The stringency of this regulation is indicated by the existence of multiple (redundant?) pH-controlling systems, such as the Na^+/H^+ antiport and the Na^+-dependent Cl^-/HCO_3^- transporter, in addition to passive and dynamic buffering systems, such as Cl^-/HCO_3^- exchange. Similarly, the concentration of free cytoplasmic Ca^{2+} ($[Ca^{2+}]_i$) is tightly regulated by multiple transport systems in the plasma membrane (the Ca^{2+} pump and the Na^+/Ca^{2+} antiport) and in intracellular compartments such as the endoplasmic reticulum and the mitochondria. However, although accurately regulated, these parameters are not invariant, and changes in pH_i and $[Ca^{2+}]_i$ are central to a variety of physiological processes including secretion, contraction, and cellular proliferation. The mechanisms underlying variations in pH_i and $[Ca^{2+}]_i$ are not entirely understood, but it is remarkable that many of the conditions that affect one of these parameters also modify the other (reviewed by Busa and Nuccitelli, 1984; see also Table I). Thus, coupling between the changes in $[Ca^{2+}]_i$ and pH_i has been suggested (Busa and Nuccitelli, 1984). The purpose of this chapter is to review recent work from our laboratory analyzing the relationship between the cytoplasmic levels of Ca^{2+} and H^+ in rat thymic lymphocytes.

Effects of Cytoplasmic Alkalinization on $[Ca^{2+}]_i$

Unlike other cell types, rat thymic lymphocytes were found to respond with a delayed increase in $[Ca^{2+}]_i$ to the addition of the tumor promoter 12-*O*-tetradecanoylphorbol-13-acetate (TPA) (Grinstein and Goetz, 1985; Fig. 1 *B*). The change in $[Ca^{2+}]_i$ was prevented by the removal of extracellular Na^+ or by the addition of amiloride or submicromolar concentrations of 5-*N*-ethyl-*N*-propylamino amiloride. At these concentrations, the latter compound is an effective blocker of the Na^+/H^+ antiport, but has little effect on either Na^+ channels or on Na^+/Ca^{2+} exchange. Therefore, the change in $[Ca^{2+}]_i$ appeared to be related to Na^+/H^+ countertransport, which is known to be activated by TPA in a variety of cells, including thymocytes (Burns and Rozengurt, 1983; Moolenaar et al., 1984; Rosoff et al., 1984; Grinstein et al., 1985; Fig. 1 *B*). This notion was supported by observations made in hypertonically treated cells. Like TPA-treated cells, osmotically shrunken thymocytes also display a pronounced activation of the Na^+/H^+ antiport (Grinstein and Rothstein, 1986; Fig. 1). This activation is also accompanied by a delayed elevation of $[Ca^{2+}]_i$ (Fig. 1 *A*), which is equally sensitive to the pyrazine diuretics and to the removal of extracellular Na^+.

The association between Na^+/H^+ countertransport and the levels of cytoplasmic Ca^{2+} was confirmed using monensin, an ionophore that can operate as an exogenous Na^+/H^+ exchanger. The addition of monensin to lymphocytes suspended in physiological (Na^+-rich) media induces the uptake of Na^+ and a pronounced cytoplasmic alkalinization, which is followed by an elevation in $[Ca^{2+}]_i$ (Fig. 1 *D*). In contrast, when added to cells in K^+-rich medium, monensin catalyzes the loss of internal Na^+, has little effect on pH_i, and does not significantly affect $[Ca^{2+}]_i$. This indicates that the effects of monensin on Ca^{2+} are due to alterations of the transmembrane gradients of monovalent cations, and not to the ionophore per se. Similar effects of monensin on $[Ca^{2+}]_i$ were subsequently reported by Rosoff and Cantley (1985) for 70Z/3 cells, and a relationship between endogenous $Na^+/$

H^+ exchange and changes in $[Ca^{2+}]_i$ has also been suggested for sea urchin sperm (Schackmann and Chock, 1986).

The change in $[Ca^{2+}]_i$ that results from stimulation of the antiport could be associated with either the increased intracellular Na^+ concentration or with the elevated cytoplasmic pH. To test these possibilities, alternative methods were used to vary these parameters independently. Changing the concentration of Na^+ could affect $[Ca^{2+}]_i$ by promoting the countertransport of external Ca^{2+} for cellular Na^+. Indeed, the presence of Na^+/Ca^{2+} exchange in a membrane-rich fraction isolated from lymphocytes was recently reported (Ueda, 1983). The concentration of intracellular Na^+ ($[Na^+]_i$) in lymphocytes can be increased rapidly by means of the

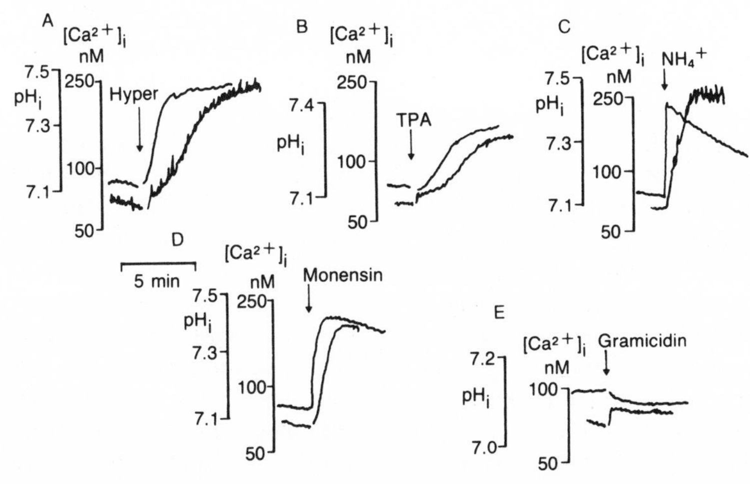

Figure 1. Changes in cytoplasmic pH (top trace of each panel) and cytoplasmic Ca^{2+} (bottom trace in each panel) in rat thymic lymphocytes suspended in Ca^{2+}-containing medium. (*A*) Where indicated, the solution was made hypertonic (550 mosM) by addition of concentrated *N*-methylglucamine Cl. (*B*) Addition of 20 nM TPA. (*C*) Addition of 20 mM NH_4Cl. (*D*) Addition of 5 μM monensin. (*E*) Addition of 1 μM gramicidin. pH_i and $[Ca^{2+}]_i$ were measured fluorimetrically using bis(carboxyethyl)carboxyfluorescein (BCECF) and quin2, respectively. (From Grinstein and Goetz, 1985.)

conductive cation channel-former gramicidin, or more slowly by inhibition of the Na^+/K^+ pump with cardiac glycosides. Incubation of rat thymocytes with 1 mM ouabain for 30 min at 37°C increased $[Na^+]_i$ from 24.9 ± 3.3 to 47.7 ± 3.7 mM (n = 4), without significantly changing $[Ca^{2+}]_i$ (89.4 ± 8.5 and 78.8 ± 4.5 nM, respectively). Incubation of the cells with 2 μM gramicidin for 6 min increased $[Na^+]_i$ to 58.1 ± 7 mM ($n = 4$). As shown in Fig. 1 *E*, under comparable conditions, pH_i was found to decline slightly (0.04 ± 0.014 units), whereas $[Ca^{2+}]_i$ increased only marginally ($\Delta[Ca^{2+}]_i = 16 \pm 3$ nM; $n = 5$). These results indicate that the level of $[Ca^{2+}]_i$ is relatively insensitive to changes in $[Na^+]_i$ and suggest that, if present, the Na^+/Ca^{2+} exchange system is not essential for $[Ca^{2+}]_i$ homeostasis. Although our data are in apparent contradiction to the results of Ueda (1983), they are in agreement with the findings of Tsien et al. (1982) for porcine mesenteric lymphocytes and of Gelfand et al. (1984) for human blood lymphocytes.

The dependence of $[Ca^{2+}]_i$ on pH_i was tested using NH_4^+ to alkalinize the cytoplasm. This cation is in equilibrium with the unprotonated weak base NH_3, which, unlike NH_4^+, readily permeates the membrane. Having reached the intracellular compartment, NH_3 can combine with cytoplasmic H^+ to form NH_4^+, thereby alkalinizing the cytoplasm. As shown in Fig. 1 *C*, the alkalinization is very rapid but is not sustained. The ensuing relaxation is thought to be due to the entry of the acidic NH_4^+, either through conductive channels or by taking the place of K^+ in the Na^+/K^+ pump. The transient alkalinization is followed by a pronounced increase in $[Ca^{2+}]_i$, which is similarly not sustained. These data indicate that the changes in $[Ca^{2+}]_i$ are secondary to the cytoplasmic alkalinization. This was further suggested by the gradual elevation of $[Ca^{2+}]_i$ detected in cells that similarly became alkaline after resuspension in a medium of $pH_o = 8.2$. Taken together, these results suggest that the elevation of $[Ca^{2+}]_i$ that accompanies the activation of Na^+/H^+ exchange is a result of the change in pH_i caused by the extrusion of H^+ equivalents.

Because quin2 is a tetracarboxylic acid, its apparent affinity for Ca^{2+} is sensitive to pH. It was therefore essential to establish that fluorescence changes like those in Fig. 1 reflected true increases in $[Ca^{2+}]_i$, as opposed to pH-dependent changes in the affinity of the probe. Four lines of evidence indicate that the measurements are reliable. (*a*) Maintaining $[Ca^{2+}]$ constant by means of Ca^{2+}-EGTA buffers, we found that quin2 fluorescence increased by $\leq 2\%$ while the pH was varied in the 7.0–7.5 range, which encompasses the range of pH_i values in Fig. 1. This small change cannot account for the increases noted in Fig. 1, which are equivalent to $\geq 15\%$ of the maximal fluorescence; the change is in keeping with the reported pK_a of the acidic groups of the probe (Tsien et al., 1982). (*b*) A comparison of the time courses of the pH_i and $[Ca^{2+}]_i$ changes indicated that the increase in quin2 fluorescence consistently lagged behind the cytoplasmic pH change. In some instances (e.g., Fig. 1, *A* and *C*), $[Ca^{2+}]_i$ attained a maximal level up to 2 min after the pH_i reached the most alkaline value. This delay is not inherent to quin2: in cell-free solutions, the response of the probe to changes in $[Ca^{2+}]$ was found to be virtually instantaneous (Grinstein and Goetz, 1985). (*c*) When measured with quin2, the pH-induced changes in $[Ca^{2+}]_i$ were greatly reduced by removal of extracellular Ca^{2+} (Fig. 2). Because removal of Ca^{2+} did not alter pH_i, these observations further rule out that the fluorescence changes merely reflect alterations in the affinity of the probe at constant $[Ca^{2+}]_i$. (*d*) The pH-induced elevation in $[Ca^{2+}]_i$ could also be detected using indo-1, a second fluorescent $[Ca^{2+}]_i$ probe with different chemical properties (Fig. 3). Even though the baseline fluorescence displayed a slight upward drift, probably caused by leakage of indo-1 from the cells, the addition of monensin (Fig. 3*A*) or NH_4^+ (not illustrated) produced an increase in $[Ca^{2+}]_i$ comparable to that recorded with quin2.

Because of its large extinction coefficient and quantum yield, indo-1 can be used at concentrations that are considerably lower than those required when using quin2. Thus, unlike quin2, indo-1 contributes only modestly to the cellular Ca^{2+}-buffering power and is more suitable for the study of small changes induced by release of finite Ca^{2+} stores, such as membrane-bound intracellular compartments. We therefore undertook studies to determine whether, in addition to increasing Ca^{2+} uptake from the medium, cytoplasmic alkalinization also led to the release of intracellular Ca^{2+} stores. Typical results are shown in Figs. 3 and 4. The addition of monensin to cells suspended in Ca^{2+}-free medium containing 1 mM EGTA produced a small but consistent increase in indo-1 fluorescence (Fig. 3*B*). This

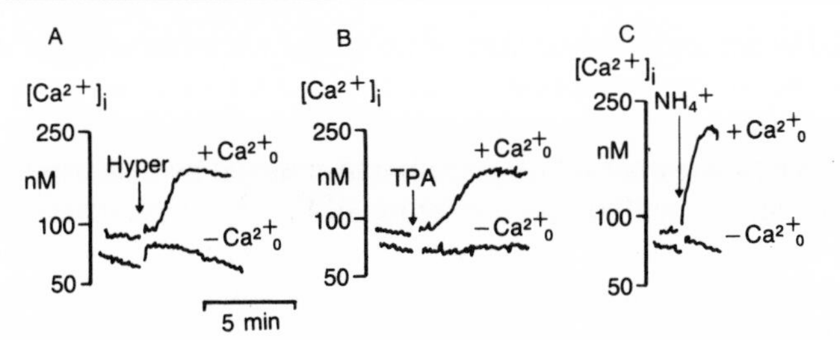

Figure 2. Comparison of the pH_i-induced changes of $[Ca^{2+}]_i$ in thymocytes suspended in Ca^{2+}-containing and Ca^{2+}-free media. (*A*) Quin2-loaded cells were suspended in medium with (top trace) or without (bottom trace) Ca^{2+}. Where indicated, cytoplasmic alkalinization was induced by hypertonic shrinking as in Fig. 1. (*B*) Cells were suspended in medium with or without Ca^{2+} as in *A* and, where indicated, 20 nM TPA was added. (*C*) Cytoplasmic alkalinization was induced at the arrow by addition of 20 mM NH_4Cl. (From Grinstein and Goetz, 1985.)

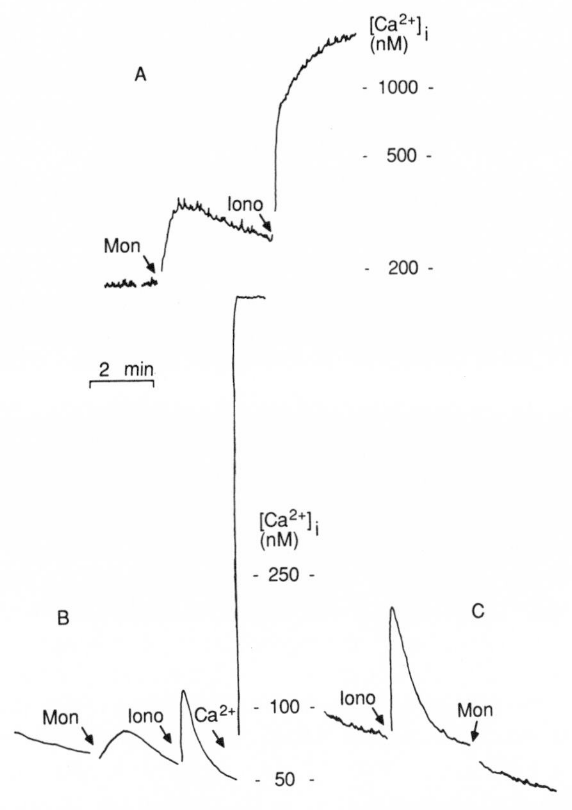

Figure 3. Effect of cytoplasmic alkalinization on $[Ca^{2+}]_i$, monitored using indo-1. Thymocytes were loaded with indo-1 by incubation with 1 μM of the parent acetoxymethyl ester. (*A*) The level of $[Ca^{2+}]_i$ was recorded continuously in Ca^{2+}-containing medium. Where indicated, 5 μM monensin was added, followed by 1 μM ionomycin. (*B*) Cells were suspended in Ca^{2+}-free medium containing 1 mM EGTA. Where indicated by the arrows, monensin (5 μM), ionomycin (1 μM), and $CaCl_2$ (2 mM) were added sequentially. (*C*) Cells were suspended in Ca^{2+}-free medium with EGTA. Ionomycin (1 μM) and monensin (5 μM) were added sequentially. Representative of three similar experiments. Temperature, 37°C.

change probably represents a release of Ca^{2+} from internal stores, rather than a nonspecific effect of the ionophore, since it failed to occur when the stores were previously depleted by pretreatment with ionomycin (Fig. 3*C*), a divalent cation ionophore. It is noteworthy that the ionomycin-induced change was smaller in cells pretreated with monensin (compare *B* and *C* of Fig. 3), as expected for a release of Ca^{2+} from overlapping pools. Similar observations were made when alkalinization was accomplished by pulsing with NH_4^+ (Fig. 4). The pH_i change was associated with a transient rise in $[Ca^{2+}]_i$, which was absent if ionomycin was added earlier. In fact, a small decrease in $[Ca^{2+}]_i$ was noted upon addition of NH_4^+ to ionomycin-treated cells, which is suggestive of Ca^{2+} binding to cellular components that may have increased their negative charge because of deprotonation. A similar observation was made after monensin addition (Fig. 3*C*). In summary, these results indicate that cytoplasmic alkalinization not only increases the net influx of extracellular Ca^{2+}, but also promotes the liberation of Ca^{2+} from internal stores.

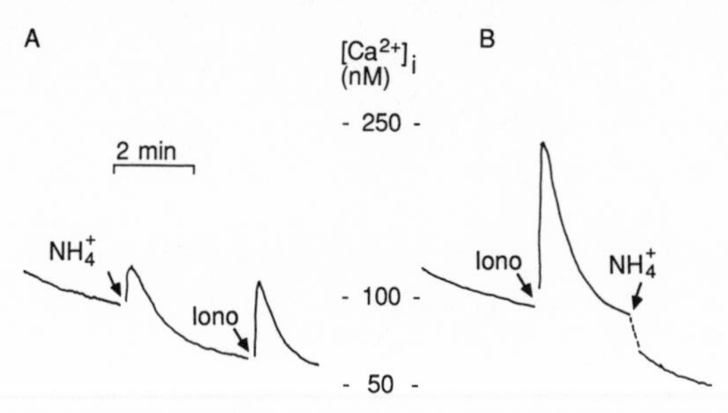

Figure 4. Release of intracellular Ca^{2+} stores by NH_4^+. Indo-1–loaded thymocytes were suspended in Ca^{2+}-free medium with EGTA. (*A*) Where indicated, 30 mM NH_4Cl was added, followed by 1 μM ionomycin. (*B*) Ionomycin (1 μM) and NH_4Cl (30 mM) were added where indicated. Other details as in Fig. 3.

Effects of Increasing $[Ca^{2+}]_i$ on Cytoplasmic pH

The relationship between $[Ca^{2+}]_i$ and the intracellular pH has been studied in a variety of cell types. A partial summary is shown in Table I. Several procedures have been used to elevate $[Ca^{2+}]_i$, including microinjection, reversal of the intrinsic Na^+/Ca^{2+} exchanger, and, more frequently, permeabilization with divalent cation ionophores such as A23187 and ionomycin. In most instances, increasing $[Ca^{2+}]_i$ lowered pH_i by 0.1–0.3 units. For instance, a distinct cytoplasmic acidification was reported by Moolenaar et al. (1983) when human foreskin fibroblasts were treated with A23187. However, using the same cell type, Muldoon et al. (1985) and Hesketh et al. (1985) found that A23187 induced an alkalinization. Thus, it appeared possible that increased $[Ca^{2+}]_i$ elicits multiple responses and that the predominant effect observed depends on the particular conditions used.

We therefore analyzed the responses of lymphocytes to Ca^{2+} ionophores under conditions where different components could be discerned. Typical results using ionomycin are illustrated in Fig. 5. At high concentrations of the ionophore (e.g., 3 μM), a rapid alkalinization was observed (Fig. 5, *D* and *F*). This effect is biologically unimportant, as it reflects the exchange of extracellular Ca^{2+} for internal H^+, catalyzed by the carboxylic ionophore. Accordingly, the alkalinization was

Table I
Effects of Elevating $[Ca^{2+}]_i$ on the pH_i

Biological system	Effect on pH_i	Reference
Fibroblasts (mammalian)	Increase	Muldoon et al. (1985)
Fibroblasts (mammalian)	Increase	Hesketh et al. (1985)
Fibroblasts (mammalian)	Decrease	Moolenaar et al. (1983)
Fibroblasts (mammalian)	Decrease	Ives and Daniel (1986)
Lymphocytes (mammalian)	No effect	Rink et al. (1982)
Lymphocytes (mammalian)	Increase	Hesketh et al. (1985)
Lymphocytes (mammalian)	Increase/decrease	Grinstein and Cohen (1987)
Synaptosomes (mammalian)	No effect	Richards et al. (1984)
Cardiac muscle (mammalian)	Decrease	Vaughan-Jones et al. (1983)
Cardiac muscle (mammalian)	Decrease	Hoerter et al. (1983)
Urinary bladder (turtle)	Decrease	Arruda et al. (1981)
Sperm (sea urchin)	Increased	Schackmann and Chock (1986)
Neuron (snail)	Decrease	Meech and Thomas (1977)

dependent on external Ca^{2+}, but was insensitive to the nature of the main monovalent cation in the medium or to the presence of amiloride. At lower ionomycin concentrations, a more complex pattern of pH_i changes emerged. In Na^+-containing solutions, a slow cytoplasmic alkalinization predominated (Fig. 5*A*). In contrast, in Na^+-free media or in Na^+ solutions containing amiloride or its 5-*N*-disubstituted analogues, treatment with ionomycin induced a distinct acidification (Fig. 5, *B* and *C*). The mechanisms underlying these pH_i changes are analyzed below.

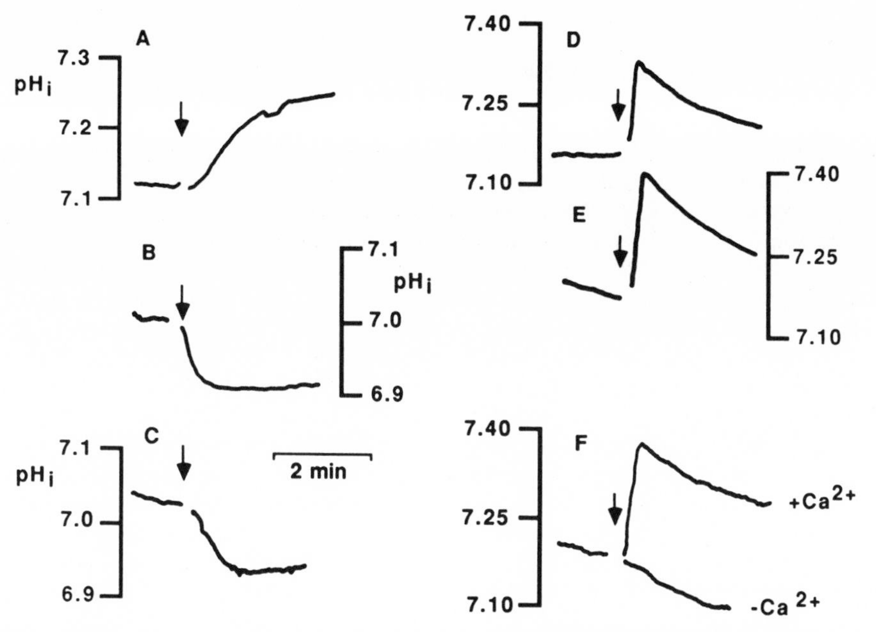

Figure 5. Effects of ionomycin on pH_i in rat thymocytes. BCECF-loaded cells were suspended in either Na^+-containing solution (*A* and *D*), Na^+-free (*N*-methylglucamine) solution (*B* and *E*), or Na^+-containing solution with 200 μM amiloride at 37°C. Where indicated by the arrows, either 0.4 μM (*A–C*) or 3 μM (*D–F*) ionomycin was added. (From Grinstein and Cohen, 1987.)

Mechanism of Intracellular Acidification

A cytoplasmic acidification can, in principle, be produced by: (*a*) increased metabolic acid production, (*b*) release of acid equivalents from intracellular stores, and (*c*) entry of acid equivalents from the medium. Evidence is presented below that, although *a* and/or *b* contribute measurably to the acidification, most of the observed changes can be accounted for by uptake of extracellular H^+ equivalents.

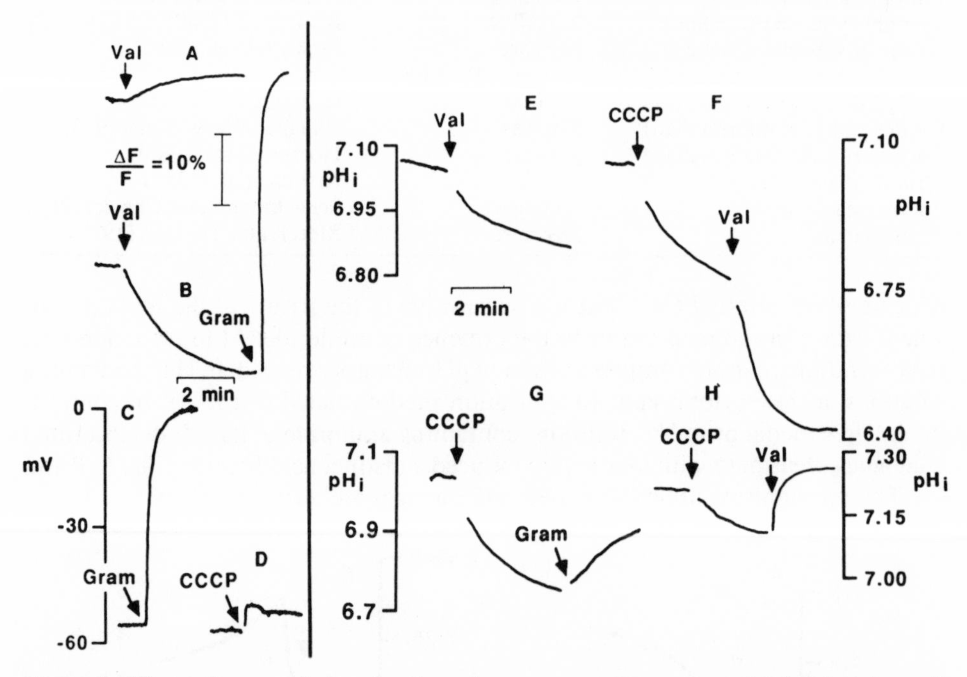

Figure 6. Effects of monovalent cation ionophores on membrane potential (left-hand panel) and pH_i (right-hand panel). Membrane potential was measured with either diS-C_3-(5) (*A* and *B*) or bis-oxonol (*C* and *D*). The former dye was used because bis-oxonol is incompatible with valinomycin. (*A*) Cells were suspended in K^+-rich medium. Where indicated, 1 μM valinomycin was added. (*B*) Cells suspended in Na^+-rich medium. Valinomycin (1 μM) and gramicidin (50 nM) were added where indicated. The fractional fluorescence change ($\Delta F/F$), relative to the fluorescence of the dye before the addition of cells, applies only to *A* and *B*. (*C*) Cells were pre-equilibrated with bis-oxonol in Na^+-rich (physiological) medium. Where indicated, 50 nM gramicidin was added. (*D*) Cells suspended in Na^+-rich medium. Where indicated, 1 μM CCCP was added. The membrane potential calibration (in millivolts) applies to *C* and *D*. (*E–H*) pH_i measurements using BCECF. (*E*) Cells in Na^+ medium with 200 μM amiloride. Valinomycin (1 μM) was added. (*F*) Cells in Na^+ medium with amiloride. First, 1 μM CCCP and then 1 μM valinomycin were added. (*G*) Cells in Na^+ medium plus amiloride. CCCP (1 μM) and gramicidin (100 nM) were added. (*H*) Cells in K^+-rich medium. CCCP (1 μM) and valinomycin (1 μM) were added. (From Grinstein and Cohen, 1987.)

Protons can be driven into the cell (or OH^- out) by the internally negative membrane potential (E_m). This phenomenon can be clearly illustrated either by increasing the H^+ conductance at constant E_m (e.g., by means of carbonyl cyanide *m*-chlorophenylhydrazone [CCCP]) or by hyperpolarizing the membrane (with valinomycin) while continuously measuring pH_i (Fig. 6). If Na^+/H^+ exchange is precluded, both procedures induce a cytoplasmic acidification, which is accentuated when the two variables are combined (Fig. 6 *F*). Conversely, the effect of CCCP can be counteracted if the cells are depolarized with gramicidin (Fig. 6 *G*). These data suggest that the acidification produced by changing $[Ca^{2+}]$ could result, at least in part, from membrane potential–driven uptake of H^+ equivalents. Indeed, a marked hyperpolarization has been observed after the addition of Ca^{2+} ionophores to lymphocytes (Rink et al., 1980; Grinstein and Cohen, 1987). This potential change is thought to be due to the increased $[Ca^{2+}]_i$, which is a major determinant of the K^+-conductive permeability.

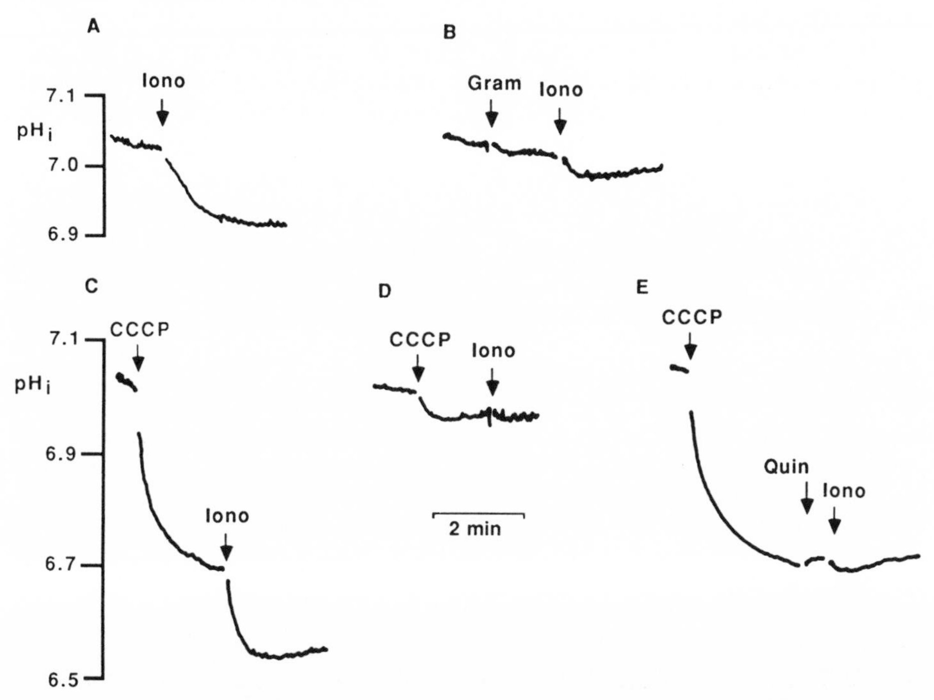

Figure 7. Changes in cation conductance affect the ionomycin-induced pH_i changes. BCECF-loaded cells were suspended in Na^+ medium with 200 μM amiloride (*A, B, C,* and *E*) or in K^+-rich solution (*D*). (*A*) Addition of 0.4 μM ionomycin. (*B*) Sequential addition of gramicidin (100 nM) and ionomycin (0.4 μM). (*C*) Sequential addition of CCCP (1 μM) and ionomycin. (*D*) Sequential addition of CCCP and ionomycin. (*E*) Sequential addition of CCCP, quinine (100 μM), and ionomycin. (From Grinstein and Cohen, 1987.)

That the effects of $[Ca^{2+}]_i$ on pH_i are associated with the hyperpolarization is suggested by two observations. First, increasing the H^+ conductance with CCCP magnified and accelerated the acidification induced by ionomycin (Fig. 7 *C*). Second, the acidification was reduced when the hyperpolarization was prevented either by elevating $[K^+]_o$ or by introducing gramicidin (Fig. 7, *B* and *D*), even when CCCP was also present. These results are consistent with an increased H^+ influx driven by the hyperpolarization, and imply that the cells have a finite conductance to H^+ (equivalents). Because the effects of ionomycin on pH_i were not entirely eliminated by depolarization, it is likely that other acid sources, such as metabolic generation or intracellular stores, also contribute to the acidification. Finally, the possible existence of a Ca^{2+}-activated K^+/H^+ exchange, such as that described in *Amphiuma* red cells by Cala (1985), has not been ruled out.

Mechanism of Cytoplasmic Alkalinization

Several lines of evidence indicate that the increase in pH_i recorded when cells are suspended in Na^+ medium with low (submicromolar) concentrations of ionomycin is due to Na^+/H^+ exchange: (*a*) the effect is accompanied by extracellular acidification; (*b*) it requires the presence of extracellular Na^+ and is associated with a gain in the cellular Na^+ content; and (*c*) all of these phenomena are prevented by micromolar concentrations of amiloride.

The consequences of increasing $[Ca^{2+}]_i$ on the rate of operation of the Na^+/H^+ antiport had been studied earlier in both intact cells and in isolated membranes. No effects were seen in renal brush border membrane vesicles (Aronson et al., 1982). In intact cells, pH_i measurements revealed little or no effect in some cell

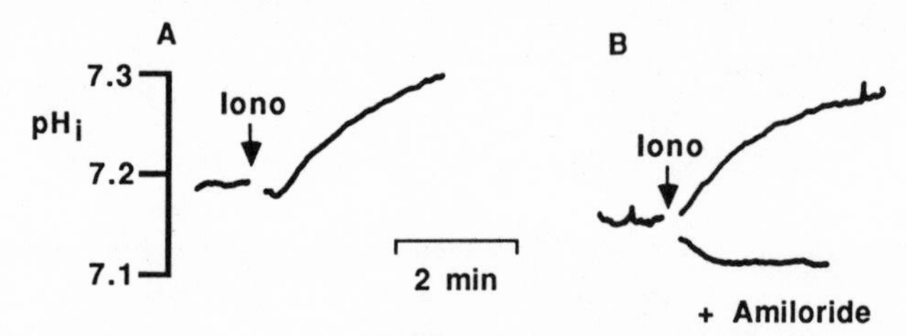

Figure 8. Effect of ionomycin on pH_i in untreated and protein kinase C–depleted thymocytes. Cells were preincubated for 24 h in the presence (*B*) or absence (*A*) of 2×10^{-7} M TPA. The cells were then washed, resuspended, and loaded with BCECF. pH_i was recorded in Na^+-containing medium and, where indicated, 0.1 μM ionomycin was added. Amiloride (200 μM) was present throughout the recording period in the bottom trace of *B*. (From Grinstein and Cohen, 1987.)

types (Moolenaar et al., 1981), but clear stimulation in others (Muldoon et al., 1985; Villereal et al., 1985). Because pH_i is affected by a variety of factors, it is conceivable that activation occurred in every instance, but was in some cases obscured by a concomitant acidifying process.

Stimulation of protein kinase C has been shown to activate Na^+/H^+ exchange in thymocytes and other cells (see Grinstein and Rothstein, 1986, for review). Because this enzyme can be activated in vitro by raising $[Ca^{2+}]$, it is conceivable that ionomycin stimulated the antiport via the Ca^{2+}-mediated activation of protein kinase C. However, the following evidence suggests that stimulation of protein kinase C did not play a significant role in the Ca^{2+}-induced Na^+/H^+ exchange: (*a*) protein phosphorylation, measured in situ, was stimulated markedly by phorbol esters, but only marginally by ionomycin; (*b*) migration of soluble C kinase to the membrane, an independent criterion of activation of this enzyme, was negligible in ionomycin-treated cells; by comparison, phorbol esters induced nearly quantitative translocation of the kinase; (*c*) ionomycin stimulated the antiport in cells pretreated with phorbol ester for 24 h (Fig. 8). The prolonged treatment with phorbol ester results in down-regulation of total protein kinase C, measurable by the loss of Ca^{2+}- and phospholipid-dependent phosphotransferase activity and by the obliteration of the pH_i response to added TPA (Grinstein et al., 1986). This

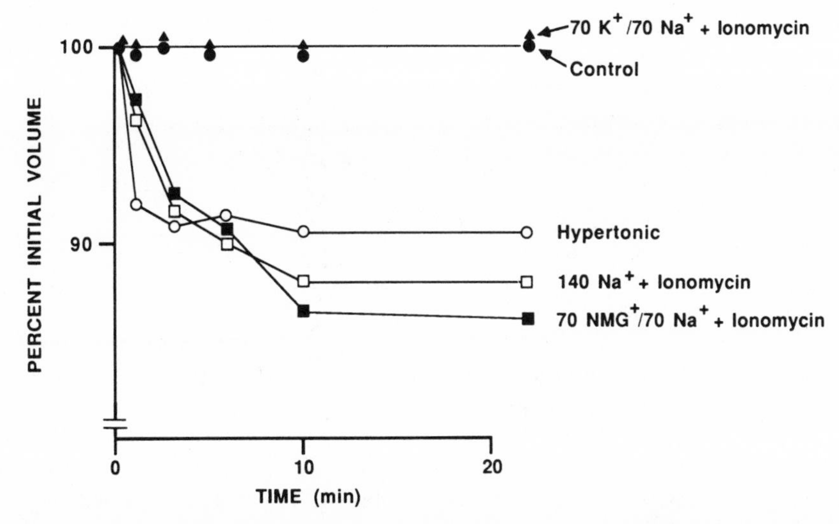

Figure 9. Volume changes in ionomycin-treated thymocytes. The cells were suspended in isotonic Na^+-containing medium (140 mM Na^+; solid circles and open squares), isotonic mixtures containing 70 mM Na^+ and 70 mM *N*-methylglucamine (solid squares), or 70 mM Na^+ and 70 mM K^+ (triangles), or in 140 mM Na^+ solution made hypertonic by addition of 25 mM *N*-methylglucamine Cl (open circles). Where indicated, 0.4 μM ionomycin was added at time zero. Cell volume was determined electronically at the indicated intervals using the Coulter-Channelyzer combination. (From Grinstein and Cohen, 1987.)

implies that the presence of protein kinase C is not essential to obtain the activation of Na^+/H^+ exchange by Ca^{2+}.

The addition of Ca^{2+} ionophores has been reported to induce cellular shrinking in other cell types (Hoffmann, 1985; Cala, 1985), resulting from the loss of KCl

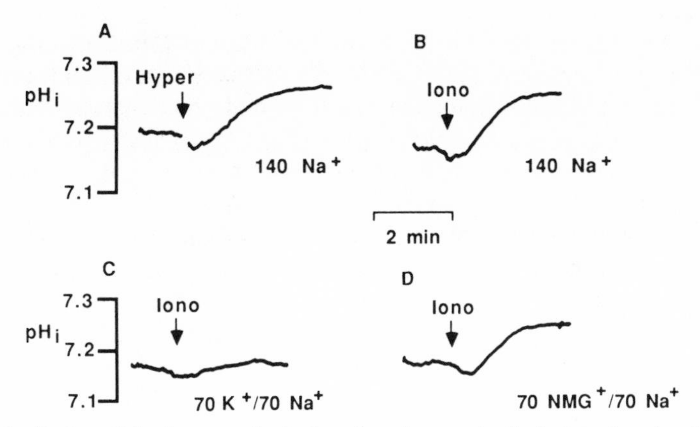

Figure 10. Similarity of the ionomycin-induced and osmotically induced activations of Na^+/H^+ exchange. Thymocytes were loaded with BCECF for the fluorimetric determination of pH_i. (*A*) Cells were suspended in isotonic Na^+ medium. Where indicated, the medium was made hypertonic by addition of 25 mM *N*-methylglucamine Cl. (*B*) Cells suspended in Na^+ medium. Ionomycin was added at the arrow. (*C*) Cells suspended in medium containing 70 mM K^+ and 70 mM Na^+. Ionomycin was added at the arrow. (*D*) Cells in medium with 70 mM *N*-methylglucamine$^+$ and 70 mM Na^+. Ionomycin was added where indicated. (From Grinstein and Cohen, 1987.)

and osmotically obliged water. Because shrinking of lymphocytes has been shown to activate Na^+/H^+ exchange, we considered the possibility that the ionomycin-induced alkalinization was secondary to a change in cell volume. Indeed, measurements of thymocyte volume by electronic sizing indicated that shrinking was occurring when cells suspended in Na^+-rich (physiological) medium were exposed to ionomycin (Fig. 9). Moreover, shrinking the cells to a comparable degree by subjecting them to mildly hypertonic media (Fig. 9) resulted in a comparable alkalinization (Fig. 10). Importantly, if the volume change induced by ionomycin was prevented by increasing $[K^+]_o$, the alkalinization was greatly reduced. This was not due to the reduction of $[Na^+]_o$, inasmuch as replacement with *N*-methyl-D-glucammonium$^+$ preserved both the shrinking and pH_i responses (Figs. 9 and 10). Taken together, these results suggest that at least part of the activation of the antiport is the result of cell shrinking.

Villereal and co-workers (Villereal et al., 1985; Muldoon et al., 1985) have suggested that a mechanism involving calmodulin is responsible for the stimulation of the Na^+/H^+ exchanger by A23187 in fibroblasts. This conclusion was based on the inhibitory effects of a variety of calmodulin antagonists. On the other hand, calmodulin blockers have been reported to block the cell volume changes induced by Ca^{2+} ionophores in peripheral blood lymphocytes (Grinstein et al., 1982). It is therefore conceivable that the stimulation of the cation antiport in fibroblasts is mediated, at least in part, by shrinking.

Concluding Remarks

A reciprocal relationship exists between pH_i and $[Ca^{2+}]_i$ in thymic lymphocytes. The cytoplasmic pH can be modified by elevating $[Ca^{2+}]_i$, whereas $[Ca^{2+}]_i$ increases when the cytoplasm becomes alkaline. The pH_i dependence of $[Ca^{2+}]_i$ could have important physiological consequences, inasmuch as pH_i is affected under a variety of circumstances. In particular, mitogenic lectins have been reported to alkalinize the cytoplasm of quiescent thymocytes within minutes (Hesketh et al., 1985), a change that is associated with increased $[Ca^{2+}]_i$. Although an initial transient phase of Ca^{2+} uptake precedes the change in pH_i, a second phase lasts several hours and could be due to the persistent increase in pH_i. Conversely, a reduction of pH_i during activation of anaerobic metabolism could be reflected in decreased $[Ca^{2+}]_i$. Similarly, the $[Ca^{2+}]_i$ sensitivity of pH_i could be manifested under physiological circumstances. Whether by changes in plasma membrane permeability and/or by release of internal stores, $[Ca^{2+}]_i$ increases when cells are stimulated with hormones, growth factors, neurotransmitters, and various physical stimuli. It will be of importance to determine whether accompanying pH_i changes occur in these instances and to define their consequences.

Acknowledgments

This work was supported by the National Cancer Institute (Canada) and the Medical Research Council (Canada). S.G. is the recipient of a Medical Research Council Scientist Award.

References

Aronson, P. S., J. Nee, and M. A. Suhm. 1982. Modifier role of internal H^+ in activating the Na^+/H^+ exchanger in renal microvillus membrane vesicles. *Nature.* 299:161–163.

Arruda, J. A., L. G. Dytko, H. Lubansky, R. Mols, R. Kleps, and C. T. Burt. 1981. Effect of calcium on intracellular pH. *Biochemical and Biophysical Research Communications.* 102:891–896.

Burns, C. P., and E. Rozengurt. 1983. Serum, platelet-derived growth factor, vasopressin and phorbol esters increase pH in Swiss 3T3 cells. *Biochemical and Biophysical Research Communications.* 116:931–938.

Busa, W. B., and R. Nuccitelli. 1984. Metabolic regulation via intracellular pH. *American Journal of Physiology.* 246:R409–R438.

Cala, P. M. 1985. Volume regulation by *Amphiuma* red blood cells: strategies of identifying alkali metal/H^+ transport. *Federation Proceedings.* 44:2500–2507.

Gelfand, E. W., R. K. Cheung, and S. Grinstein. 1984. Role for membrane potential in the regulation of lectin-induced calcium uptake. *Journal of Cellular Physiology.* 121:533–539.

Grinstein, S., and S. Cohen. 1987. Cytoplasmic [Ca^{2+}] and intracellular pH in lymphocytes. Role of membrane potential and volume-activated Na^+/H^+ exchange. *Journal of General Physiology.* 89:185–213.

Grinstein, S., S. Cohen, J. D. Goetz, A. Rothstein, and E. W. Gelfand. 1985. Characterization of the activation of Na^+/H^+ exchange in lymphocytes by phorbol esters. Change in the cytoplasmic pH-dependence of the antiport. *Proceedings of the National Academy of Sciences.* 82:1429–1433.

Grinstein, S., A. DuPre, and A. Rothstein. 1982. Volume regulation by human lymphocytes. Role of Ca^{2+}. *Journal of General Physiology.* 79:849–868.

Grinstein, S., and J. D. Goetz. 1985. Control of free cytoplasmic calcium by intracellular pH in rat lymphocytes. *Biochimica et Biophysica Acta.* 819:267–270.

Grinstein, S., E. Mack, and G. B. Mills. 1986. Osmotic activation of the Na^+/H^+ antiport in protein kinase C-depleted lymphocytes. *Biochemical and Biophysical Research Communications.* 134:8–13.

Grinstein, S., and A. Rothstein. 1986. Mechanisms of regulation of the Na^+/H^+ exchanger. *Journal of Membrane Biology.* 90:1–12.

Hesketh, T. R., J. P. Moore, J. D. H. Morris, M. V. Taylor, J. Rogers, G. A. Smith, and J. C. Metcalfe. 1985. A common sequence of calcium and pH signals in the mitogenic stimulation of eukaryotic cells. *Nature.* 313:481–484.

Hoerter, J. A., M. V. Miceli, D. G. Rendlund, E. G. Lakatta, and W. E. Jacobus. 1983. Intracellular calcium alters high energy phosphates and intracellular pH. *Biophysical Journal.* 41:249*a*. (Abstr.)

Hoffmann, E. K. 1985. Role of separate K and Cl channels and of Na/Cl cotransport in volume regulation in Ehrlich cells. *Federation Proceedings.* 44:2513–2519.

Ives, H. E., and T. O. Daniel. 1986. Growth factor-induced transient acidification is caused by increased intracelllar Ca^{2+}. *Journal of General Physiology.* 88:30*a*. (Abstr.)

Meech, R. W., and R. C. Thomas. 1977. The effect of calcium injection on the intracellular sodium and pH of snail neurons. *Journal of Physiology.* 265:867–879.

Moolenaar, W. H., C. L. Mummery, P. Van der saag, and S. W. DeLaat. 1981. Rapid ionic events of the initiation of growth in serum-stimulated neuroblastoma cells. *Cell.* 23:789–798.

Moolenaar, W. H., L. G. J. Tertoolen, and S. W. de Laat. 1984. Phorbol ester and diacylglycerol mimic growth factors in raising cytoplasmic pH. *Nature.* 312:371–374.

Moolenaar, W. H., R. Y. Tsien, P. T. van der Saag, and S. W. de Laat. 1983. Na^+/H^+ exchange and cytoplasmic pH in the action of growth factors in human fibroblasts. *Nature.* 304:645–648.

Muldoon, L. L., R. J. Dinerstein, and M. L. Villereal. 1985. Intracellular pH in human fibroblasts: effect of mitogens, A23187 and phospholipase activation. *American Journal of Physiology.* 249:C140–C148.

Richards, C. D., J. C. Metcalfe, G. A. Smith, and T. R. Hesketh. 1984. Changes in free-calcium levels and pH in synaptosomes during transmitter release. *Biochimica et Biophysica Acta.* 803:215–220.

Rink, T. J., C. Montecucco, T. R. Hesketh, and R. Y. Tsien. 1980. Lymphocyte membrane potential assessed with fluorescent probes. *Biochimica et Biophysica Acta.* 595:15–30.

Rink, T. J., R. Y. Tsien, and T. Pozzan. 1982. Cytoplasmic pH and free Mg in lymphocytes. *Journal of Cell Biology.* 95:189–196.

Rosoff, P. M., and L. C. Cantley. 1985. Stimulation of the T3-T receptor-associated Ca influx enhances the activity of the Na^+/H^+ exchanger in leukemic human T cell line. *Journal of Biological Chemistry.* 260:14053–14059.

Rosoff, P. M., L. F. Stein, and L. C. Cantley. 1984. Phorbol esters induce differentiation in a pre-B lymphocyte cell line by enhancing Na^+/H^+ exchange. *Journal of Biological Chemistry.* 259:7056–7060.

Schackmann, R. W., and P. B. Chock. 1986. Alteration of intracellular [Ca^{2+}] in sea urchin sperm by the egg peptide speract. Evidence that increased intracellular Ca^{2+} is coupled to Na^+ entry and increased intracellular pH. *Journal of Biological Chemistry.* 261:8719–8728.

Tsien, R. Y., T. Pozzan, and T. J. Rink. 1982. Calcium homeostasis in intact lymphocytes: cytoplasmic free calcium monitored with a new, intracellularly trapped fluorescent indicator. *Journal of Cell Biology.* 94:325–334.

Ueda, T. 1983. Na^+/Ca^{2+} exchange activity in rabbit lymphocyte plasma membrane. *Biochimica et Biophysica Acta.* 734:342–346.

Vaughn-Jones, R. D., W. J. Lederer, and D. A. Eisner. 1983. Calcium ions can affect intracellular pH in mammalian cardiac muscle. *Nature.* 301:522–524.

Villereal, M. L., N. E. Owen, L. M. Vicentini, L. L. Mix-Muldoon, and G. A. Jamieson, Jr. 1985. Mechanism for growth factor-induced increase in Na^+/H^+ exchange and rise in Ca^{2+} activity in cultured human fibroblasts. *Cancer Cells.* 3:417–424.

Calcium Involvement in Intracellular Events

Chapter 15

Mechanism of Activation of Protein Kinase C: Role of Diacylglycerol and Calcium Second Messengers

Yusuf A. Hannun and Robert M. Bell

Departments of Medicine and Biochemistry, Duke University Medical Center, Durham, North Carolina

Introduction

Protein kinase C, a Ca^{2+}- and phospholipid-dependent protein kinase, is now recognized as a pivotal regulatory element in signal transduction, tumor promotion, and cell regulation (Nishizuka, 1984). Nishizuka and co-workers discovered protein kinase C in rat brain cytosol (Takai et al., 1977). Later, protein kinase C was found to require phospholipid and supraphysiologic concentrations of Ca^{2+} for optimal activity. In a subsequent evaluation, it was found that neutral lipid fractions, specifically diacylglycerol (DG), increased the affinity of the enzyme for Ca^{2+}: in the presence of DG, low micromolar concentrations of Ca^{2+} were sufficient for full activation (Kishimoto et al., 1980). These results led Nishizuka to postulate a link between phosphatidylinositol turnover and activation of protein kinase C (for review, see Nishizuka, 1984). This mechanism of transmembrane signaling is thought to involve receptor-mediated phosphatidylinositol (PI) turnover through a GTP-dependent activation of phospholipase C. PI breakdown results in the formation of two second messengers, DG and inositol trisphosphate (IP_3). DG then activates protein kinase C and IP_3 leads to the mobilization of intracellular Ca^{2+} (Berridge, 1984). This increase in intracellular Ca^{2+} acts synergistically with DG in activating protein kinase C.

Further interest in protein kinase C was generated with the finding that phorbol esters, potent tumor promoters, activate protein kinase C (Castagna et al., 1982). Protein kinase C was also shown to be the phorbol ester receptor (Niedel et al., 1983). These findings implicated protein kinase C in tumor promotion, regulation of cell proliferation, and oncogenesis (Nishizuka, 1983, 1984).

DGs as Second Messengers

With the discovery that DGs modulate protein kinase C activity in vitro, it became necessary to demonstrate that DGs also induce biological responses in intact cells. This was particularly important because of the involvement of DG as a common intermediate in glycerolipid metabolism, a role that initially appears to be at variance with a second-messenger function of these molecules. The usual DGs, such as *sn*-1-stearyl-2-arachidonylglycerol and *sn*-1,2-dioleoylglycerol, have a very low monomer solubility in water, which precludes their use in biological systems. To overcome this problem, *sn*-1-oleoyl-2-acetylglycerol was synthesized and then tested in vitro and in intact cells such as human platelets. It activated protein kinase C in both situations (Kaibuchi et al., 1983). Additional studies were performed with a "cell-permeable" DG, *sn*-1,2-dioctanoylglycerol (diC_8). diC_8 was able to enter cells and activate protein kinase C (for review, see Ganong et al., 1986*a*). These studies clearly demonstrated that exogenous DG was biologically active and capable of activating protein kinase C, whereas closely related analogues were not. In certain situations, such as with platelets, DG by itself was insufficient to generate a full biologic response. Under these circumstances, DG and Ca^{2+} acted synergistically to activate platelets (Kaibuchi et al., 1983).

In addition to the demonstration that exogenously delivered DGs were biologically active, experiments were performed to study the changes in DG levels upon stimulation. These experiments took two general forms. Labeling cellular lipids with radiolabeled arachidonic acid or other fatty acids allowed the examination of the DG, phosphatidic acid, and phosphatidylinositide changes that occurred after stimulation in numerous cell systems (Nishizuka, 1983). The other approach to this problem was based on the measurement of DG mass in crude lipid extracts of tissues by the quantitative conversion of DG to phosphatidic acid through the action of DG kinase from *Escherichia coli* (Loomis et al., 1985). These studies confirmed the increase in DG levels in thrombin-stimulated platelets and in vasopressin-stimulated hepatocytes. Since K-*ras*- and *sis*-transformed normal rat kidney cells had elevated levels of DG, the possibility that DG activation of protein kinase C may play a role in cellular transformation was raised (Priess et al., 1986). These mass measurements extended the work of Fleishmann et al. (1986) in *ras*-transformed cells.

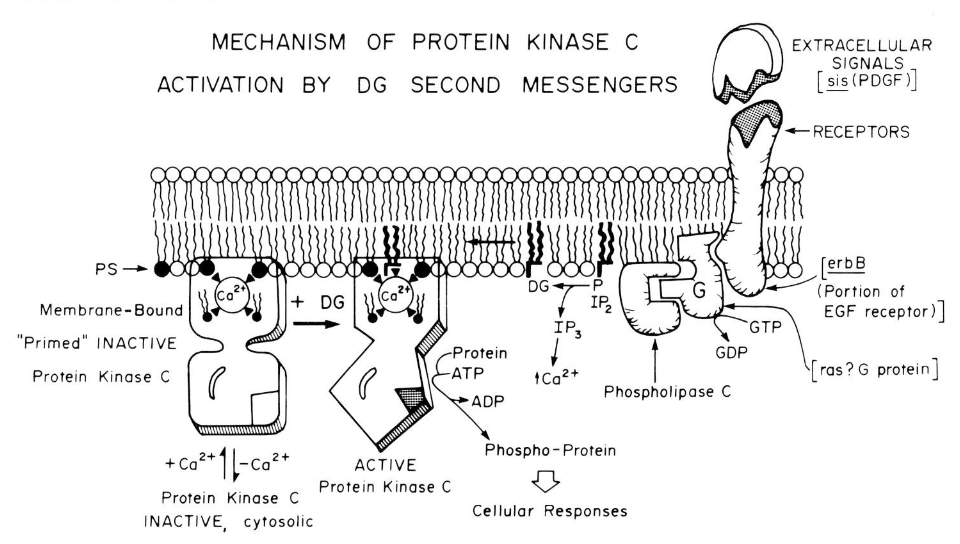

Figure 1. Model of protein kinase C activation by DG second messengers. (Reproduced from Bell, 1986, with permission from *Cell.*)

In Vitro Analysis of Lipid Cofactor Activation of Protein Kinase C

With the establishment of a second-messenger function of DGs through the activation of protein kinase C (Fig. 1), it became necessary to study the molecular mechanisms of the interaction of protein kinase C with its lipid cofactors. Nishizuka (1983) had established the requirement for Ca^{2+} and phospholipid, especially phosphatidylserine (PS), for protein kinase C activity. In vitro modulation of protein kinase C activity by DG dramatically increased the affinity of the enzyme for Ca^{2+} (Kishimoto et al., 1980). Further efforts at analyzing the activation of protein kinase C by its lipid cofactors were hampered by the physical properties of sonicated lipid dispersions that were generally used. These vesicles are heterogeneous in size and exist as multilamellar or unilamellar forms. Moreover, such vesicles aggregate in the presence of Ca^{2+}. To overcome these difficulties, our laboratory developed a mixed-micelle assay for protein kinase C activity (Hannun et al., 1985) wherein the lipid cofactors are dispersed in detergent micelles. Triton X-100 was used as the detergent of choice because of the extensive physical characterization of Triton X-100 micelles and Triton X-100/phospholipid mixed micelles and because of the suitable kinetics of micelle formation with Triton X-

100. Since Triton X-100/phospholipid mixed micelles are homogeneous in size and composition and do not coexist with pure Triton X-100 micelles, they allow the systematic and independent variation of lipid cofactors. Therefore, the number of lipid cofactors required for activation could be estimated. In the presence of 100 μM Ca^{2+}, enzyme activity could be fully reconstituted by Triton X-100 mixed micelles containing 8 mol % PS and 2 mol % DG. Triton X-100 provides an inert surface (with respect to protein kinase C) into which lipid cofactors are partitioned. Other non-ionic detergents, as well as some ionic detergents, were also able to support protein kinase C activity in the presence of Ca^{2+} and the lipid cofactors (Hannun et al., 1986*a*). Furthermore, enzyme activity was primarily determined by the surface concentration of the lipid cofactors, i.e., by the lipid:detergent mole percent rather than by the bulk concentration of lipids. Protein kinase C enzyme activity was lost when the surface concentration of both lipid cofactors (DG and PS) was diluted by the addition of detergent (Hannun et al., 1986*a*). Activity was insensitive to the Triton X-100 concentration (as long as the critical micellar concentration was greatly exceeded) if the mole ratio of PS and DG was kept unchanged. That is, enzyme activity was independent of the micelle number, but dependent on the micelle composition. This behavior is expected with amphipathic molecules that preferentially partition into micelles. These results illustrate the important point that protein kinase C is sensitive to the microenvironment of the surface (membranes, vesicles, or mixed micelles).

Next, the stoichiometry of protein kinase C activation by lipid cofactors was studied using the mixed-micelle methods (Hannun et al., 1985). The enzyme was found to have a requirement for PS, DG, and Ca^{2+}. At 8 mol % PS and in the presence of Ca^{2+}, activity was strongly dependent on DG. Maximal activity was observed at 1 mol % DG (Fig. 2), which corresponds to an average of 1.4 DG molecules in each Triton X-100 mixed micelle. When the PS dependence was studied in the presence of Ca^{2+} and 2.5 mol % DG (Fig. 3), there was an initial lag in activity until 3–4 mol % PS was reached, after which activation by PS occurred in a highly cooperative manner, with an apparent Hill number near 4.8. Next, the interrelationships between Ca^{2+}, PS, and DG were studied. DG increased the affinity of the enzyme for both Ca^{2+} and PS (Fig. 4). Similarly, Ca^{2+} modulated the affinity of the enzyme to DG and to PS. Double-reciprocal plots showed changes in K_m (apparent) with little effect on V_{max}. At all concentrations of Ca^{2+} and DG used, PS activation remained highly cooperative, with Hill numbers ranging between 4 and 5. These data suggested that protein kinase C activation requires Ca^{2+}, one molecule of DG, and at least four molecules of PS in each micelle. To further elucidate the stoichiometry of the activation and to study the number of lipid molecules interacting with the enzyme, it became necessary to study the interaction of protein kinase C with the detergent/lipid mixed micelles. Molecular sieve chromatography of protein kinase C and Triton X-100/lipid mixed micelles was used to analyze these interactions (Hannun et al., 1985; Hannun and Bell, 1986). These studies showed that monomeric protein kinase C interacted with single Triton X-100/lipid mixed micelles. The results imply that protein kinase C interacts with the lipid cofactors present in each micelle. The stoichiometry of the interaction between the enzyme and its lipid cofactors could then be deduced to be one DG and four PS molecules interacting with monomeric protein.

The molecular sieve studies also proved to be a model for the analysis of the

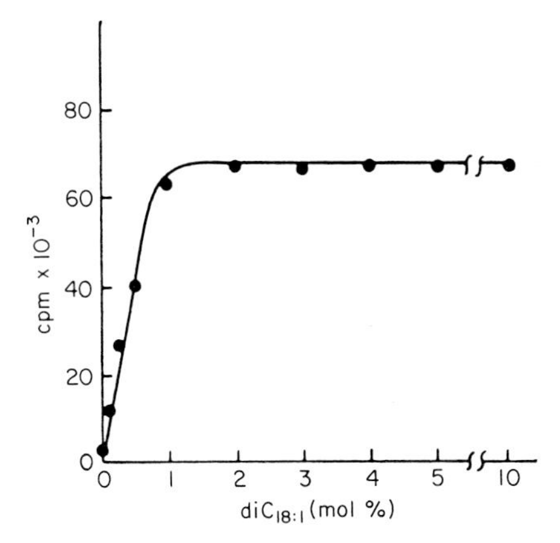

Figure 2. DG dependence of protein kinase C activation by mixed micelles. Mixed micelles were formed with 8.0 mol % PS. $CaCl_2$ was added to a final concentration of 100 μM. 1 mol % of $diC_{18:1}$ corresponds to 43 μM. (Reproduced from Hannun et al., 1985, with permission of *The Journal of Biological Chemistry.*)

interaction of protein kinase C with surfaces. It was found that the association of protein kinase C with mixed micelles required Ca^{2+} and PS only. In this state, the enzyme was bound to a surface (mixed micelles) but was inactive. The subsequent addition of DG resulted in activation. These findings suggested a two-step mechanism for protein kinase C activation, where in the first step the enzyme associates with a surface, and in the second step it becomes activated by the generation of DG. This two-step mechanism and the deduced stoichiometry closely reflect the in vivo situation. Plasma membranes contain 8–20 mol % PS. In the presence of Ca^{2+}, protein kinase C would be found as a membrane-associated, but inactive, protein. The breakdown of PI would then lead to the generation of DG, which would activate protein kinase C.

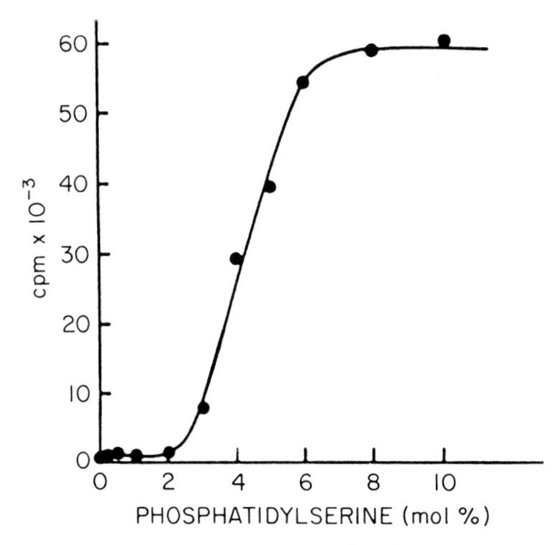

Figure 3. PS dependence of protein kinase C activation by mixed micelles. $diC_{18:1}$ was present at 2.5 mol % (107 μM) and $CaCl_2$ at 100 μM. 1 mol % PS corresponds to 43 μM. (Reproduced from Hannun et al., 1985, with permission of *The Journal of Biological Chemistry.*)

The mixed-micelle methods were also applied for the study of activation and binding of protein kinase C by phorbol esters (Hannun and Bell, 1986). The results showed that monomeric protein kinase C binds phorbol dibutyrate in the presence of Ca^{2+} and PS. Again, the PS dependence displayed the same degree of cooperativity that was seen when DG was used as the activator. Phorbol esters also modulated the affinity of the enzyme for PS and Ca^{2+} in a manner analogous to that of DG.

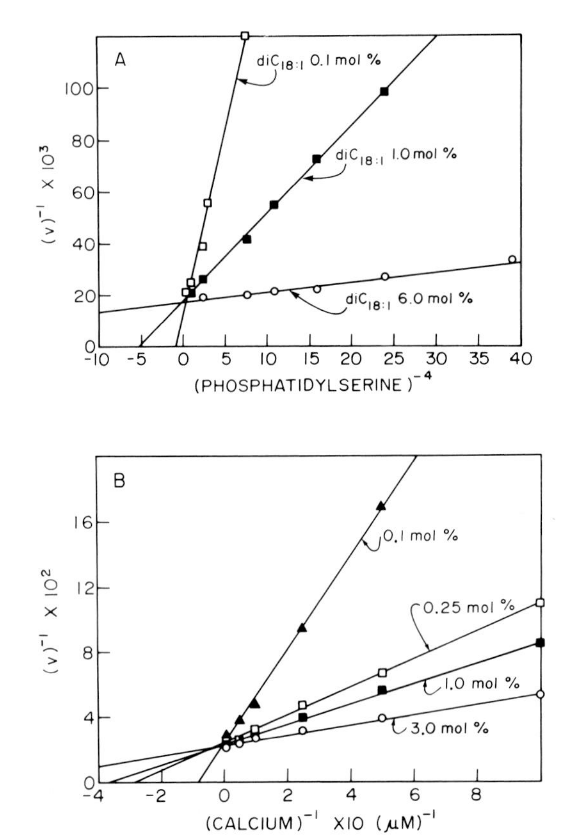

Figure 4. (*A*) PS and DG interdependence of protein kinase C activation by mixed micelles. Double-reciprocal plots showing a predominant effect of DG on the K_m (apparent) of PS. Activity was measured in the presence of $diC_{18:1}$ at 0.1 (□), 1.0 (■), and 6.0 (○) mol %. $CaCl_2$ was present at 100 μM. (*B*) Double-reciprocal plots of the Ca dependence as modulated by DG. DG was present at 0.1 (▲), 0.25 (□), 1.0 (■), and 3.0 (○) mol %. (Reproduced from Hannun et al., 1986*a*, with permission of *The Journal of Biological Chemistry*.)

A study was undertaken of the structural features of DG required for activation of protein kinase C. The chain length of the fatty acids present in *sn*-1,2-DG was varied. The resulting DGs fully activated protein kinase C, except for those DGs with the shortest acyl chains and the lowest hydrophobicity (*sn*-1,2-dibutyrylglycerol and *sn*-1,2-dipropanoylglycerol) (Hannun et al., 1986*a*). Another series of DG analogues with substitutions at the *sn*-1, -2, and -3 positions were synthesized and evaluated for their effects on protein kinase C activation using the mixed-micelle methods. In brief, these studies established that both oxygen esters were essential, as was the *sn*-3 hydroxyl, and activation was stereospecific (Ganong et al., 1986*b*; Hannun et al., 1986*a*; Boni and Rando, 1985).

Model for Protein Kinase C Activation by PS, DG, and Ca^{2+}

The mechanistic studies using the mixed-micelle methods and the structure-function studies conducted with the DG analogues have allowed the formulation of a molecular model for protein kinase C activation (Ganong et al., 1986*b*). A complex forms between Ca^{2+}, PS, and protein kinase C (Fig. 5), whereby the protein becomes associated with the micellar surface through a Ca^{2+} bridge binding the protein and four PS molecules. In this state, protein kinase C is "primed" but inactive. DG (or phorbol ester) then activates the enzyme by specifically binding to protein kinase C and Ca^{2+} through its *sn*-1 and -2 oxygen esters and the *sn*-3 hydroxyl. Ca^{2+} plays a central role in the formation of the active complex by simultaneously binding four PS molecules, one DG molecule, and protein kinase C. This model of protein kinase C activation by lipid cofactors accounts for the observed lipid dependences, the role of Ca^{2+} and PS in enzyme-micelle interaction, the observed stoichiometry with DG and PS, and the interdependence of Ca^{2+}, PS, and DG, where each one of these molecules modulates the affinity of the enzyme for the other two. This model is also consistent with phorbol esters activating protein kinase C by a mechanism similar to that of DG. Additional interactions may occur between phorbol esters and protein kinase C to account for their greater affinity for the enzyme. This model remains speculative at the present time; it may help in the design of further experiments to study the molecular mechanisms of protein kinase C activation and regulation.

Inhibitors of Protein Kinase C: Mixed-Micelle Analysis

A number of inhibitors of protein kinase C have been identified. These include the phenothiazines (Mori et al., 1980), adriamycin (Wise et al., 1982), dibucaine (Mori et al., 1980), alkyllysophospholipid (Helfman et al., 1983), cp-46,665, an antineoplastic lipoidal amine (Shoji et al., 1985), the antiestrogen tamoxifen (O'Brien et al., 1985), verapamil (Mori et al., 1980), and H-7 (Hidaka et al., 1984). Except for the last, which interacts at the ATP site of the enzyme, the analysis of the mechanisms of inhibition has suggested that these compounds interfere with the lipid cofactor regulation of protein kinase C. Detailed analysis of the mechanism of inhibition by these inhibitors has been fraught with technical problems. First, these molecules are amphiphiles that partition, in part, into membranes or vesicles. Their "effective" surface concentrations are therefore unknown, while their bulk concentrations will be misleading. Second, many of these compounds have detergent or "membrane-perturbing" effects, and since they are effective at concentrations equivalent to those of PS used in the assays (10–100 μM), their physical effects on vesicles may obscure specific interactions with the enzyme. Third, the nonhomogeneous nature of sonic dispersions of lipid vesicles precludes molecular analysis. The mixed-micelle methods, on the other hand, circumvent many of these problems. Triton X-100/phospholipid mixed micelles are homogeneous in size and composition. Lipid cofactors are dispersed uniformly and in a low mole fraction; therefore, the physical characteristics of micelles are determined principally by Triton X-100. Similarly, the addition of low mole fractions of "membrane-active"

Figure 5. Model of protein kinase C activation by PS, DG, and Ca^{2+}. (*A*) Overhead view of four space-filling models of PS. The four carboxyl groups are shown in close proximity to bind Ca^{2+}. (*B*) A view in the transverse plane of the membrane bilayer showing the complex of Ca^{2+} and four molecules of PS. An available bond to Ca^{2+} is shown unoccupied by ligand (arrow). (*C*) DG is shown to ligate directly to Ca^{2+} present in the PS/Ca^{2+}/protein kinase C complex. DG may then have two additional contact points with protein kinase C. (*D*) Protein kinase C is envisioned to bind to the complex shown in *B* on the membrane surface. In this state, the enzyme is inactive. (*E*) Protein kinase C is activated when DG becomes associated with the four-PS/Ca^{2+}/protein kinase C complex. (Fig. 5 of Ganong et al., 1986*b*.)

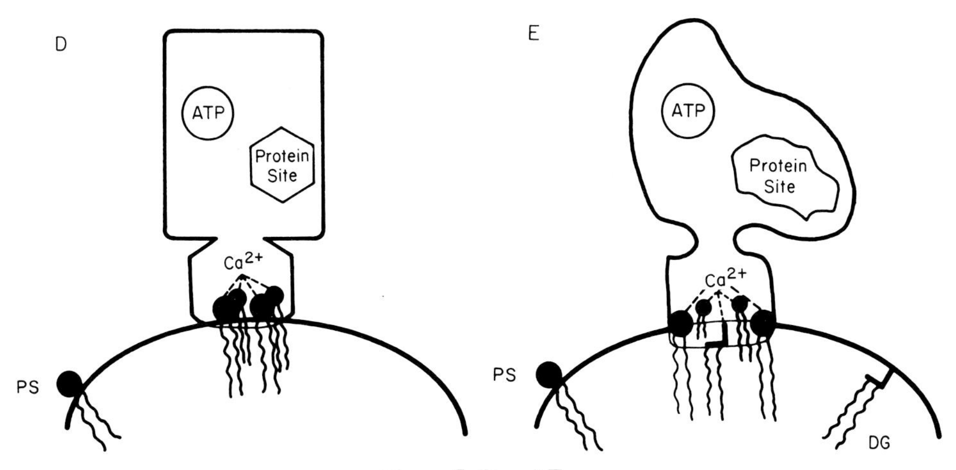

Figure 5, D and E.

inhibitors would not change the basic structure of mixed micelles. Also, these amphiphilic inhibitors are soluble in detergent solutions, and surface dilution kinetics can be analyzed to evaluate the relative importance of surface (mole percent) concentrations as opposed to bulk (molar) concentrations. Finally, since the assay is quantitative, it lends itself to kinetic analysis. Using these methods, we evaluated the mechanism of action of a number of compounds (Hannun, Y. A., and R. M. Bell, unpublished observations). The phenothiazines were found to be competitive inhibitors of protein kinase C with respect to PS in concentrations equivalent to those of PS required for activation. The mechanism involved inhibition of the DG–phorbol ester interaction with protein kinase C. Also, these inhibitors were found to display surface dilution kinetics. An increase in the number of Triton X-100/PS/DG mixed micelles, while it does not affect baseline enzyme activity, diminishes the potency of these inhibitors. This is because the surface concentration of the inhibitors decreases with the addition of more micelles. Keeping the inhibitors at the same mole percent (with respect to Triton X-100) produces the same degree of inhibition, regardless of the molar bulk concentration of the inhibitors. Similar analyses have been conducted on other inhibitors, such as adriamycin and verapamil, as well as some novel inhibitors, such as acridine dyes (Fig. 6) and related compounds. Preliminary analysis of inhibition by acridine orange suggests that one molecule of this inhibitor interacts with two molecules of PS within the four-PS/Ca^{2+}/protein kinase C complex, thus preventing the interaction of DG (phorbol esters) with the mixed-micelle/protein kinase C complex (Hannun, Y. A., and R. M. Bell, manuscript in preparation). The enzyme remains bound to the micelle, but in an inactive state.

Sphingosine Inhibition of Protein Kinase C

Sphingosine was found to be a potent and reversible inhibitor of protein kinase C (Hannun et al., 1986*b*). It was almost as effective an inhibitor as DG was an activator (Fig. 6). Mechanistic studies showed sphingosine to interact with the regulatory domain of protein kinase C, with little effect on the catalytic domain. Kinetic analysis also showed that sphingosine interacted competitively with Ca^{2+}, PS, and DG. Sphingosine also inhibited phorbol ester binding to protein kinase C, but did not cause the dissociation of the enzyme from the mixed-micelle surface. Since sphingosine is a naturally occurring compound with a role in the intermediary metabolism of sphingolipids, we were interested in the possible physiologic role sphingosine may have as a negative effector of protein kinase C. To explore this hypothesis, the effects of sphingosine on protein kinase C activation were tested in a number of systems. In human platelets (Hannun et al., 1986*b*), sphingosine was found to potently inhibit 40-kD phosphorylation induced by thrombin, diC_8, or phorbol ester. Sphingosine also inhibited [^{3}H]phorbol dibutyrate binding to human platelets. In neutrophils (Wilson et al., 1986), sphingosine inhibited the effects of phorbol ester, DG, the chemotactic peptide f-met-leu-phe, opsonized zymosan, and arachidonate on the activation of the oxidative burst. Finally, sphingosine was found to inhibit the DG- and phorbol ester–induced differentiation of the human promyelocytic cell line HL-60 (Merrill et al., 1986). These experiments established that sphingosine inhibits the activation of protein kinase C in different cell types

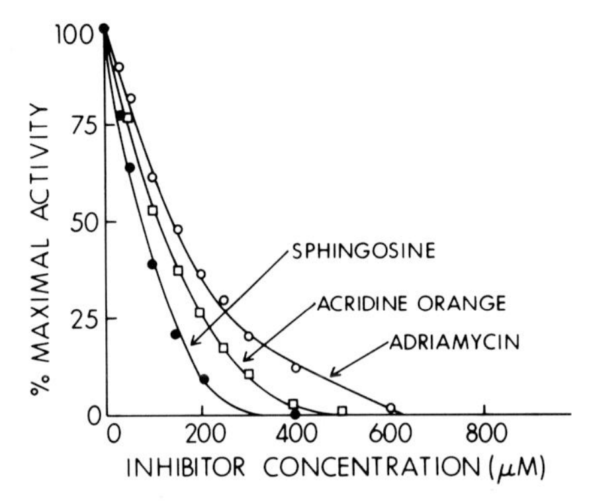

Figure 6. Inhibition of protein kinase C by adriamycin, acridine orange, and sphingosine. Triton X-100 mixed micelles contained PS at 6 mol % and $diC_{18:1}$ at 2 mol %. Ca^{2+} was at 100 μM. Acridine organge and adriamycin were added from fresh aqueous solutions. Sphingosine was added with the lipid cofactors.

and by different agonists. They also suggest that sphingosine may be generally useful as an inhibitor for the investigation of protein kinase C function in cell systems. The hypothesis that sphingosine modulates protein kinase C activity physiologically could explain a number of observations. First, resting intracellular levels of DG do not activate protein kinase C. This could be explained if protein kinase C activity in the resting state is modulated by both positive (DG) effectors and negative (sphingosine) effectors. Second, previous studies have revealed that widely varying levels of exogenous DG are required to activate protein kinase C in different cell systems (Ganong et al., 1986*a*). This may be a result of differences in the levels of free sphingosine exerting variable inhibition of protein kinase C. Finally, the generation of sphingosine in response to extracellular signals may play a negative feedback role on protein kinase C activation by agonists. Further investigation of these intriguing possibilities requires quantitative measurements of sphingosine levels in cells, as well as measurements of sphingolipid turnover.

Acknowledgments

This work was supported by National Institutes of Health grant AM-20205 and by American Cancer Society grant BC-511.

References

Bell, R. 1986. Protein kinase C activation by diacylglycerol second messengers. *Cell.* 45:631–632.

Berridge, M. J. 1984. Inositol trisphosphate and diacylglycerol as second messengers. *Biochemical Journal.* 220:345–360.

Boni, L. T., and R. R. Rando. 1985. The nature of protein kinase C activation by physically defined phospholipid vesicles and diacylglycerols. *Journal of Biological Chemistry.* 260:10819–10825.

Castagna, M., Y. Takai, K. Kaibuchi, K. Sano, U. Kikkawa, and Y. Nishizuka. 1982. Direct activation of calcium-activated, phospholipid-dependent protein kinase by tumor-promoting phorbol esters. *Journal of Biological Chemistry.* 257:7847–7851.

Fleishmann, L. F., S. B. Chahwala, and L. Cantley. 1986. *Ras*-transformed cells: altered levels of phosphatidylinositol-4,5-bisphosphate and catabolites. *Science.* 231:407–410.

Ganong, B. R., C. R. Loomis, Y. A. Hannun, and R. M. Bell. 1986*a*. Regulation of protein kinase C by lipid cofactors. *In* Cell Membranes: Methods and Reviews. E. Elson, W. Frazier, and L. Glaser, editors. Plenum Publishing Corp., New York.

Ganong, B. R., C. R. Loomis, Y. A. Hannun, and R. M. Bell. 1986*b*. Specificity and mechanism of protein kinase C activation by *sn*-1,2-diacylglycerols. *Proceedings of the National Academy of Sciences.* 83:1184–1188.

Hannun, Y. A., and R. M. Bell. 1986. Phorbol ester binding and activation of protein kinase C on Triton X-100 mixed micelles containing phosphatidylserine. *Journal of Biological Chemistry.* 261:9341–9347.

Hannun, Y. A., C. R. Loomis, and R. M. Bell. 1985. Activation of protein kinase C by Triton X-100 mixed micelles containing diacylglycerol and phosphatidylserine. *Journal of Biological Chemistry.* 260:10039–10043.

Hannun, Y. A., C. R. Loomis, and R. M. Bell. 1986*a*. Protein kinase C activation in mixed micelles. Mechanistic implications of phospholipid, diacylglycerol and calcium interdependencies. *Journal of Biological Chemistry.* 261:7184–7190.

Hannun, Y. A., C. R. Loomis, A. H. Merrill, and R. M. Bell. 1986*b*. Sphingosine inhibition of protein kinase C activity and of phorbol dibutyrate binding *in vitro* and in human platelets. *Journal of Biological Chemistry.* 261:12604–12609.

Helfman, D. M., K. C. Barnes, J. M. Kinkade, W. R. Volger, M. Shoji, and J. F. Kuo. 1983. Phospholipid-sensitive Ca^{2+}-dependent protein phosphorylation system in various types of leukemic cells from human patients and in human leukemic cell lines HL60 and K562, and its inhibition by alkyl-lysophospholipid. *Cancer Research.* 43:2955–2961.

Hidaka, H., M. Inapaki, S. Kawamoto, and Y. Sasaki. 1984. Isoquinoline sulfonamides, novel and potent inhibitors of cyclic nucleolide dependent protein kinase and protein kinase C. *Biochemistry.* 23:5036–5041.

Kaibuchi, K., Y. Takai, M. Sawamura, M. Hoshijima, T. Fujikura, and Y. Nishizuka. 1983. Synergistic functions of protein phosphorylation and calcium mobilization in platelet activation. *Journal of Biological Chemistry.* 258:2010–2013.

Kishimoto, A., Y. Takai, T. Mori, U. Kikkawa, and Y. Nishizuka. 1980. Activation of calcium and phospholipid-dependent protein kinase by diacylglycerol, its possible relation to phosphatidylinositol turnover. *Journal of Biological Chemistry.* 255:2273–2276.

Loomis, C. R., J. P. Walsh, and R. M. Bell. 1985. *sn*-1,2-Diacylglycerol kinase of *Escherichia coli*, purification, reconstitution and partial amino- and carboxyl-terminal analysis. *Journal of Biological Chemistry*. 260:4091–4097.

Merrill, A. H., A. Sereni, W. Steven, Y. Hannun, R. M. Bell, and J. Kinkade. 1986. Inhibition of phorbol ester dependent differentiation of human promyelocytic leukemic (HL60) cells by sphinganine and other long-chain bases. *Journal of Biological Chemistry*. 261:12610–12615.

Mori, T., Y. Takai, R. Minakuchi, B. Yu, and Y. Nishizuka. 1980. Inhibitory action of chlorpromazine, dibucaine, and other phospholipid-interacting drugs on calcium-activated, phospholipid-dependent protein kinase. *Journal of Biological Chemistry*. 255:8378–8380.

Niedel, J. E., L. J. Kuhn, and G. R. Vandenbark. 1983. Phorbol diester receptor copurifies with protein kinase C. *Proceedings of the National Academy of Sciences*. 80:36–40.

Nishizuka, Y. 1983. Calcium, phospholipid turnover and transmembrane signalling. *Philosophical Transactions of the Royal Society of London, Series B*. 302:101–112.

Nishizuka, Y. 1984. The role of protein kinase C in cell surface signal transduction and tumor promotion. *Nature*. 308:693–698.

O'Brien, C. A., R. M. Liskamp, D. H. Solomon, and I. B. Weinstein. 1985. Inhibition of protein kinase C by tamoxifen. *Cancer Research*. 45:2462–2465.

Preiss, J., C. R. Loomis, W. R. Bishop, R. Stein, J. E. Niedel, and R. M. Bell. 1986. Quantitative measurement of *sn*-1,2-diacylglycerols present in platelets, hepatocytes and *ras*- and *sis*-transformed normal rat kidney cells. *Journal of Biological Chemistry*. 261:8597–8600.

Shoji, M., W. R. Vogler, and J. F. Kuo. 1985. Inhibition of phospholipid/Ca^{2+}-dependent protein kinase and phosphorylation of leukemic cell proteins by CP-46,665-1, a novel antineoplastic lipoidal amine. *Biochemical and Biophysical Research Communications*. 127:590–595.

Takai, Y., A. Kishimoto, M. Inoue, and Y. Nishizuka. 1977. Studies on a cyclic nucleotide-independent protein kinase and its proenzyme in mammalian tissues. *Journal of Biological Chemistry*. 252:7603–7609.

Wilson, E., M. C. Olcott, R. M. Bell, A. H. Merrill, and J. D. Lambeth. 1986. Inhibition of the oxidative burst in human neutrophils by sphingoid long-chain bases: role of protein kinase C in activation of the burst. *Journal of Biological Chemistry*. 261:12616–12623.

Wise, B. C., D. B. Glass, C. H. Jen Chou, R. L. Raynor, N. Katoh, R. C. Schatzman, R. S. Turner, R. F. Kilber, and J. F. Kuo. 1982. Phospholipid-sensitive Ca^{2+}-dependent protein kinase from heart. *Journal of Biological Chemistry*. 257:8489–8495.

Chapter 16

The Relationship Between the Cytosolic Free Calcium Ion Concentration and the Control of Pyruvate Dehydrogenase

Richard G. Hansford and James M. Staddon

Energy Metabolism and Bioenergetics Section, Laboratory of Cardiovascular Science, Gerontology Research Center, National Institute on Aging, National Institutes of Health, Baltimore, Maryland

Introduction

It is well established that the Ca ion plays a central role as a mediator of excitation-contraction and stimulus-secretion coupling (see, e.g., Katz, 1970; Rasmussen, 1981). Perhaps less well known is the role that this ion also plays in activating mitochondrial oxidation and thereby making available ATP at an increased rate to match the increased energy demands associated with contraction and secretion. This activation has been most clearly established for pyruvate dehydrogenase phosphatase (Denton et al., 1972; Pettit et al., 1972), glycerol 3-phosphate dehydrogenase (Hansford and Chappell, 1967), NAD-isocitrate dehydrogenase (Denton et al., 1978), and 2-oxoglutarate dehydrogenase (McCormack and Denton, 1979). In the case of pyruvate dehydrogenase phosphatase, activation by micromolar concentrations of Ca ions leads to the generation of an increased amount of the active, dephospho form of the pyruvate dehydrogenase complex (PDH_A) (for reviews, see Hansford, 1980; Wieland, 1983). In the case of the other three dehydrogenases, activation is of an allosteric nature and results in a decreased K_m for the substrate.

Further, recent work by Moreno-Sánchez (1985*a*, *b*) has also raised the possibility that low, physiologically appropriate concentrations of Ca^{2+} activate the mitochondrial adenine nucleotide translocase, or the ATP-synthase, or both. This would mean that Ca^{2+} is involved not only in stimulating the reactions that give rise to the generation of the mitochondrial proton electrochemical gradient ($\Delta\bar{\mu}_{H^+}$) (see Mitchell, 1966, 1979), but also in the reaction that couples $\Delta\bar{\mu}_{H^+}$ to the synthesis of ATP. Aspects of this will be discussed later in this chapter.

It is now established not only that purified pyruvate dehydrogenase phosphatase responds to Ca ions (Denton et al., 1972; Pettit et al., 1972), but also that intact, respiring heart mitochondria are sensitive to variations in the Ca^{2+} concentration of the surrounding medium ($[Ca^{2+}]_o$), such that a minority of the total pyruvate dehydrogenase is in the form of PDH_A when $[Ca^{2+}]_o$ is 10^{-7} M, but a large majority is in this form when $[Ca^{2+}]_o$ is 10^{-6} M. The results for the purified phosphatase are presented in Fig. 1*A*, and the results for intact rat heart mitochondria are given in Fig. 1*B*. It is noted that whereas the sensitivity of enzyme interconversion to the extramitochondrial free Ca^{2+} concentration can be manipulated by changing the Na^+ concentration (Fig. 1*B*; Denton et al., 1980) or the Mg^{2+} concentration (Denton et al., 1980; Hansford, 1981)—procedures that change the magnitude of the mitochondrial membrane Ca^{2+} gradient—50% activation of pyruvate dehydrogenase occurs at a Ca^{2+} concentration of 0.41 μM when the ionic composition of the medium mimics that of the cytosol. This value is somewhere between resting and stimulated values for the heart cytosolic Ca^{2+} concentration ($[Ca^{2+}]_c$; see below).

Thus, control by Ca^{2+} is plausible for the intact cell. But can it be demonstrated to occur? This chapter presents recent work from our laboratory in which we follow the response of the PDH_A content to the elevation of $[Ca^{2+}]_c$ to determine whether this is consistent with a cause-and-effect relationship. The cell types employed are freshly isolated, Ca^{2+}-tolerant cardiac myocytes from adult animals, in which we raise $[Ca^{2+}]_c$ by plasma membrane depolarization, and freshly isolated hepatocytes, in which case we use the Ca^{2+}-mobilizing hormones phenylephrine, vasopressin, and glucagon.

Methods

The procedures for the isolation of Ca^{2+}-tolerant cardiac myocytes from adult rats, the loading of these cells with quin2, and the studies of quin2 fluorescence and of the measurement of pyruvate dehydrogenase have all been described in Hansford (1987).

The procedures for the isolation of hepatocytes, the loading of these cells with quin2, and the measurement of the fluorescence of intracellular quin2 have been described by Staddon and Hansford (1986). Procedures for the measurements

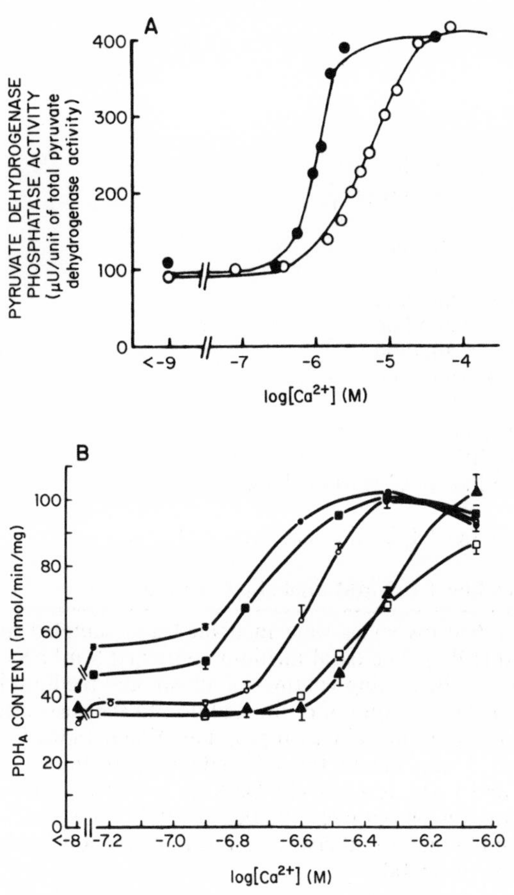

Figure 1. (*A*) Activation of pyruvate dehydrogenase phosphatase by Ca^{2+}. (*B*) Effect of increasing $[Ca^{2+}]_o$ on the PDH_A content of respiring, coupled rat heart mitochondria. (*A*) Enzyme activity was measured in mitochondrial extracts and is presented as a function of the concentration of Ca^{2+} (filled circles) or Sr^{2+} (open circles). (Reprinted with permission from McCormack and Denton, 1980.) (*B*) Mitochondria were incubated in media containing the concentration of Ca^{2+} indicated, stabilized with EGTA buffers, and sampled for PDH_A content when enzyme interconversion had attained a steady state. The respiratory substrate was glutamate plus malate; full details are given in Hansford (1981). The concentrations of Na^+ in the incubation were: none (filled circles), 1 mM (open circles), 10 mM (open squares), 20 mM (filled triangles). All incubations contained 1 mM Mg^{2+}.

of the fluorescence of intracellular indo-1 and of reduced nicotinamide adenine dinucleotide and reduced nicotinamide adenine dinucleotide phosphate [NAD(P)H] and for the extraction and measurement of hepatocyte pyruvate dehydrogenase have been described by Staddon and Hansford (1987).

Table I
Effect of Ruthenium Red and Ryanodine on the Increase in PDH_A Content Induced by KCl and Veratridine Plus Ouabain

Condition	PDH_A (% of total)
(*A*) 5 mM KCl	31.7±1.4 (32)
20 mM KCl	39.5±0.6 (4)
40 mM KCl	49.5±4.6 (6)
55 mM KCl	49.3±3.6 (8)
80 mM KCl	60.6±3.9 (11)
(*B*) 5 mM KCl + ruthenium red	31.5±2.7 (3)
40 mM KCl + ruthenium red	38.7±2.3 (3)
55 mM KCl + ruthenium red	35.2±3.0 (5)*
80 mM KCl + ruthenium red	37.8±3.3 (7)‡
(*C*) 20 mM KCl + ryanodine	39.8±4.1 (3)
40 mM KCl + ryanodine	56.0±6.9 (8)
55 mM KCl + ryanodine	62.3±7.0 (8)
80 mM KCl + ryanodine	72.7±9.2 (8)
(*D*) 5 μM veratridine	37.6±4.5 (6)
25 μM veratridine	51.0±1.9 (10)
25 μM veratridine + 0.2 mM ouabain	57.6±3.9 (8)
0.2 mM ouabain	45.6±5.0 (5)
25 μM veratridine + 0.2 mM ouabain + ruthenium red	35.8±2.0 (10)§
(*E*) 25 μM veratridine + 0.2 mM ouabain + ryanodine	52.6±4.1 (3)

Suspensions of rat cardiac myocytes were incubated and sampled for PDH_A content as described in Hansford (1987). The basal medium contained 5 mM KCl: higher concentrations of K^+ were achieved by mixing volumes of an isotonic medium in which K replaced Na ions. Sampling was done 10 min after the addition to the suspension of K^+-containing medium or of veratridine plus ouabain, as appropriate. Where indicated, ruthenium red and ryanodine were added 3 min before the K^+-containing medium or veratridine, to give concentrations of 12 and 1 μM, respectively. Data are presented as means ± SEM, with the number of preparations in parentheses. Values of PDH_A are significantly lower in the presence of ruthenium red than in otherwise identical incubations omitting ruthenium red. * $p < 0.05$. ‡ $p < 0.005$. § $p < 0.001$.

Results and Discussion

Response of PDH_A Content and of $[Ca^{2+}]_c$ of Cardiac Myocytes to Plasma Membrane Depolarization

It can be seen from Table I that the elevation of the extracellular K^+ concentration from 5 to 80 mM, a procedure that leads to a graded depolarization and increase in $[Ca^{2+}]_c$ (Powell et al., 1984; Sheu et al., 1986), also gives rise to a progressive increase in PDH_A content. Equally, the treatment of the myocytes with veratridine plus ouabain, a procedure that might be expected to lead to depolarization and

Na^+ overload (Escueta and Appel, 1969; Ohta et al., 1973), gives rise to a large increase in PDH_A.

The response of $[Ca^{2+}]_c$ to these interventions is shown in Fig. 2. Raising $[K^+]$ to 40 mM gives a larger increase (*A*) than does raising $[K^+]$ to 20 mM (*B*). Ouabain alone, added to inhibit the Na^+-K^+-ATPase (Escueta and Appel, 1969), gives a very modest increase in $[Ca^{2+}]_c$, which is consistent with the known insensitivity of the rat to cardiac glycosides (*C*). However, ouabain does potentiate the effect of

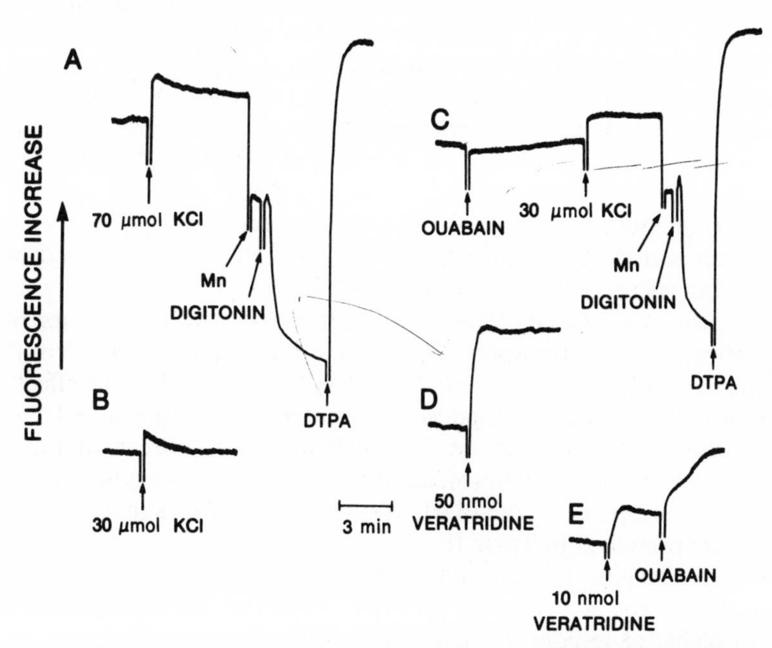

Figure 2. The effect of plasma membrane depolarization on the fluorescence of suspensions of cardiac myocytes loaded with quin2. Cells were incubated at 3.5 mg of protein/ml in medium containing 7.5 μM quin2 AM, for 30 min at 37°C. This generated cells containing 0.73 nmol of quin2/mg protein. After this loading step, the cells were washed by centrifugation and used for the fluorescence studies shown. Where indicated, millimolar Cl_2 was added to 0.1 mM to quench the fluorescence of extracellular quin2, digitonin was added to 5 μM to allow the Mn^{2+} to react with intracellular quin2, and diethylenetriaminepentaacetic anhydride (DTPA) was added to 1 mM to remove Mn^{2+} from the quin2/Mn^{2+} chelate and therefore generate maximal fluorescence. Full details are given in Hansford (1987).

a nonsaturating concentration of veratridine (*E*), which acts by holding open Na^+ channels (Ohta et al., 1973; Blaustein and Goldring, 1975). Veratridine alone gives a large, rapid increase in $[Ca^{2+}]_c$ (*D*) when added to the high concentration (25 μM) used for the pyruvate dehydrogenase studies.

It is clear that there is a relation between the decrease of increase in $[Ca^{2+}]_c$ shown in Fig. 2 and the degree of increase in PDH_A content shown in Table I. It is not possible, in our opinion, to define the changes in $[Ca^{2+}]_c$ more quantitatively, as the calibration of the fluorescence signal from quin2 is marred by the apparent presence of more than one compartment for the dye within the cell (see Hansford, 1987, and Staddon and Hansford, 1986, for more discussion).

Whenever $[Ca^{2+}]_c$ is sufficiently elevated, a muscle will undergo contraction and the availability of ADP to the mitochondria will increase. Pyruvate dehydrogenase interconversion is sensitive to changes in the mitochondrial ATP/ADP ratio, as ADP inhibits the pyruvate dehydrogenase kinase (Hucho et al., 1972; see also Hansford, 1976), and so the question arises as to how important are changes in $[Ca^{2+}]_c$ per se in signaling metabolic demand to this enzyme system, vis-à-vis changes in the ATP/ADP ratio that occur because of muscle contraction. The following arguments bear on this point.

(*a*) The presence of ruthenium red, an inhibitor of the uptake of Ca^{2+} by the mitochondria (Moore, 1971), largely prevents any increase in PDH_A content due to high concentrations of K^+ or to treatment with veratridine plus ouabain (Table I). However, ruthenium red does not attenuate the response of $[Ca^{2+}]_c$ to these interventions. (This is presented not only because of the degree of fluorescence quenching due to ruthenium red: the percentage change of the quin2 signal in response to KCl or veratridine is identical in the presence and absence of the inhibitor.) Ruthenium red also does not alter the frequency of spontaneous "waving" that occurs in a significant fraction of these cell populations and has been taken to reflect values of $[Ca^{2+}]_c$ (Kort et al., 1985).

(*b*) Treatment of the cell suspensions with ryanodine, an inhibitor of sarcoplasmic reticulum Ca^{2+} transport (Sutko and Kenyon, 1983), renders the cells totally quiescent when viewed under the microscope, but has no effect on the elevation of PDH_A content in response to the interventions shown in Table I.

(*c*) Loading of the cells with the Ca^{2+}-chelating agent quin2, to an intracellular content of 0.5–0.7 nmol/mg of protein, makes the cells totally quiescent, but does not diminish the response of the PDH_A content to procedures that elevate $[Ca^{2+}]_c$. These data are presented in Table II.

From these results, it seems quite clear that increased mechanical work is not a prerequisite for the activation of pyruvate dehydrogenase and that the enzyme system is capable of responding to changes in $[Ca^{2+}]_c$ alone. However, that does not rule out a supplementary effect of a decreased mitochondrial ATP/ADP ratio in actively contracting heart muscle, and it is noted that the degree of mechanical work performed by the "waving" cells in this study is likely to be very small compared with their workload in the animal. Nevertheless, the efficacy of the Ca^{2+} signal is established.

We regard the use of quin2-loaded cells as very informative and note that major changes in steady state values of $[Ca^{2+}]_c$ can be demonstrated (Fig. 2), concomitant with enzyme activation (Table II), but without any evident contraction. It is apparent that at ~0.5 mM intracellular quin2, the degree of Ca^{2+} buffering is such that Ca^{2+}-induced Ca^{2+} release from the sarcoplasmic reticulum (Fabiato and Fabiato, 1975; Fabiato, 1983) is inhibited.

Response of PDH_A Content and of $[Ca^{2+}]_c$ of Hepatocytes to Glucagon, Vasopressin, and Phenylephrine

It is known that α_1-adrenergic stimulation and the hormone vasopressin raise $[Ca^{2+}]_c$ in hepatocytes (Charest et al., 1983; Berthon et al., 1984; Thomas et al., 1984) as well as stimulating the pathway of gluconeogenesis. Recently, it has been found that the hormone glucagon, which also stimulates gluconeogenesis, but which had formerly been considered to act solely through the elevation of cyclic AMP, also gives rise to an increase in $[Ca^{2+}]_c$ (Charest et al., 1983; Sistare et al.,

Table II
Effect of Addition of KCl and of Veratridine Plus Ouabain on the PDH_A Content of Quin2-loaded Cardiac Myocytes

Condition	PDH_A (% of total)
5 mM KCl	34.2±4.6 (6)
20 mM KCl	41.5 (2)
40 mM KCl	45.3±3.1 (4)
55 mM KCl	44.7±5.1 (5)
80 mM KCl	57.3±5.5 (3)
25 μM veratridine + 0.2 mM ouabain	62.5±2.8 (4)

The experiment was conducted as described in Table I, with the exception that the myocytes were loaded with quin2 by a prior 30-min incubation with 7.5 μM quin2 AM. The data are derived from the number of experiments indicated but only three preparations of cells were used. (From Hansford, 1987.)

1985; Staddon and Hansford, 1986). For this reason, we chose to investigate whether the PDH_A content of hepatocytes responded to these agents and, if so, whether there was a quantitative relationship between changes in $[Ca^{2+}]_c$ and changes in PDH_A content. Early in this investigation, we discovered that the activation of protein kinase C by phorbol esters results in an attenuation of the increase in $[Ca^{2+}]_c$ that normally occurs upon exposure to glucagon (Staddon and Hansford, 1986). The site of action of protein kinase C is distal to the glucagon receptor, as phorbol ester can also effectively antagonize the action of dibutyryl cAMP and forskolin in raising $[Ca^{2+}]_c$ (Staddon and Hansford, 1986). There is a distinction here between the effects of phorbol esters on the response to glucagon and that to the α_1-adrenergic agonist phenylephrine: in the latter case, the locus of action is thought to be the α_1-receptor (Cooper et al., 1985; Lynch et al., 1985).

For the purposes of the present study, glucagon, vasopressin, and phenylephrine provided tools allowing the elevation of $[Ca^{2+}]_c$, and the phorbol ester 4β-phorbol 12-myristate 13-acetate (PMA) provided a tool for reversing this effect, when stimulation was by either glucagon or phenylephrine. Table III shows that phenylephrine, vasopressin, and glucagon each gave an increase in PDH_A content,

Table III
Activation of Pyruvate Dehydrogenase by Phenylephrine, Vasopressin, and Glucagon in Hepatocytes and the Influence of PMA

Hormone	PDH activity	Plus PMA (500 nM)
	nmol/min/mg cell protein	
—	1.01±0.05 (8)	0.98±0.06 (8)
Phenylephrine (25 μM)	1.20±0.09 (6)*	1.01±0.08 (6)
Vasopressin (25 nM)	1.69±0.13 (8)‡	1.58±0.11 (8)‡
Glucagon (10 nM)	2.27±0.22 (8)‡	1.25±0.13 (8)§

Pyruvate dehydrogenase activity was measured as described in Staddon and Hansford (1987). Values shown are means ± SEM, with the number of cell preparations indicated in parentheses. The statistical significance for the difference between values in each vertical column was determined by a paired *t* test: * $p < 0.01$; ‡ $p < 0.001$; § $p < 0.05$. In comparison between columns: PMA was without effect on unstimulated PDH_A content; PMA was without effect on the vasopressin-induced and phenylephrine-induced increase in PDH_A content; PMA significantly ($p < 0.001$) diminished the activation caused by glucagon.

with increasing effectiveness in that order. The response to phenylephrine was abolished by pretreatment with PMA, whereas the response to glucagon was severely attenuated and that to vasopressin was unaffected (Table III).

When $[Ca^{2+}]_c$ was monitored in quin2-loaded hepatocytes (Fig. 3), it was found that glucagon, vasopressin, and phenylephrine each gave a large increase in $[Ca^{2+}]_c$, which was sustained, provided that the cells were incubated in a medium containing physiologically appropriate levels of Ca^{2+}. Prior treatment with PMA abolished the effect of phenylephrine, as had previously been shown by Cooper et al. (1985) and Lynch et al. (1985), and severely attenuated the response to glucagon. The response to vasopressin was unaffected.

The general correspondence between changes in $[Ca^{2+}]_c$ and changes in PDH_A content is consistent with the thesis that the former is a determinant of the latter,

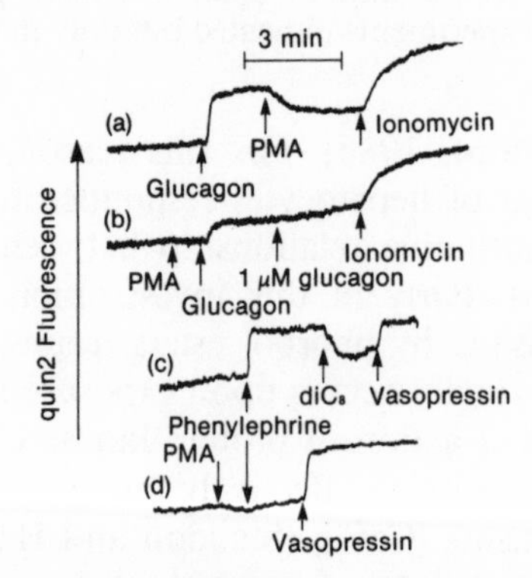

Figure 3. The response of hepatocyte $[Ca^{2+}]_c$ to treatment with glucagon, phenylephrine, and vasopressin. Hepatocytes were loaded with quin2 by incubation of the cells (2.5 mg protein/ml) with 100 μM quin2 AM for 15 min at 37°C. This procedure gave cells containing 1 nmol/mg of cell protein, or ~1 mM intracellular quin2. Fluorescence measurements were made as described by Staddon and Hansford (1986). In trace *a*, glucagon was added to 10 nM, PMA to 500 nM, and ionomycin to 10 μM. In trace *b*, PMA was added before glucagon: the second addition of glucagon gave 1 μM. In trace *c*, phenylephrine was added to 25 μM, and L-1-α-1,2-dioctanoyl glycerol (diC_8) was added to 10 μM; this unnatural diacylglycerol is used to activate protein kinase C, in analogy with the use of PMA. A subsequent addition of vasopressin, to 25 nM, was effective in elevating $[Ca^{2+}]_c$. In trace *d*, treatment of the hepatocytes with PMA (500 nM) abolished the response to phenylephrine (25 μM), but allowed an undiminished response to vasopressin (25 nM). The medium contained 2.5 mM Ca^{2+} in these studies.

in the response of hepatocytes to these hormones. An apparent anomaly, which may be informative, is that vasopressin gives as large an increase in $[Ca^{2+}]_c$ as glucagon, but less of an activation of pyruvate dehydrogenase (Table III). Conceivably, glucagon has the additional effect of resetting the mitochondrial Ca^{2+} transport cycle, such that uptake is favored; alternatively, vasopressin might reset the cycle in the favor of efflux (see Nicholls and Åkerman, 1982, and Hansford, 1985, for reviews on the mitochondrial Ca^{2+} transport cycle). More experiments seem to be warranted on this matter.

There is no doubt that the presence of 1 mM quin2 within the hepatocytes in studies of the sort shown in Fig. 3 leads to a significant damping of changes in $[Ca^{2+}]_c$. As one of our goals was to relate changes in $[Ca^{2+}]_c$ to changes in mitochondrial NADH, as an index of mitochondrial dehydrogenase activity in general (see below), and we were interested in temporal relationships, it seemed preferable to use less highly buffered cell preparations. Fig. 4 presents the results of an experiment using hepatocytes loaded with ~0.1 mM indo-1, a fluorescent Ca^{2+}-chelating agent (Grynkiewicz et al., 1985). At this relatively low level of intracellular Ca^{2+} buffering, it is apparent that phenylephrine and glucagon both cause very rapid increases in $[Ca^{2+}]_c$, and moreover that the subsequent addition of PMA leads to a rapid reversal. As the action of vasopressin is not sensitive to PMA, it is possible again to raise $[Ca^{2+}]_c$ by adding this hormone.

The Relationship Between Energy Demand and Energy Supply

Activation of pyruvate dehydrogenase, NAD-linked isocitrate dehydrogenase, and 2-oxoglutarate dehydrogenase by Ca^{2+} would be expected to increase the rate of provision of NADH to the respiratory chain and, in the absence of an activation of the chain itself or of phosphorylation reactions (which is, however, a possibility;

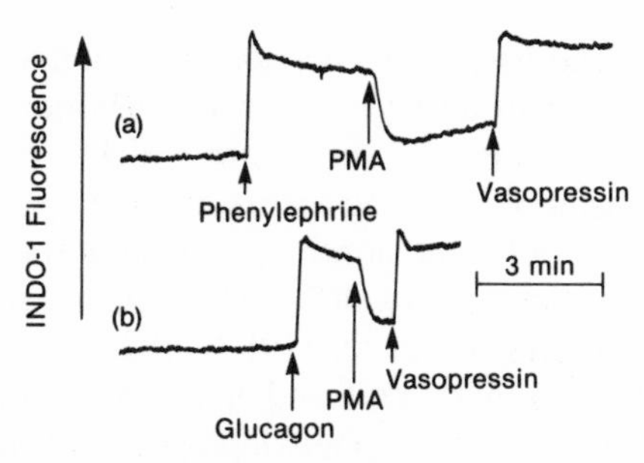

Figure 4. Demonstration of the reversal by PMA of increases in $[Ca^{2+}]_c$ induced by phenylephrine and glucagon. Hepatocytes were loaded with indo-1, to an intracellular concentration of ~0.1 mM. Fluorescence was excited at 365 nm and emission at 400 nm was recorded. Phenylephrine was added to 25 μM, glucagon to 10 nM, vasopressin to 25 nM, and PMA to 500 nM. The medium contained 2.5 mM Ca^{2+}.

see Moreno-Sánchez, 1985*a*, *b*), to lead to an increased steady state content of mitochondrial NADH. Activation of energy-utilizing processes in the cytosol, by contrast, might be expected to lower the mitochondrial NADH content by increasing ADP availability to the mitochondrion, according to the classic theory of Chance and Williams (1956). Therefore, we were curious to see what the response of mitochondrial NADH content would be to the perturbations described above.

In work with cardiac myocytes (Fig. 5), the addition of KCl to 30 or 45 mM was found to give rise to a negligible redox change, when this was monitored by whole-cell fluorescence. The response varied between no change and the 6% change in the direction of oxidation that is shown in the figure. This is quantitated in terms of pyridine nucleotide that can be oxidized by the proton ionophore FCCP and reduced by the respiratory chain inhibitor rotenone. A similar oxidation of mitochondrial nicotinamide nucleotide has been reported for perfused heart, when the work load is increased (Illingworth et al., 1975). It is likely that the activation of actomyosin ATPase by Ca^{2+} is a dominant process under these conditions.

By contrast, Ca^{2+}-mobilizing hormones lead to a pronounced increase in fluorescence in suspensions of hepatocytes (Fig. 6). Glucagon, phenylephrine, and

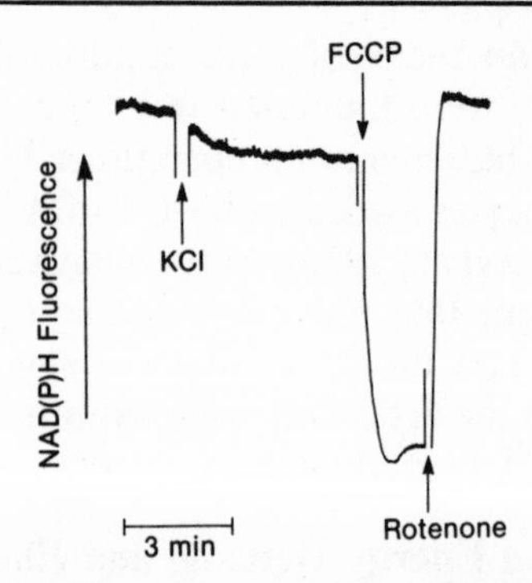

Figure 5. Effect of plasma membrane depolarization on the NAD(P)H fluorescence of a suspension of cardiac myocytes. Fluorescence of a suspension of myocytes (2.5 mg protein/ml) was excited at 365 nm and collected at wavelengths >460 nm. Where indicated, KCl was added to raise the concentration in the medium from 5 to 45 mM. The uncoupling agent FCCP was added to 2.5 μM; rotenone was then added to 2.5 μM.

vasopressin (separately) lead to a rapid increase in fluorescence, which is not fully sustained. This is attributed to an increase in mitochondrial NADH content on the basis of control experiments (not shown) in which cytosolic NAD is reduced with ethanol and the transfer of reducing agents into the mitochondria is blockaded with amino-oxyacetate. Under these conditions, results similar to those shown in Fig. 6 are obtained. An increase in NAD(P)H fluorescence in response to glucagon has also been reported by Sugano et al. (1980), Balaban and Blum (1982), and Sistare et al. (1985) in work with perfused liver or isolated hepatocytes. PMA prevents the rise in NADH due to phenylephrine and attenuates that due to glucagon, in striking analogy to the results for $[Ca^{2+}]_c$ (Fig. 3) and for PDH_A content (Table III).

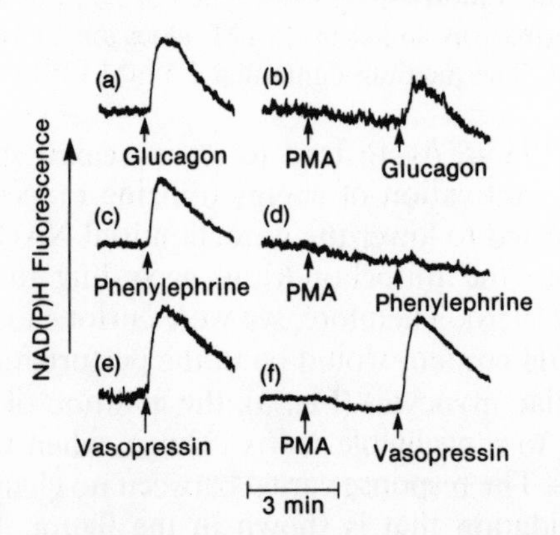

Figure 6. Response of the NAD(P)H fluorescence of a suspension of hepatocytes to Ca^{2+}-mobilizing hormones. The fluorescence of a suspension of hepatocytes was monitored as described for Fig. 5. The final concentrations of added compounds were: 10 nM glucagon, 500 nM PMA, 25 μM phenylephrine, and 25 nM vasopressin. The Ca^{2+} concentration of the medium was 2.5 mM. The maximum increases in fluorescence caused by the agonists, expressed as a percentage of the increase in fluorescence obtained by adding 5 μM rotenone plus 20 mM DL-3-hydroxybutyrate, were 37% glucagon (*a*), 17% glucagon (*b*), 51% phenylephrine (*c*), and 44% vasopressin (*e*).

That the elevation of $[Ca^{2+}]_c$ per se is sufficient to increase dehydrogenase activity, and therefore mitochondrial NADH content, emerges clearly from Fig. 7. At low concentrations of ionomycin, which are sufficient materially to increase $[Ca^{2+}]_c$ (not shown), there is a large, time-dependent increase in fluorescence. However, at higher concentrations of ionomycin, the response is biphasic, with an initial reduction succeeded by an oxidation. Presumably, the energy demand imposed by Ca^{2+} cycling at the mitochondrial membrane, the endoplasmic reticulum, and the plasma membrane begins to overwhelm the ability of the mitochondria to maintain the $\Delta\bar{\mu}_{H^+}$.

When the Ca^{2+} concentration of the surrounding medium is rapidly decreased to ~0.2 μM by the addition of EGTA, a suspension of hepatocytes can still respond to Ca^{2+}-mobilizing hormones (Fig. 8). However, the increase in $[Ca^{2+}]_c$ is quite transient: this is especially clear in this experiment, owing to the use of 0.1 mM indo-1 as the indicator. Under these conditions, it is reasonable to presume that

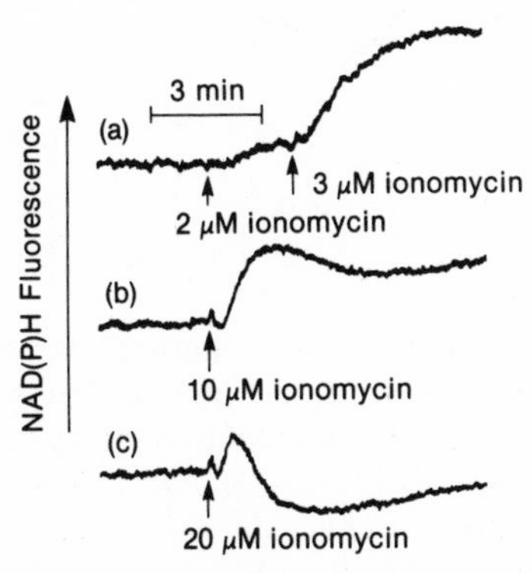

Figure 7. Effect of ionomycin on the NAD(P)H fluorescence of a suspension of hepatocytes. The fluorescence of a suspension of hepatocytes was monitored as given for Fig. 5. The incubation medium contained 2.5 mM Ca^{2+}. In trace *a*, sequential additions of ionomycin were made, giving final concentrations of 2 and 5 μM. In traces *b* and *c*, ionomycin was added to 10 and 20 μM, respectively.

Ca^{2+} is being released from an intracellular pool, and the consensus now is that this is located in the endoplasmic reticulum (Kleineke and Söling, 1985; Somlyo et al., 1985). After release by glucagon, little remains of this pool to be released by vasopressin (Fig. 8). However, when the glucagon-induced release has been opposed by PMA, more Ca^{2+} remains to be released by vasopressin. Fig. 8 shows that the addition of glucagon still causes an increase in mitochondrial NAD(P)H fluorescence, in the absence of extracellular Ca^{2+}. Our result here contrasts with the finding of Balaban and Blum (1982) that incubation in a Ca^{2+}-free medium abolished the response of NADH to glucagon: cells were exposed to Ca^{2+}-free medium for a greater length of time in this earlier study, with the likelihood that the reticular store of Ca^{2+} had become depleted before the hormone was added. In general, there is found to be a correspondence between changes in $[Ca^{2+}]_c$ and changes in NAD(P)H fluorescence (Fig. 8).

However, the addition of vasopressin after the depletion of intracellular Ca^{2+} pools with glucagon is found to give essentially no increase in NADH content, despite an appreciable increase in $[Ca^{2+}]_c$. This apparent anomaly is interesting.

One possible mechanism is that glucagon enhances the activity of the respiratory chain (Quinlan and Halestrap, 1986), as well as that of the dehydrogenases being discussed here. If the activation of the respiratory chain is slower, this could account for the failure of NADH content to rise in response to the second hormonal stimulation. A second possibility is that these hepatocytes are substrate-poor and that there is an insufficient supply of pyruvate and acetyl coenzyme A to the tricarboxylate cycle to give high flux, despite the activation of the dehydrogenases by Ca^{2+}. There is some evidence in favor of this notion, as the mitochondrial $NADH/NAD^+$ ratios of isolated hepatocytes are normally lower than those of the perfused liver. One could presumably discriminate between these two different

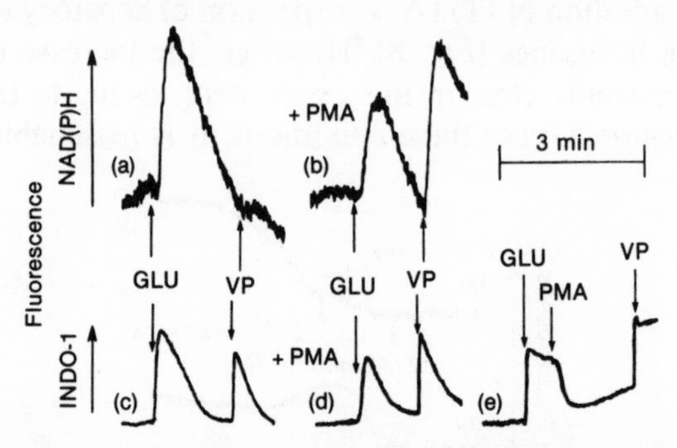

Figure 8. Effect of PMA on the response of an intracellular pool of Ca^{2+} to stimulation of hepatocytes by glucagon. Parallel recordings are presented of the fluorescence of suspensions loaded with indo-1 (traces *c–e*) and of the autofluorescence of suspensions of "nonloaded" cells, which reflects mainly NAD(P)H (traces *a* and *b*). Cells were preincubated in a medium containing 0.5 mM $CaCl_2$. 3 min before the addition of hormone, EGTA was added to 0.75 mM, giving a final concentration of Ca^{2+} in the medium of ~0.2 μM; where appropriate, 500 nM PMA was added at the same time as the EGTA. In trace *e*, alone, the $CaCl_2$ concentration was 2.5 mM, and there was no addition of EGTA. Glucagon was added to 10 nM and vasopressin to 25 nM. The maximum increases in NAD(P)H fluorescence caused by the agonists, expressed as a percentage of the increase in fluorescence obtained by adding 5 μM rotenone plus 20 mM D,L-3-hydroxybutyrate, were 35% glucagon (*a*) and 19% glucagon (*b*). Full details are given in Staddon and Hansford (1987).

models on the grounds of rates of O_2 uptake: if NADH content is low owing to substrate limitation, O_2 uptake should be slow; if NADH content is low owing to increased respiratory chain activity, then O_2 uptake should be more rapid.

In general, it is clear that the increase of $[Ca^{2+}]_c$ in hepatocytes increases the mitochondrial $NADH/NAD^+$ ratio, via an activation of dehydrogenases. The significance of this for oxidative phosphorylation is presented in Fig. 9, which is schematic in nature. On the basis of the chemiosmotic theory of energy transduction (Mitchell, 1966, 1979), the rate of hydrogen flow down the respiratory chain is driven by the drop in free energy between the ΔEh of the span of the respiratory chain being considered and the proton electrochemical gradient $\Delta\bar{\mu}_{H^+}$. Equally, the rate of phosphoryl group transfer to ADP is driven by the free-energy drop between the $\Delta\bar{\mu}_{H^+}$ and the free energy of ATP synthesis, under prevailing cellular conditions (ΔG_{ATP}). The increase in $NADH/NAD^+$ owing to dehydrogenase activation necessarily increases the ΔEh component for the respiratory chain as a whole, as the terminal acceptor couple 1/2 O_2/H_2O is unchanged. Given near-equilibrium at sites 1 and 2, and nonequilibrium at site 3 (see Hansford, 1980, for a review), the increase in the $NADH/NAD^+$ ratio would be expected to raise the cytochrome *c* Fe^{2+}/Fe^{3+} ratio, and thus the flux through site 3. The enhanced rate of respiration would be expected to lead to elevated values of $\Delta\bar{\mu}_{H^+}$. The latter change in turn will make possible higher rates of phosphorylation of ADP and the generation of elevated values of the $ATP/ADP \times P_i$ ratio in the steady state. There is indeed evidence that glucagon treatment of liver leads to an enhanced $\Delta\bar{\mu}_{H^+}$ and ATP/ $ADP \times P_i$ ratio (Bryla et al., 1977; Halestrap, 1978; Titheradge et al., 1979; Titheradge and Haynes, 1980).

In excitable tissues, the situation is different in that activation is normally associated with increased availability to the mitochondrion of both Ca^{2+} and ADP.

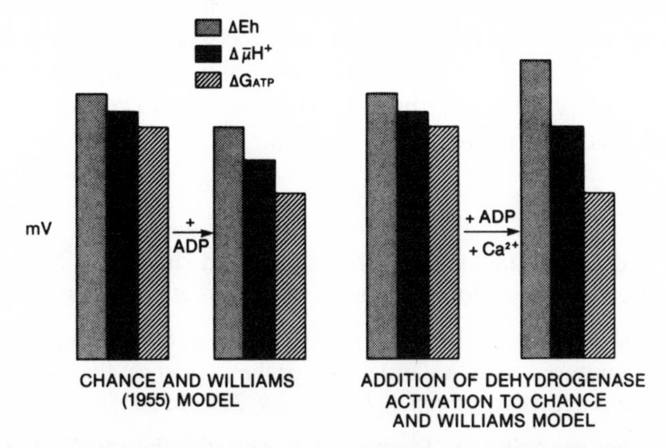

Figure 9. Energetics of oxidative phosphorylation: the effect of superimposing the activation of dehydrogenases by Ca^{2+} and ADP on the classic model. The height of each bar represents one of the thermodynamic quantities involved in oxidative phosphorylation. ΔEh is the difference in observed reduction potentials between the redox couple acting as an electron donor and the couple acting as an acceptor for a given redox loop, or loops, of the respiratory chain. In the experiments of Johnson and Hansford (1977) and Hansford and Castro (1981), two spans were considered, from $NADH/NAD^+$ to cytochrome *c* Fe^{2+}/Fe^{3+}. $\Delta\bar{\mu}_{H^+}$ is the proton electrochemical gradient, or protonmotive force of Mitchell (1966, 1979). The parameter measured is multiplied by an assumed stoichiometry of protons per site to allow the comparison with the other thermodynamic parameters. ΔG_{ATP} is the free energy of synthesis of ATP, under prevailing conditions of concentration of reactants and products and Mg^{2+}. The purpose of this plot is didactic, and it is accordingly stylized. For real results, the reader is referred to Johnson and Hansford (1977).

Increased availability of ADP per se leads to a drop in $\Delta\bar{\mu}_{H^+}$, because it acts as a substrate for the proton-transporting ATP-synthase (Mitchell, 1966, 1979). In turn, respiration is activated and the steady state mitochondrial $NADH/NAD^+$ ratio tends to fall. This is a restatement in chemiosmotic terms of the classic Chance and Williams (1956) mechanism of respiratory control. Superimposed on this, we believe, is activation of dehydrogenases by Ca^{2+} and, indeed, by ADP (Denton et al., 1972, 1978; Pettit et al., 1972; Hansford, 1972; Hucho et al., 1972; McCormack and Denton, 1979). The net effect of activation of both ATP-synthase (by ADP) and of pyruvate, NAD-isocitrate, and 2-oxoglutarate dehydrogenases (by Ca^{2+} and by ADP) is that mitochondrial $NADH/NAD^+$ ratios decreases little, if at all, on activation: certainly they decrease less than they would in the Chance and Williams

(1956) model. In heart mitochondria oxidizing 2-oxoglutarate, the $NADH/NAD^+$ ratio does not change upon adding ADP and Ca^{2+}, to 0.5–1 μM (Hansford and Castro, 1981): with suspensions of cardiac myocytes, the mitochondrial $NADH/NAD^+$ ratio changes very little on depolarization of the cells with elevated concentrations of KCl (Fig. 5). In mitochondria from fly flight muscle, a tissue that is exquisitely well adapted to energy transduction and shows dehydrogenase level control to a very pronounced degree, the addition of ADP plus Ca^{2+} (0.5–1 μM) results in an increase in the $NADH/NAD^+$ ratio, when the substrate is pyruvate plus glycerol 3-phosphate (Hansford and Sacktor, 1971). At the same time, the rates of O_2 uptake are increased by 30-fold.

The consequence of this preservation, or enhancement, of the $NADH/NAD^+$ ratio in the face of increased energy demand upon the mitochondria is that the free-energy drop between ΔEh and $\Delta\bar{\mu}_{H^+}$ and between $\Delta\bar{\mu}_{H^+}$ and ΔG_{ATP} can be increased, allowing higher rates of oxidative phosphorylation, without ΔG_{ATP} being allowed to fall to the point where the ATP/ADP × P_i ratio could no longer support the ATP-driven reactions of the rest of the cell.

References

Balaban, R. S., and J. J. Blum. 1982. Hormone-induced changes in NADH fluorescence and O_2 consumption of rat hepatocytes. *American Journal of Physiology.* 242:C172–C177.

Blaustein, M. P., and J. M. Goldring. 1975. Membrane potentials in pinched-off presynaptic nerve terminals monitored with a fluorescent probe: evidence that synaptosomes have potassium diffusion potentials. *Journal of Physiology.* 247:589–615.

Bryla, J. E., E. J. Harris, and J. A. Plumb. 1977. The stimulatory effect of glucagon and dibutyryl cyclic AMP on ureogenesis and gluconeogenesis in relation to the mitochondrial ATP content. *FEBS Letters.* 80:443–448.

Chance, B., and G. R. Williams. 1956. The respiratory chain and oxidation phosphorylation. *Advances in Enzymology.* 17:65–134.

Charest, R., P. F. Blackmore, B. Berthon, and J. H. Exton. 1983. Changes in free cytosolic Ca^{2+} in hepatocytes following α_1-adrenergic stimulation. Studies on Quin-2-loaded hepatocytes. *Journal of Biological Chemistry.* 258:8769–8773.

Cooper, R. H., K. E. Coll, and J. R. Williamson. 1985. Differential effects of phorbol ester on phenylephrine and vasopressin-induced Ca^{2+} mobilization in isolated hepatocytes. *Journal of Biological Chemistry.* 260:3281–3288.

Denton, R. M., J. G. McCormack, and N. J. Edgell. 1980. Role of calcium ions in the regulation of intramitochondrial metabolism. Effects of Na^+, Mg^{2+} and ruthenium red on the Ca^{2+}-stimulated oxidation of oxoglutarate and on pyruvate dehydrogenase activity in intact rat heart mitochondria. *Biochemical Journal.* 190:107–117.

Denton, R. M., P. J. Randle, and B. R. Martin. 1972. Stimulation by Ca^{2+} of pyruvate dehydrogenase phosphate phosphatase. *Biochemical Journal.* 128:161–163.

Denton, R. M., D. A. Richards, and J. G. Chin. 1978. Calcium ions and the regulation of NAD^+-linked isocitrate dehydrogenase from the mitochondria of rat heart and other tissues. *Biochemical Journal.* 176:899–906.

Escueta, A. V., and S. H. Appel. 1969. Biochemical studies of synapses in vitro. II. Potassium transport. *Biochemistry.* 8:725–733.

Fabiato, A. 1983. Calcium-induced release of calcium from the cardiac sarcoplasmic reticulum. *American Journal of Physiology.* 245:C1–C14.

Fabiato, A., and F. Fabiato. 1975. Contractions induced by a calcium-triggered release of calcium from the sarcoplasmic reticulum of single skinned cardiac cells. *Journal of Physiology.* 249:469–495.

Grynkiewicz, G., M. Poenie, and R. Y. Tsien. 1985. A new generation of Ca^{2+} indicators with greatly improved fluorescence properties. *Journal of Biological Chemistry.* 260:3440–3450.

Halestrap, A. P. 1978. Stimulation of pyruvate transport in metabolizing mitochondria through changes in the transmembrane pH gradient induced by glucagon treatment of rats. *Biochemical Journal.* 172:389–398.

Hansford, R. G. 1972. The effect of adenine nucleotides upon the 2-oxoglutarate dehydrogenase of blowfly flight muscle. *FEBS Letters.* 21:139–141.

Hansford, R. G. 1976. Studies on the effects of coenzyme A-SH: acetyl coenzyme A, nicotinamide adenine dinucleotide: reduced nicotinamide adenine dinucleotide and adenosine diphosphate: adenosine triphosphate ratios on the interconversion of active and inactive pyruvate dehydrogenase in isolated rat heart mitochondria. *Journal of Biological Chemistry.* 251:5483–5489.

Hansford, R. G. 1980. Control of mitochondrial substrate oxidation. *Current Topics in Bioenergetics.* 10:217–278.

Hansford, R. G. 1981. Effect of micromolar concentrations of free Ca^{2+} ions on pyruvate dehydrogenase interconversion in intact rat heart mitochondria. *Biochemical Journal.* 194:721–732.

Hansford, R. G. 1985. Relation between mitochondrial calcium transport and energy metabolism. *Reviews of Physiology, Biochemistry and Pharmacology.* 102:1–72.

Hansford, R. G. 1987. Relation between cytosolic free Ca^{2+} and the control of pyruvate dehydrogenase in isolated cardiac myocytes. *Biochemical Journal.* 241:145–151.

Hansford, R. G., and F. Castro. 1981. Effect of micromolar concentrations of free calcium ions on the reduction of heart mitochondrial NAD (P) by 2-oxoglutarate. *Biochemical Journal.* 198:525–533.

Hansford, R. G., and J. B. Chappell. 1967. The effect of Ca^{2+} on the oxidation of glycerol phosphate by blowfly flight muscle mitochondria. *Biochemical and Biophysical Research Communications.* 27:686–692.

Hansford, R. G., and B. Sacktor. 1971. Oxidative metabolism of Insecta. *In* Chemical Zoology. M. Florkin and B. T. Scheer, editors. Academic Press, Inc., New York and London. 6(Pt. B):213–247.

Hucho, F., D. D. Randall, T. E. Roche, M. W. Burgett, J. W. Pelley, and L. J. Reed. 1972. α-Keto acid dehydrogenase complexes. XVII. Kinetic and regulatory properties of pyruvate dehydrogenase kinase and pyruvate dehydrogenase phosphatase from bovine kidney and heart. *Archives of Biochemistry and Biophysics.* 151:328–340.

Illingworth, J. A., W. C. L. Ford, K. Kobayashi, and J. R. Williamson. 1975. Regulation of myocardial energy metabolism. *In* Recent Advances in Studies on Cardiac Structure and Metabolism. P.-E. Roy and P. Harris, editors. University Park Press, Baltimore, MD. 8:271–290.

Johnson, R. N., and R. G. Hansford. 1977. The nature of controlled respiration and its relationship to protonmotive force and proton conductance in blowfly flight-muscle mitochondria. *Biochemical Journal.* 164:305–322.

Katz, A. M. 1970. Contractile proteins of the heart. *Physiological Reviews.* 50:63–158.

Kleineke, J., and H.-D. Söling. 1985. Mitochondrial and extramitochondrial Ca^{2+} pools in the perfused rat liver. Mitochondria are not the origin of calcium mobilized by vasopressin. *Journal of Biological Chemistry.* 260:1040–1045.

Kort, A. A., M. C. Capogrossi, and E. G. Lakatta. 1985. Frequency, amplitude, and propagation velocity of spontaneous Ca^{++}-dependent contractile waves in intact adult rat cardiac muscle and isolated myocytes. *Circulation Research.* 57:844–855.

Lynch, C. J., R. Charest, S. B. Bocckino, J. H. Exton, and P. F. Blackmore. 1985. Inhibition of hepatic α_1-adrenergic effects and binding by phorbol myristate acetate. *Journal of Biological Chemistry.* 260:2844–2851.

McCormack, J. G., and R. M. Denton. 1979. The effects of calcium ions and adenine nucleotides on the activity of pig heart 2-oxoglutarate dehydrogenase complex. *Biochemical Journal.* 180:533–544.

Mitchell, P. 1966. Chemiosmotic Coupling in Oxidative and Photosynthetic Phosphorylation. Glynn Research, Bodmin. 192 pp.

Mitchell, P. 1979. Compartmentation and communication in living systems. Ligand conduction: a general catalytic principle in chemical, osmotic and chemiosmotic reaction systems. *European Journal of Biochemistry.* 95:1–20.

Moore, C. L. 1971. Specific inhibition of mitochondrial Ca^{2+} transport by ruthenium red. *Biochemical and Biophysical Research Communications.* 42:298–305.

Moreno-Sánchez, R. 1985*a*. Regulation of oxidative phosphorylation in mitochondria by external free Ca^{2+} concentrations. *Journal of Biological Chemistry.* 260:4028–4034.

Moreno-Sánchez, R. 1985*b*. Contribution of the translocator of adenine nucleotides and the ATP synthase to the control of oxidative phosphorylation and arsenylation in liver mitochondria. *Journal of Biological Chemistry.* 260:12554–12560.

Nicholls, D., and K. Åkerman. 1982. Mitochondrial calcium transport. *Biochimica et Biophysica Acta.* 683:57–88.

Ohta, M., T. Narahashi, and R. F. Keeler. 1973. Effect of veratrum alkaloids on membrane potential and conductance of squid and crayfish giant axons. *Journal of Pharmacology and Experimental Therapeutics.* 184:143–154.

Pettit, F. H., T. E. Roche, and L. J. Reed. 1972. Function of calcium ions in pyruvate dehydrogenase phosphatase activity. *Biochemical and Biophysical Research Communications.* 49:563–571.

Powell, T., P. E. R. Tatham, and V. W. Twist. 1984. Cytoplasmic free calcium measured by Quin 2 fluorescence in isolated ventricular myocytes at rest and during potassium-depolarization. *Biochemical and Biophysical Research Communications.* 122:1012–1020.

Quinlan, P. T., and A. P. Halestrap. 1986. The mechanism of the hormonal activation of respiration in isolated hepatocytes and its importance in the regulation of gluconeogenesis. *Biochemical Journal.* 236:789–800.

Rasmussen, H. 1981. Calcium and cAMP as Synarchic Messengers. John Wiley & Sons, New York.

Sheu, S.-S., V. K. Sharma, and A. Uglesity. 1986. Na^{+}-Ca^{2+} exchange contributes to increase of cytosolic Ca^{2+} concentration during depolarization in heart muscle. *American Journal of Physiology.* 250:C651–C656.

Sistare, F. D., R. A. Picking, and R. C. Haynes, Jr. 1985. Sensitivity of the response of cytosolic calcium in Quin-2-loaded rat hepatocytes to glucagon, adenine nucleosides and adenine nucleotides. *Journal of Biological Chemistry.* 260:12744–12747.

Somlyo, A. P., M. Bond, and A. V. Somlyo. 1985. Calcium content of mitochondria and endoplasmic reticulum in liver frozen rapidly in vivo. *Nature.* 314:622–625.

Staddon, J. M., and R. G. Hansford. 1986. 4-β-Phorbol 12-myristate 13-acetate attenuates the glucagon-induced increase in cytoplasmic free Ca^{2+} concentration in isolated rat hepatocytes. *Biochemical Journal.* 238:737–743.

Staddon, J. M., and R. G. Hansford. 1987. The glucagon-induced activation of pyruvate dehydrogenase in hepatocytes is diminished by 4β-phorbol 12-myristate 13-acetate: a role for cytoplasmic Ca^{2+} in dehydrogenase regulation. *Biochemical Journal.* 241:729–735.

Sugano, T., M. Shiota, H. Khono, M. Shimada, and N. Oshino. 1980. Effects of calcium ions on the activation of gluconeogenesis by norepinephrine in perfused rat liver. *Journal of Biochemistry.* 87:465–472.

Sutko, J. L., and J. L. Kenyon. 1983. Ryanodine modification of cardiac muscle responses to potassium-free solutions. Evidence for inhibition of sarcoplasmic reticulum calcium release. *Journal of General Physiology.* 82:385–404.

Titheradge, M. A., and R. C. Haynes. 1980. The hormonal stimulation of ureogenesis in isolated hepatocytes though increases in mitochondrial ATP production. *Archives of Biochemistry and Biophysics.* 201:44–55.

Titheradge, M. A., J. L. Stringer, and R. C. Haynes. 1979. The stimulation of the mitochondrial uncoupler-dependent-ATPase in isolated hepatocytes by catecholamines and glucagon and its relationship to gluconeogenesis. *European Journal of Biochemistry.* 102:117–127.

Wieland, O. H. 1983. The mammalian pyruvate dehydrogenase complex: structure and regulation. *Reviews of Physiology, Biochemistry and Pharmacology.* 96:124–170.

Chapter 17

Membrane and Microfilament Organization and Vasopressin Action in Transporting Epithelia

Dennis A. Ausiello, John Hartwig, and Dennis Brown

Renal and Hematology-Oncology Units, Massachusetts General Hospital, and Harvard Medical School, Boston, Massachusetts

Introduction

Vasopressin is the major hormone involved in the regulation of water balance by mammals. Its mechanism of action has been studied in the key target epithelium, the renal collecting duct, as well as in in vitro models: the amphibian skin and bladder. In general, vasopressin-sensitive epithelia are remarkably impermeable to water in the resting state, in contrast to the usual high permeability of luminal membranes of epithelial cells in general. When stimulated by vasopressin, the permeability of the renal collecting duct or amphibian bladder increases to a level comparable to that of most other biological membranes. Vasopressin induces this transformation in water permeability in its target cells by first binding to its receptor on the basolateral surface of the epithelial cell, stimulating adenylate cyclase, and increasing the cytoplasmic concentration of the second messenger, cyclic AMP. While the molecular events distal to cAMP production are obscure, an understanding of the resulting morphological events has been defined in the toad bladder and, to a lesser extent, in the kidney collecting duct. Alterations in luminal membrane structure were first demonstrated by freeze-fracture electron microscopy by Chevalier et al. (1974). They showed that, within minutes after stimulation by antidiuretic hormone, aggregates of small intramembranous particles (IMPs) appear in the luminal membrane of the frog bladder granular cells. This observation has been confirmed in a number of laboratories (for review, see Hays, 1983, and Hays et al., 1985) and has been extended to the collecting duct of the mammalian kidney (Harmanci et al., 1978) and to toad epidermis (Brown et al., 1983). A further understanding of this process came from the work of Wade (1978) and Humbert et al. (1977), who reported that thin, tubular structures in the apical cytoplasm of toad bladder granular cells contain the IMP aggregates in a helical array, and appear to deliver them to the luminal membrane after vasopressin-induced fusion. Wade et al. (1981), in the "shuttle hypothesis," proposed that the cytoplasmic tubular structures travel from sites in the apical cytoplasm to the luminal membrane. In support of this, they demonstrated that the frequency of these structures in the cytoplasm decreased after vasopressin stimulation, which suggested that they had migrated to fusion sites in the membrane. Muller et al. (1980) developed the idea of vasopressin-induced fusion further by showing that fusion events were rare in the absence of vasopressin and increased sharply in the presence of hormone. Thus, the final steps in the action of vasopressin seem to be a special case of fusion by exocytosis, entirely in keeping with a hormonal response that must appear and disappear rapidly as the state of hydration of the subject changes.

In this chapter, we will review the phenomena of endocytosis and exocytosis as they apply to vasopressin action and attempt to integrate this hormonal response to alterations in membrane and microfilament organization.

Endocytosis and Exocytosis: General Considerations

One of the more intriguing problems in cell biology today is the mechanism by which cells can initiate a continuous and highly controlled movement of membrane-embedded proteins from one cell organelle to another. Our current understanding of this process has come, in large part, from the elegant work of Brown and Goldstein (1986), who have coined the term "receptor-mediated endocytosis" to define the selective uptake of specific membrane proteins from the cell surface.

This process is common to virtually all eukaryotic cells, which use receptor-mediated endocytosis for a wide variety of functions, including cellular activities involved with nutrition, host defense, and transport and processing of extracellular molecules (Stahl and Schwartz, 1986). In general, molecular ligands bind to specific cell surface receptors that move to and aggregate within specialized regions of the plasma membrane termed "coated pits" (Brown and Goldstein, 1986). When viewed in the transmission electron microscope, these invaginations of the plasma membrane demonstrate a characteristic fuzzy cytoplasmic border that is composed of a family of proteins with a major species, clathrin, having a molecular weight of 180,000 daltons (Pearse, 1976), and other polypeptides, including a 100,000-dalton species (Zaremba and Keen, 1983) that has been implicated in both coat assembly and in the binding of clathrin to the membrane. When viewed in rapidly frozen, deeply etched images under the electron microscope, the coated pits and their progeny, coated vesicles, appear as a cage-like structure composed of a polygonal lattice of hexagons and pentagons (Heuser, 1980). Ungewickell and Branton (1981) have demonstrated that this polyhedral surface lattice is constructed from three-legged, hexameric protein complexes that are formed of three clathrin heavy chains and three light chains, and have the contours of triskelions. In general, coated pits comprise 1–2% of the plasma membrane surface area of most cells, and in some cell types they may be much larger than in others. Once a ligand-receptor complex is localized to a coated pit, the pits pinch off from the plasma membrane and become coated vesicles (Fine and Ockleford, 1984). These coated vesicles, however, appear as evanescent structures; once the coated pits have severed their connection with the extracellular milieu, there is a rapid disassembly of the clathrin coat (Goud et al., 1985). The resultant smooth-surfaced, uncoated vesicles, together with their contents of ligands and receptors, are then delivered to the endosomal compartment (Helenius et al., 1983). This endosomal compartment is composed of a network of tubules and vesicles that form a reticulum within the peripheral cytoplasm of the cell. This intracellular compartment is similar, but not identical, in various cell types, and many of the characteristics of these vesicles have not yet been fully defined. Most, if not all, of these vesicles share a common H^+-ATPase that appears to be responsible for the acidic nature of the endosomal compartment in comparison with the cell cytoplasm. The fate of the receptor-ligand complex, once entering this intracellular compartment, is variable. In many instances, however, the ligand is dissociated from the receptor and is degraded in lysosomes while the receptor is recycled, through exocytosis of smooth vesicles. The signals that govern receptor cycling lie both within the ligand and within the receptor. Some receptors internalize and recycle in the absence of their ligands, but often at a slower rate (Brown and Goldstein, 1986). Ligand binding to receptors may trigger one of many regulatory events, including phosphorylation, that may be signals for the endocytotic process (Stahl and Schwartz, 1986).

While the pattern described above appears to be well established in many instances of receptor-mediated endocytosis, the processing and movement of nonreceptor proteins is less well defined. There is evidence for the participation of clathrin-coated vesicles in exocytotic transport (Fine and Ockleford, 1984; Rothman and Fine, 1980), and in liver, separation of endocytotic and exocytotic coated vesicles has been achieved using a novel cholinesterase-mediated density shift technique (Helmy et al., 1986). While these vesicles did not appear to differ in their protein constituents, there were significant differences in their lipid makeup, which may be responsible for the selective targeting of molecules to the respective type of coated vesicle. However, the complete pathway for the cycling of endogenous proteins still remains to be fully defined. The picture that seems to be emerging is that clathrin-coated vesicles are restricted to routes connecting the Golgi with some terminal destinations of biosynthetic traffic, but perhaps not the plasma membrane (Rothman, 1986). While the normal enzymes and secretory proteins destined for storage granules may be selectively removed from the Golgi in clathrin-coated membranes (Orci et al., 1984), it is possible that others, including cell surface proteins, are transported to the plasma membrane by other types of vesicles whose content may not be nearly so selective. Indeed, in kidney intercalated cells, large coated vesicles have been observed by Brown and Orci (1986) that do not contain clathrin but do contain an enzyme, H^+-ATPase, which is known to be cycled to and from the plasma membrane surface (Brown D., S. Gluck, and J. Hartwig, unpublished observations). In addition, non–clathrin-coated vesicles may also be involved in transport processes within the Golgi complex (Orci et al., 1986). The fact that some cells can survive without clathrin further supports the variability and multiplicity of the vesicle cycling process. Payne and Schekman (1985) deleted the single clathrin gene from yeast; these cells grew quite well and were able to secrete the enzyme invertase at essentially normal rates. Thus, in yeast, intracellular transport and secretion of this enzyme does not require clathrin-coated vesicles. The effect of deletion of clathrin on endocytosis was not reported in this study. For any given endocytotic/exocytotic process, the nature and pattern of vesicle translocation may be different from cell type to cell type and from process to process. This has particular bearing on the process of membrane insertion and recycling of water channels in vasopressin-sensitive epithelia (see below).

Membrane Cycling of IMPs

As described above, an increased transepithelial flow of water induced by vasopressin is associated with the appearance of IMP aggregates seen by freeze-fracture electron microscopy within apical plasma membranes of sensitive cells in amphibian urinary bladder, amphibian epidermis, and the kidney collecting duct. In the toad bladder and the toad skin, a correlation has been shown between vasopressin-induced osmotic water flow and cluster frequency in the apical membrane of vasopressin-sensitive cells (Brown et al., 1983; Kachadorian et al., 1977). In the kidney collecting duct, a dose-response relationship has been shown between the vasopressin-induced increase in urine osmolality and cluster frequency in the apical membrane of vasopressin-sensitive cells in Brattleboro rats (Harmanci et al., 1980). Although direct evidence is lacking, it is presumed that the IMPs within the clusters serve as the water channels across the previously impermeable membrane of these epithelial cells. We have begun to investigate the mechanisms of water channel insertion and removal from the apical plasma membranes of collecting duct principal cells in response to vasopressin. In the collecting duct (Brown and Orci, 1983), as well as in the toad bladder (Orci et al., 1980), we found that the IMP aggregates were not labeled with filipin-sterol complexes, in contrast to the heavy labeling of the rest of the plasma membrane (Figs. 1–3). Filipin is a sterol-specific antibiotic that can be used as a cytochemical probe for membrane cholesterol and related β-hydroxy sterols, and it produces deformations within the membrane that are easily visible in freeze-fracture electron microscopy, and which perturb the membrane when viewed by thin-section electron microscopy. Vasopressin-induced IMP aggregates are shown in Fig. 2 in the apical membrane of a Brattleboro rat principal cell treated with vasopressin. In Fig. 3, a similar membrane region is shown, but it is from a sample of tissue that had been preincubated with filipin before freeze-fracturing. The small bumps characteristic of filipin-sterol complexes are visible throughout the membrane, but are absent from many circular or oval regions, marked with arrowheads, which correspond to the IMP aggregates. This absence of filipin-sterol complexes enabled the aggregates to be identified in thin sections of the collecting duct, since they retained a trilaminar appearance while the rest of the membrane was disrupted by the presence of the filipin-sterol complexes (Fig. 1). In this way, the particle aggregates in the collecting duct were found to correspond to coated pits, structures that are involved in receptor-mediated endocytosis in other cell types (Brown and Goldstein, 1986). When Brattleboro rats, which genetically lack vasopressin, were treated with this hormone, a 10-fold increase in coated pits was measured in collecting duct principal cells, concomitant with the appearance of freeze-fracture aggregates (Brown and Orci, 1983). The question now remaining to be resolved is, how are coated pits involved in the increase of transmembrane water permeability? One intriguing possibility is that coated membrane segments are inherently more permeable to water than the rest of the membrane, but a more likely possibility is that the membrane domains containing the IMP aggregates are selectively removed from the apical plasma membrane by internalization via coated pits. Indeed, we have recently shown (Brown D., P. Weyer, and L. Orci, unpublished results) that vasopressin stimulates a large increase in the endocytotic uptake of horseradish peroxidase by principal cells in Brattleboro rats, and that this endocytotic uptake occurs via coated pits. It seems, therefore, that in collecting duct principal cells, IMP aggregates are removed from the apical plasma membrane via an endocytotic mechanism involving coated pits. How the aggregates are recycled back to the apical membrane has not yet been determined in the collecting duct.

Membrane and Microfilament Organization

Most of the experimental efforts undertaken to understand vasopressin-induced water flow have devoted their attention to the changes induced in the apical membrane, presumably secondary to the direct insertion of the water channels into this membrane as described above (Chevalier et al., 1974; Harmanci et al., 1978; Wade, 1978; Muller et al., 1980; Brown and Orci, 1983). The aspect of transepithelial water transport dealing with the water pathway through the cytoplasmic compartments of the kidney collecting duct cells or other vasopressin target cells has, with a few exceptions discussed below, been largely ignored. While it is appropriate that so much attention has been directed at plasma membranes as the rate-limiting barriers in transepithelial transport processes, the large wealth of new information on the organization of the complex structure of cytoplasm and membrane-cytoplasm interactions has made it necessary to understand the role of the cytoplasm in any transepithelial transport event. This point of view has been most eloquently described in a review by DiBona (1983). In it, he states that

"... the cell as a whole must be considered as instrumental to transepithelial osmotic water movements." Using toad bladder granular cells as the major focus for their study of vasopressin-sensitive water transport, DiBona and colleagues

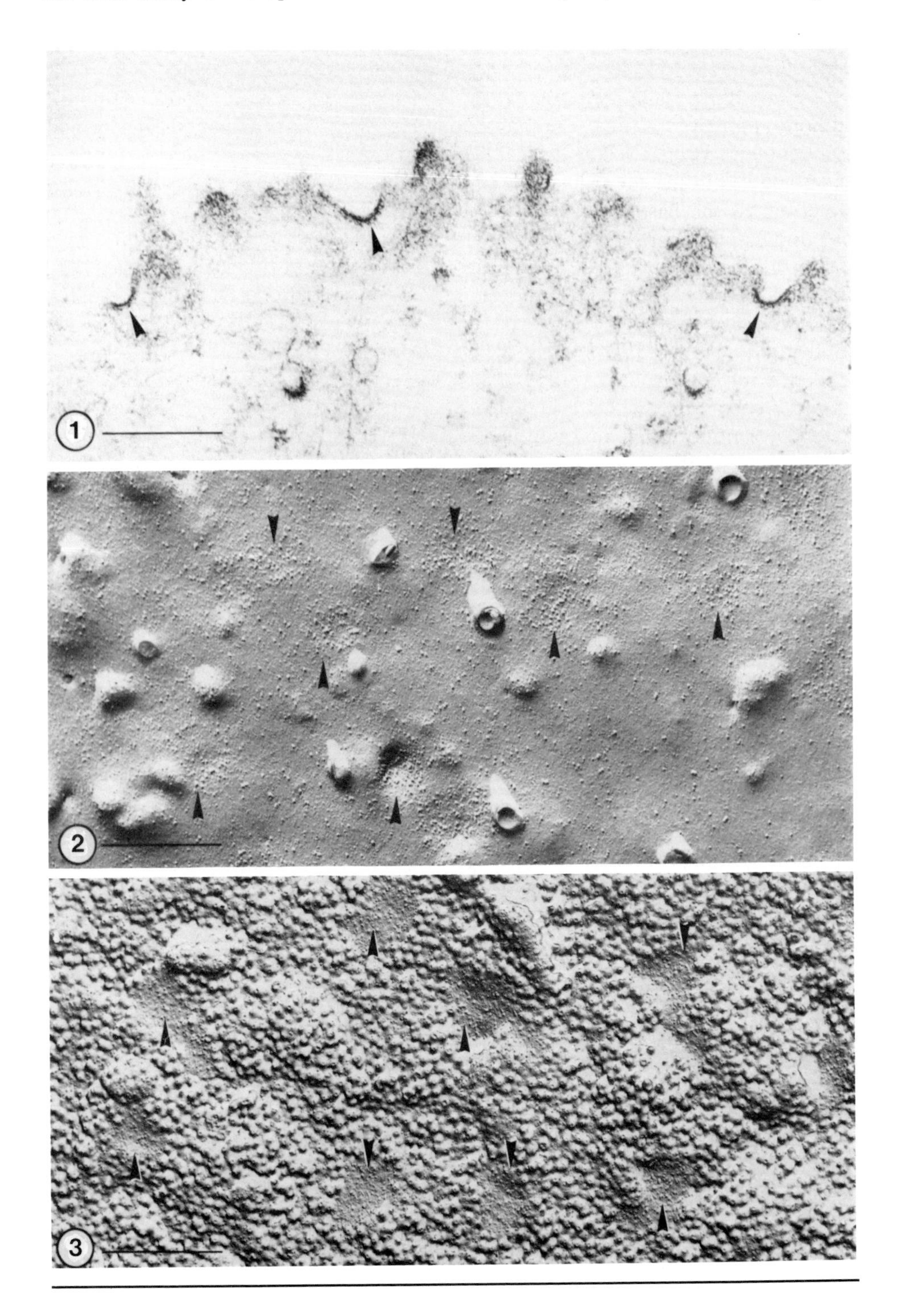

have generated data supporting this point of view (DiBona et al., 1969; DiBona, 1978, 1979, 1981, 1983; Hardy and DiBona, 1982; Kirk et al., 1984). Drawing largely on this work and some earlier studies of others (Eggena, 1972; Davis et al., 1974; DeSousa et al., 1974; Kachadorian et al., 1979; LeFurgey and Tisher, 1981; Pearl and Taylor, 1983), the following paragraph describes vasopressin-induced water flow in toad bladder granular cells.

When the granular cells are exposed to vasopressin in the presence of a large osmotic gradient, with the mucosal surface dilute, significant alterations in granular cell geometry occur. Water moves across the apical aspect of the granular cell membrane, diluting the cell interior so that an osmotic force develops to move water across the basolateral membrane to the intercellular spaces, which are in osmotic equilibrium with the serosal bathing solutions. These spaces have been demonstrated to be widely distended during hormone-stimulated water flow (Pak Poy and Bentley, 1960; DiBona et al., 1969). The flow into the serosal bathing solution is achieved by the development of excess hydrostatic pressure in the moderately compliant intercellular spaces. An examination of this process yields evidence for dramatic changes in surface and cytoskeletal morphology in the granular cells. First, when a significant osmotic gradient is present, vasopressin-induced osmotic water flow results in an approximate doubling of granular cell volume, presumably with an associated reduction in intracellular osmolality. Remarkably, this dramatic dilution is nondestructive to intracellular organelles and metabolic processes. This is in direct contrast to when cells are swollen to the same extent by dilution of the serosal bathing media, which results in serious damage to intracellular organelle integrity (DiBona et al., 1969; DiBona, 1981). There thus appears to be something very different in the manner of swelling or in the handling of water between hormone-stimulated apical to basolateral water movement vs. intracellular swelling owing to basolateral hypotonicity. The control of this process may very well rest in the nature of hormone- and/or water flow–induced changes in the cytoskeletal makeup of the subapical membrane regions of granular cells.

When one views this process using differential interference contrast microscopy of living tissue, vasopressin-induced water flow produces a change in the granular cell from a large, flat apical cell into an "ice cream cone"–shaped cell with a bulging apical membrane (DiBona, 1978, 1979, 1983). These changes are consistent with our hypothesis that the cytoskeletal elements that constitute the granular cell

---

Figures 1–3. Figs. 1–3 show luminal plasma membranes of collecting duct principal cells from Brattleboro rats that had received 1 U of vasopressin tannate in peanut oil, subcutaneously, for four consecutive days. Fig. 1 is a thin section, whereas Figs. 2 and 3 are freeze-fracture replicas. Vasopressin treatment induces the appearance of IMP clusters (arrowheads, Fig. 2) on these membranes, in parallel with an increase in urine osmolality. When vasopressin-treated kidneys are exposed to filipin before fracture, the typical filipin-sterol complexes are abundant in the apical membrane, but are absent from the IMP clusters (arrowheads, Fig. 3). This property can be used to identify the IMP clusters in thin sections, since most of the apical membrane is disrupted by the filipin (Fig. 1), whereas the regions corresponding to the clusters retain a trilaminar membrane structure. In Fig. 1, three such regions are present (arrowheads) and they all correspond to coated pits. In this way, the vasopressin-induced IMP clusters were identified as being present in coated pits. Bar, 0.5 μm.

equivalent of a terminal web are relaxed or depolymerized to effect a local gel-sol transformation after hormonal stimulation (Ausiello et al., 1984; Ausiello and Hartwig, 1985). There are a number of indications that vasopressin action includes a component of cytoskeletal involvement that changes the membrane conformation at a macroscopic level. A striking feature of the granular cell mucosal surface, after vasopressin-induced water flow, is that the coral array of surface ridges is transformed into straight, finger-like projections or microvilli (Spinelli et al., 1975; Mills and Malick, 1978; DiBona, 1979; LeFurgey and Tisher, 1981; Kirk et al., 1984). Although it is controversial whether or not the ridge-to-villus transformation requires water flow, DiBona (1983) has reported that vasopressin application to toad bladder epithelial cells in the absence of an osmotic gradient (and thus no water flow) can cause this conformational change. Using the techniques of rapid freezing and freeze drying, we have confirmed that the ridge-to-villus transformation occurs even in the absence of osmotic water flow (Figs. 4 and 5).

Some attempts have been made to define the precise architecture of the cytoplasmic structure immediately subjacent to the apical membrane. This region is rich in microfilaments, as expected (Pearl and Taylor, 1983), and microtubules and intermediate filaments are also abundant in these cells (Taylor, 1977; Kraehenbuhl et al., 1979). DiBona (1983) has studied, by fluorescence microscopy, vasopressin-induced alterations in actin structure using the probe NBD-phallacidin. This probe binds only to actin filaments (Wulf et al., 1979). Actin distribution of granular cells from nonstimulated tissue appears to predominate around the cell margins: after vasopressin stimulation, the granular cells showed a spongy reticular cytoplasm with bright fluorescent rings girdling vacuoles accompanying the transition to the "ice cream cone" cell shape. DiBona (1983) hypothesized that this shape change induced by vasopressin might very well contribute to the nondestructive nature of the water flow process that dilutes cytosolic content. He postulated that the more basally distributed mitochondria, for example, might be subject to a less severe dilution of their surroundings if a gradient of water activity existed from apex to base in the granular cell cytoplasm. This condition is expected if the influx of water across the apical membrane is sufficient to offset the upstream diffusion of mobile solutes toward an intracellular equilibrium distribution. A nonisomorphous distention of the cytomatrix would favor an asymmetric distribution of the solute by generating gradients of mobile cation distribution, at least to the extent that the fiber lattice contains fixed negative charges, as would be expected for actin filaments.

Other studies also implicate a role for alterations in cytoplasmic filament

Figures 4 and 5. Electron micrographs of rapidly frozen and freeze-dried toad bladder epithelium showing modifications of the apical surface of granular cells induced by vasopressin. Bladders were incubated for 30 min at room temperature with a hydroosmotic gradient across the cells (serosal side bathed in toad Ringer's, mucosal side bathed in distilled water) in the absence (Fig. 4) and presence (Fig. 5) of vasopressin. Bladders were then fixed for 10 min in 1% glutaraldehyde/0.2% tannic acid, washed with distilled water, rapidly frozen in 15% methanol, freeze-dried at −80°C, and metal-coated. In both micrographs, the high resolution of this technique reveals that the surface glycocalyx is arranged into a dense filamentous network. The ridge-like surface of the granular cell (Fig. 4) is transformed into discrete microvilli in the presence of vasopressin (Fig. 5). A similar observation was made in the absence of an osmotic gradient. Bar, 1 μm.

organization, specifically microfilament organization, in transepithelial water flow in vasopressin-sensitive tissue. It had been shown a number of years ago (Davis et al., 1979; Pearl and Taylor, 1983) that the microfilament disrupter cytochalasin B inhibits vasopressin-induced water flow across toad bladder epithelia. More recent studies (Hardy and DiBona, 1982) have demonstrated that cytochalasin B application inhibits the vasopressin-induced effect on osmotic water flow (P_f) by 40–60%, but not its action on measured tritiated water diffusion (P_d). In the absence

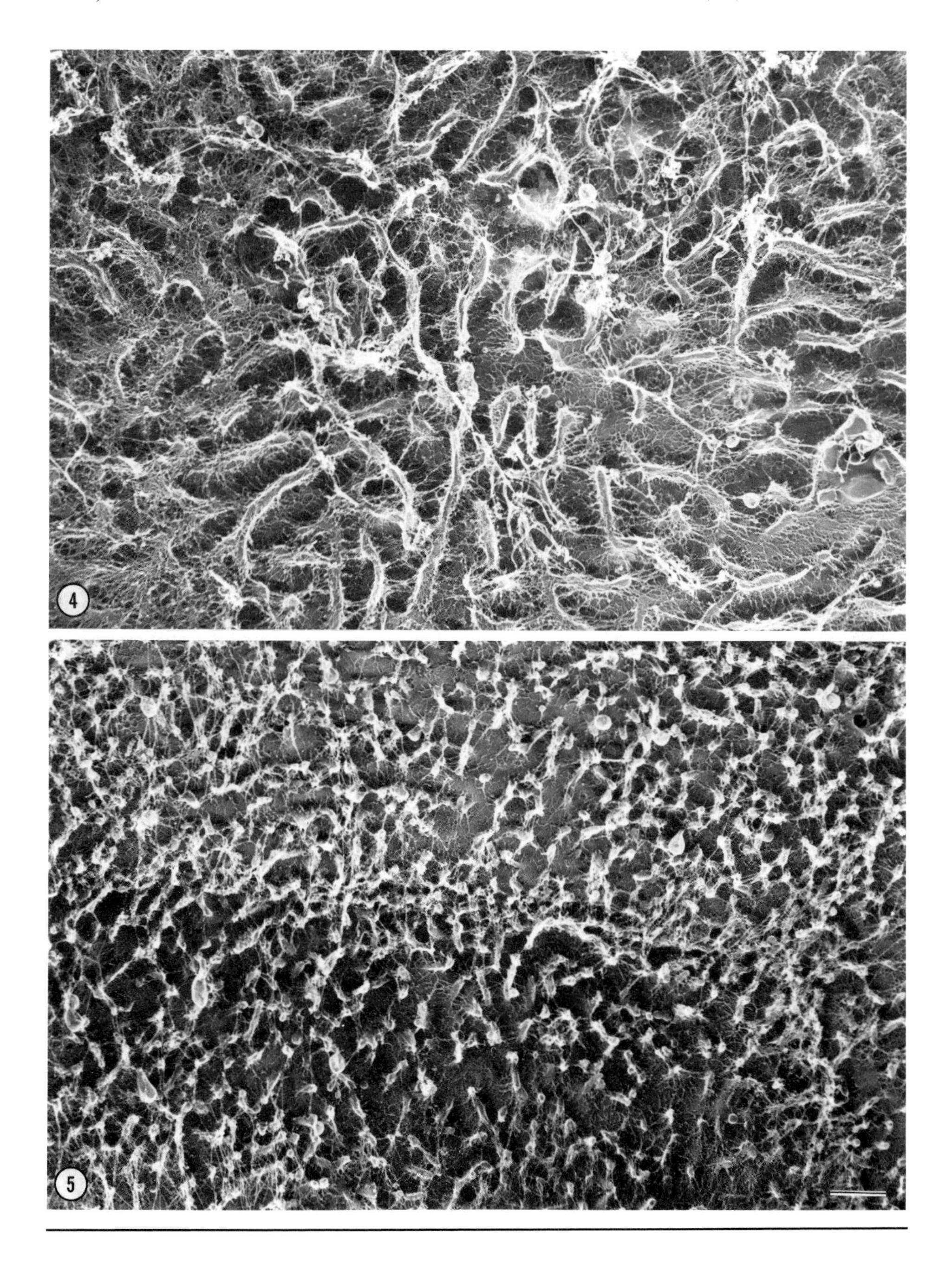

of an osmotic gradient, cytochalasin B produces no obvious effects on cytoplasm geometry, but during brisk osmotic flow, the granular cell interior is full of large-sized vacuoles. If glutaraldehyde fixation of tissue is conducted before gradient application, cytochalasin B does not inhibit vasopressin-induced water flow (Eggena, 1972). Thus, precluding the geometric changes induced by cytochalasin (in this case, vacuolation of the cytoplasm) successfully removes the apparent inhibition. Although cytochalasin B has also been shown to alter the presence of the presumed water channel aggregates in the apical membrane (Kachadorian et al., 1979), these observations have only been made in the presence of an osmotic gradient. Thus, in the experiment just described, the nature of the water channels is not known. However, it strongly suggests that the integrity of water flow is preserved with cytochalasin B if cytoskeletal disruption does not occur. While this does not imply that a transcytoplasmic route by itself is the major modulator of water flow in vasopressin-sensitive tissue, it does support the concept that there is considerable interaction between the transepithelial water route and the state of membrane permeability per se.

It is now known that the cytochalasins bind to only one end of the actin filament to inhibit actin monomer addition at this end (Brenner and Korn, 1979; Brown and Spudich, 1979; Hartwig and Stossel, 1979; Flanagan and Lin, 1980; MacLean-Fletcher and Pollard, 1980). Since this end of the actin filament has a 10-fold-higher affinity for monomeric actin than the other end, binding causes net filament disassembly, thereby producing a population of shortened filaments (Hartwig and Stossel, 1979; MacLean-Fletcher and Pollard, 1980). The addition of cytochalasin to filaments also causes some filament fragmentation (Hartwig and Stossel, 1979). Its interaction with actin in vitro is most dramatic during filament assembly, where, by blocking monomer addition at the end growing at the most rapid rate, it depresses the overall rate of filament assembly. In cells, this effect of cytochalasin is even more potent because the bulk of filament assembly is believed to occur selectively at this high-affinity filament end (Korn, 1982; Weeds, 1982). Therefore, its addition to cells will inhibit new filament assembly and shorten the weight average filament length by causing net filament disassembly and by fragmenting filaments. Such decreases in actin filament length will have large effects on the mechanical properties of actin networks.

Glutaraldehyde fixation does prevent cell swelling during osmotic water flow (Eggena, 1972). Thus, the observation that fixation of the permeability state of the plasma membranes does not alter the water permeability induced by vasopressin, but does prevent the “ice cream cone” shape distensibility of the cell, suggests that cell swelling is not necessary for the high water permeability of the cell. Why, then, do cells swell in response to vasopressin? DiBona (1983) raises the interesting possibility that swelling involves a cellular redistribution of cytoskeletal elements that delineate low-resistance hydraulic pathways. Although direct observations of living tissues during water flow (DiBona, 1978, 1979, 1981; Kirk et al., 1984) have not disclosed any visual evidence of selective water pathways across the cell, DiBona has concluded from quantitative estimates that these might be too small to be resolved with such techniques. The pathway structure for such transcytoplasmic water routes must be described in a great deal more detail before speculation can lead to a reasonable hypothesis for transepithelial water flow. It seems reasonable to conclude, however, that there is specificity to the cytoplasmic changes that occur in vasopressin-induced water flow from apical to basolateral membrane, in contrast to the indiscriminate swelling with intracellular damage that is observed when the basolateral medium is diluted. The most likely source of the regulation of changes in shape geometry is the cytoskeleton, in particular the microfilaments, since they comprise a large percentage of the fibrous network below the apical membrane. An inverse correlation exists between the extent of actin-filament cross-linking and the ability of actin networks to imbibe water, and there is a direct relationship between the extent of cross-linking and the amount of pressure a filament network can withstand before undergoing a phase transition, i.e., gel to sol transition (Zaner, K., unpublished observations). These data imply that vasopressin-sensitive epithelial cells may be able to create domains in the cytoplasm that retard water (i.e., resist swelling) by cross-linking actin filaments into a tight network. These would have the capacity to selectively exclude water that has penetrated into the cell through the apical membrane. Conversely, local decreases in the extent of filament cross-linking or the activation of proteins that function to fragment and shorten filaments could establish cytoplasmic domains where swelling and water movement could occur (i.e., in subapical membrane areas). The ability of cytochalasin to inhibit water flow may thus be related to its disruptive effect on apical actin networks functioning in this exclusion capacity or in the unregulated solation of apical actin networks.

It has been shown that actin is a major protein in toad bladder epithelia and that actin filaments reside in the cortical cytoplasm (Pearl and Taylor, 1983). However, the three-dimensional organization of actin filaments in toad bladder granular cells has not been studied. In the last few years, we have begun to define the nature of microfilament organization in vasopressin-sensitive cells. Our data have revealed that toad bladder epithelial cells contain the actin-associated regulatory proteins, actin-binding protein, and villin (Ausiello et al., 1984). Actin-binding protein is a large molecule of mammalian cells that binds to and cross-links actin filaments into isotropic networks (Weeds, 1982; Ausiello et al., 1984; Hartwig and Shevlin, 1986). Villin, like the related protein gelsolin, is a calcium-activated protein that severs and binds the high-affinity end of actin filaments (Bretscher and Weber, 1980; Yin et al., 1981; Corwin and Hartwig, 1983). In addition, villin has the ability to bundle actin filaments in the absence of calcium (Bretscher and Weber, 1980; Corwin and Hartwig, 1983). Using a technique for rapid freezing of tissue for examination in the electron microscope (Heuser and Kirschner, 1980), we have been able to obtain a detailed three-dimensional view of the apical membrane-filament interactions in toad bladder granular cells (Figs. 6 and 7). The dense array of actin filaments (decorated with the S1 subfragment of myosin in Fig. 6) and nonactin filaments demonstrates the complex interaction between cytoskeletal elements and both the membrane surface (ridges and villi) and cytoplasmic organelles, particularly vesicles (Fig. 7).

In addition, we have also demonstrated that, using an immuno-gold localization procedure, actin, actin-binding protein, gelsolin, and myosin are concentrated in the subapical areas of rabbit renal collecting duct principal cells (Brown et al., 1986). This is the first direct demonstration of proteins intimately involved in microfilament organization in the mammalian kidney. While the cytoplasmic changes induced by vasopressin have not been extensively studied in the kidney, in comparison with those described above in the toad bladder, Kirk et al. (1984)

have demonstrated that during the onset of vasopressin-stimulated volume flow, there is an apparent increase in the volume of the epithelium. This is accompanied by apical rounding of the cells, dilation of the lateral intercellular spaces, and cytoplasmic vacuolization. Thus, it appears likely that geometric changes similar to those described for the toad bladder also occur in the kidney, and that alterations

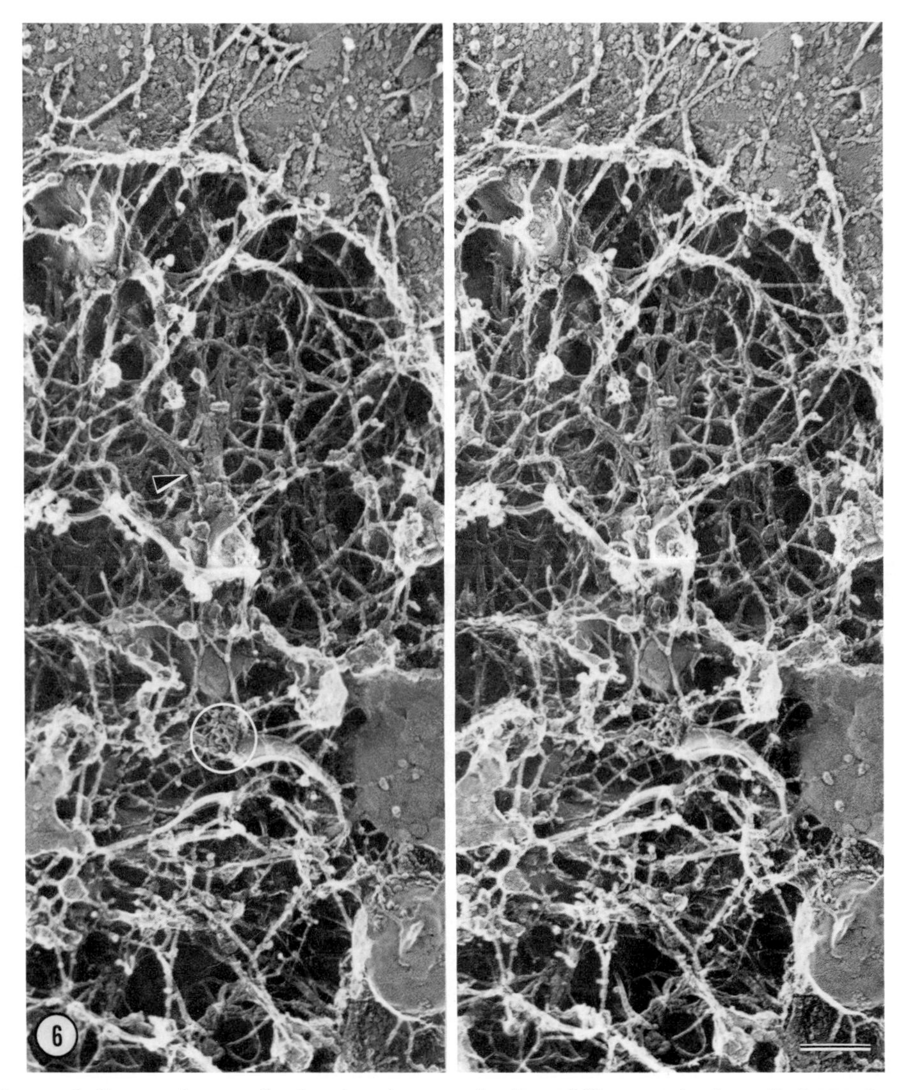

Figure 6. Stereomicrographs showing the organization of filaments in the apical cytoplasm of a granular cell incubated in the presence of vasopressin and an osmotic gradient. The apical cytoplasm was revealed by ripping off the apical membrane by attaching and removing a polylysine-coated coverslip to the bladder epithelium. The bladder was then fixed, frozen, and prepared for the microscope as in Fig. 4. Viewing in stereo reveals many points where cytoplasmic filaments appear to attach to the underside of the plasma membrane. The insertion of filamentous glycocalyx on the apical surface is also apparent. The arrow indicates a region where 10-nm filaments have coalesced into a bundle forming the core of a microvillus. A small clathrin-coated vesicle is circled. Bar, 0.2 μm.

in microfilament organization might play an important role in these morphological changes. It is clear that the demonstration of actin-associated proteins in these cell types provides toad bladder epithelium and kidney collecting duct cells with the capacity to undergo a calcium-regulated, actin-based, gel-sol transition, as well as the ability to form surface microvilli during water flow, as seen in the toad bladder (Spinelli et al., 1975; Mills and Malick, 1978; LeFurgey and Tisher, 1981). While

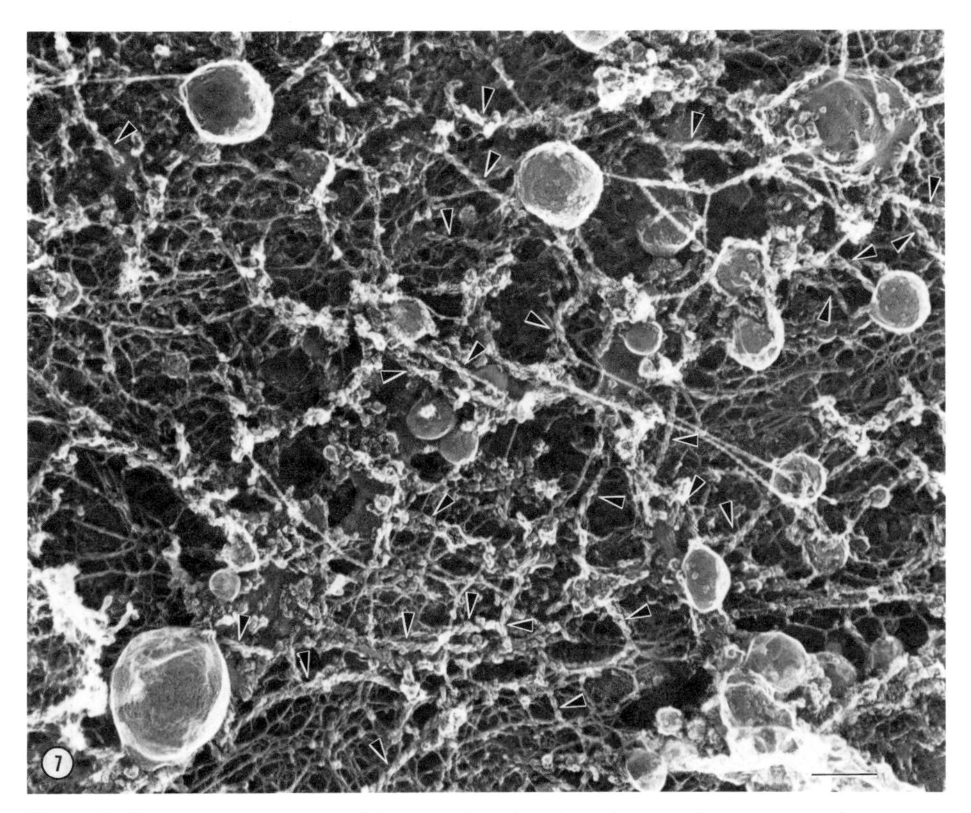

Figure 7. Electron micrograph of the cytoplasmic side of the apical membrane of a granular cell. This micrograph is a companion view to Fig. 6 in that it shows a membrane fragment and associated filaments that were ripped from a granular cell. The cytoplasmic side of the apical membrane is coated with a dense, three-dimensional network of filaments. Actin filaments have been identified by incubating the specimen with the S1 subfragment of skeletal muscle myosin. This binds to the filaments, giving them a thickened, cable-like appearance (arrows). In many cases, the actin filaments are found to be attached to small membrane-bounded vesicles. Bar, 0.2 μm.

we have concentrated on these proteins, it is likely that many more actin-associated molecules are functional components of the cytoskeleton of these cells (Korn, 1982; Weeds, 1982). The regulation by such proteins of microfilament gel-sol transformation is likely to play an important role in the pathways involved in vasopressin-mediated transepithelial water transport. These include both the exocytotic and endocytotic pathways for the insertion and removal of water channels from the apical membrane, as well as the proposed alteration in cytoplasmic cytoskeletal geometry, which may be important in dictating the transcellular pathway for water movement across the epithelium. It appears likely that a common alteration in microfilament organization will facilitate both processes.

References

Ausiello, D. A., H. L. Corwin, and J. H. Hartwig. 1984. Identification of actin-binding protein and villin in toad bladder epithelia. *American Journal of Physiology.* 246:F101–F104.

Ausiello, D. A., and J. H. Hartwig. 1985. Microfilament organization and vasopressin action. *In* Vasopressin. R. W. Schrier, editor. Raven Press, New York. 89–96.

Brenner, S. L., and E. D. Korn. 1979. Substoichiometric concentrations of cytochalasin B inhibit actin polymerization. Additional evidence for an f-actin treadmill. *Journal of Biological Chemistry.* 254:9982–9985.

Bretscher, A., and K. Weber. 1980. Villin is a major protein of the microvillus cytoskeleton which binds both G and F actin in a calcium-dependent manner. *Cell.* 20:839–847.

Brown, D., D. Ausiello, J. Hartwig, T. Stossel, and L. Orci. 1986. Actin-regulatory proteins locate predominantly in the apical cytoplasm of kidney collecting duct. *Kidney International.* 29:413. (Abstr.)

Brown, D., A. Grosso, and R. C. DeSousa. 1983. Correlation between water flow and intramembrane particle aggregates in toad epidermis. *American Journal of Physiology.* 245:C334–C342.

Brown, D., and L. Orci. 1983. Vasopressin stimulates formation of coated pits in rat kidney collecting ducts. *Nature.* 302:253–255.

Brown, D., and L. Orci. 1986. The "coat" of kidney intercalated cell tubulovesicles does not contain clathrin. *American Journal of Physiology.* 250:C605–C608.

Brown, M. S., and J. L. Goldstein. 1986. A receptor-mediated pathway for cholesterol homeostasis. *Science.* 232:34–47.

Brown, S. S., and J. A. Spudich. 1979. Cytochalasin inhibits the rate of elongation of actin filament fragments. *Journal of Cell Biology.* 83:657–662.

Chevalier, J., J. Bourguet, and J. S. Hugon. 1974. Membrane associated particles: distribution in frog urinary bladder epithelium at rest and after oxytocin treatment. *Cell Tissue Research.* 152:129–140.

Corwin, H. L., and J. H. Hartwig. 1983. Isolation of actin-binding protein and villin from toad oocytes. *Developmental Biology.* 99:61–74.

Davis, W. L., D. B. P. Goodman, R. J. Schuster, H. Rasmussen, and J. H. Martin. 1974. Effects of cytochalasin B on the response of toad urinary bladder to vasopressin. *Journal of Cell Biology.* 63:986–997.

De Sousa, R. C., A. Grosso, and C. Rufener. 1974. Blockade of the hydroosmotic effect of vasopressin by cytochalasin B. *Experientia.* 30:175–177.

DiBona, D. R. 1978. Direct visualization of epithelial morphology in the living amphibian urinary bladder. *Journal of Membrane Biology.* 40:45–70.

DiBona, D. R. 1979. Direct visualization of ADH-mediated transepithelial osmotic flow. *In* Hormonal Control of Epithelial Transport. J. Bourguet, editor. INSERM, Paris. 195–206.

DiBona, D. R. 1981. Vasopressin action on the conformational state of the granular cell in the amphibian bladder. *In* Epithelial Ion and Water Transport. A. D. C. Macknight and J. P. Leader, editors. Raven Press, New York. 241–255.

DiBona, D. R. 1983. Cytoplasmic involvement in ADH-mediated osmosis across toad urinary bladder. *American Journal of Physiology.* 245:C297–C307.

DiBona, D. R., M. M. Civan, and A. Leaf. 1969. The cellular specificity of the effect of vasopressin on toad urinary bladder. *Journal of Membrane Biology.* 1:79–91.

Eggena, P. 1972. Glutaraldehyde-fixation method for determining the permeability to water of the toad urinary bladder. *Endocrinology.* 91:240–246.

Fine, R., and C. D. Ockleford. 1984. Supramolecular cytology of coated vesicles. *International Review of Cytology.* 91:1–43.

Flanagan, M. D., and S. Lin. 1980. Cytochalasins block actin filament elongation by binding to high affinity sites associated with F-actin. *Journal of Biological Chemistry.* 255:835–838.

Goud, B., C. Huet, and D. Louvard. 1985. Assembled and unassembled pools of clathrin: a quantitative study using an enzyme immunoassay. *Journal of Cell Biology.* 100:521–527.

Hardy, M. A., and D. R. DiBona. 1982. Microfilaments and the hydrosmotic action of vasopressin in toad urinary bladder. *American Journal of Physiology.* 243:C200–C204.

Harmanci, M. C., P. Stern, W. A. Kachadorian, H. Valtin, and V. A. DiScala. 1978. Antidiuretic hormone-induced intramembranous alterations in mammalian collecting duct. *American Journal of Physiology.* 235:F440–F443.

Harmanci, M. C., P. Stern, W. A. Kachadorian, H. Valtin, and V. A. DiScala. 1980. Vasopressin and collecting duct intramembranous particle clusters: a dose response relationship. *American Journal of Physiology.* 239:F560–F564.

Hartwig, J. H., and P. Shevlin. 1986. The architecture of actin filaments and the ultrastructural location of actin-binding protein in the periphery of lung macrophages. *Journal of Cell Biology.* 103:1007–1020.

Hartwig, J. H., and T. P. Stossel. 1979. Cytochalasin B and the structure of actin gels. *Journal of Molecular Biology.* 134:539–554.

Hays, R. 1983. Alteration of luminal membrane structure by antidiuretic hormone. *American Journal of Physiology.* 245:C289–C296.

Hays, R. M., J. Sasaki, S. M. Tilles, L. Meiteles, and N. Franki. 1985. Morphological aspects of the cellular action of antidiuretic hormone. *In* Vasopressin. R. W. S. Schrier, editor. Raven Press, New York. 79–87.

Helenius, A., I. Mellman, D. Wall, and A. Hubbard. 1983. Endosomes. *Trends in Biochemical Sciences.* 8:245–250.

Helmy, S., K. Porter-Jordan, E. A. Dawidowicz, P. Pilch, A. L. Schwartz, and R. E. Fine. 1986. Separation of endocytic from exocytic coated vesicles using a novel cholinesterase mediated density shift technique. *Cell.* 44:497–506.

Heuser, J. 1980. Three dimensional visualization of coated vesicle formation in fibroblasts. *Journal of Cell Biology.* 84:560–583.

Heuser, J. E., and M. W. Kirschner. 1980. Filament organization revealed in platinum replicas of freeze-dried cytoskeletons. *Journal of Cell Biology.* 86:212–234.

Humbert, F., R. Montesano, A. Grosso, R. C. DeSousa, and L. Orci. 1977. Particle aggregates in plasma and intracellular membranes of toad bladder granular cell. *Experientia.* 33:1364–1367.

Kachadorian, W. A., S. J. Ellis, and J. Muller. 1979. Possible roles of microtubules and microfilaments in ADH action on toad urinary bladder. *American Journal of Physiology.* 236:F14–F20.

Kachadorian, W. A., J. B. Wade, C. C. Uiterwyk, and V. A. DiScala. 1977. Membrane structural and functional responses to vasopressin in toad bladder. *Journal of Membrane Biology.* 30:381–401.

Kirk, K. L., J. A. Schafer, and D. R. DiBona. 1984. Quantitative analysis of the structural events associated with antidiuretic hormone-induced volume reabsorption in the rabbit cortical collecting tubule. *Journal of Membrane Biology.* 79:65–74.

Korn, E. D. 1982. Actin polymerization and its regulation by proteins from non-muscle cells. *Physiological Reviews.* 62:672–737.

Kraehenbuhl, J. P., J. Pfeiffer, M. Rossier, and B. C. Rossier. 1979. Microfilament-rich cells in the toad bladder epithelium. *Journal of Membrane Biology.* 48:167–180.

LeFurgey, A., and C. C. Tisher. 1981. Time course of vasopressin-induced formation of microvilli in granular cells of toad urinary bladder. *Journal of Membrane Biology.* 61:13–19.

MacLean-Fletcher, S., and T. D. Pollard. 1980. Mechanism of action of cytochalasin B on actin. *Cell.* 20:329–341.

Mills, J. W., and L. E. Malick. 1978. Mucosal surface morphology of the toad urinary bladder. Scanning electron microscope study of the natriferic and hydro-osmotic response to vasopressin. *Journal of Cell Biology.* 77:598–610.

Muller, J., W. A. Kachadorian, and V. A. DiScala. 1980. Evidence that ADH-stimulated intramembrane particle aggregates are transferred from cytoplasmic to luminal membranes in toad bladder epithelial cells. *Journal of Cell Biology.* 85:83–95.

Orci, L., B. S. Glick, and J. E. Rothman. 1986. A new type of coated vesicular carrier that appears not to contain clathrin: its possible role in protein transport within the Golgi stack. *Cell.* 46:171–184.

Orci, L., P. Halban, M. Amherdt, M. Ravazzola, J.-D. Vassalli, and A. Perrelet. 1984. A clathrin-coated, Golgi-related compartment of the insulin secreting cell accumulates proinsulin in the presence of monensin. *Cell.* 39:39–47.

Orci, L., R. Montesano, and D. Brown. 1980. Heterogeneity of toad bladder granular cell luminal membrane: distribution of filipin-sterol complexes in freeze-fracture. *Biochimica et Biophysica Acta.* 601:443–452.

Pak Poy, R. F. K., and P. Bentley. 1960. Fine structure of the toad urinary bladder. *Experimental Cell Research.* 20:235–237.

Payne, G. S., and R. Schekman. 1985. A test of clathrin function in protein secretion and cell growth. *Science.* 230:1009–1014.

Pearl, M., and A. Taylor. 1983. Actin filaments and vasopressin-stimulated water flow in toad urinary bladder. *American Journal of Physiology.* 245:C28–C39.

Pearse, B. M. F. 1976. Clathrin: a unique protein associated with intracellular transfer of membrane by coated vesicles. *Proceedings of the National Academy of Sciences.* 73:1255–1259.

Rothman, J. E. 1986. Life without clathrin. *Nature.* 319:96–97.

Rothman, J. E., and R. E. Fine. 1980. Coated vesicles transport newly synthesized membrane glycoproteins from endoplasmic reticulum to plasma membrane in two successive stages. *Proceedings of the National Academy of Sciences.* 77:780–784.

Spinelli, F., A. Grosso, and R. C. DeSousa. 1975. The hydroosmotic effect of vasopressin: a scanning electron-microscope study. *Journal of Membrane Biology.* 23:139–156.

Stahl, P., and A. L. Schwartz. 1986. Receptor-mediated endocytosis. *Journal of Clinical Investigation.* 77:657–662.

Taylor, A. 1977. Role of microtubules and microfilaments in the action of vasopressin. *In* Disturbances in Body Fluid Osmolality. T. E. Andreoli, J. J. Grantham, and F. C. Rector, editors. American Physiological Society, Washington, DC. 97–124.

Ungewickell, E., and D. Branton. 1981. Assembly units of clathrin coats. *Nature.* 289:420–422.

Wade, J. B. 1978. Membrane structural specialization of the toad urinary bladder revealed by the freeze-fracture technique. III. Location, structure and vasopressin dependence of intramembranous particle arrays. *Journal of Membrane Biology.* 40:(Special Suppl.):281–296.

Wade, J. B., D. L. Stetson, and S. A. Lewis. 1981. ADH action: evidence for a membrane shuttle mechanism. *Annals of the New York Academy of Sciences.* 372:106–117.

Weeds, A. 1982. Actin-binding proteins: regulators of cell architecture and motility. *Nature.* 296:811–816.

Wulf, E. A., A. Deboben, F. A. Bautz, H. Faulstich, and T. H. Wieland. 1979. Fluorescent phallotoxin, a tool for the visualization of cellular actin. *Proceedings of the National Academy of Sciences.* 76:4498–4502.

Yin, H. L., J. H. Hartwig, K. Maruyama, and T. P. Stossel. 1981. Ca^{2+} control of actin polymerization. Interaction of macrophage gelsolin with actin monomers and effects on actin polymerization. *Journal of Biological Chemistry.* 256:9693–9697.

Zaremba, S., and J. H. Keen. 1983. Assembly polypeptides from coated vesicles mediate reassembly of unique clathrin coats. *Journal of Cell Biology.* 97:1339–1347.

Chapter 18

Multiple Roles of Calcium in Anoxic-induced Injury in Renal Proximal Tubules

Lazaro J. Mandel, Takehito Takano, Stephen P. Soltoff, William R. Jacobs, Ann LeFurgey, and Peter Ingram

Department of Physiology, Duke University Medical Center, Durham, North Carolina

Introduction

Oxygen deprivation produces a series of complex changes in mammalian cells that are reversible or irreversible, depending on the duration of this deprivation. Proximal tubules from the kidney lose all exchangeable K (Harris et al., 1981; Soltoff and Mandel, 1986), gain Na (Mason et al., 1981), and decrease their ATP levels by 60–90% (Gerlach et al., 1963; Balaban et al., 1980*a*) within 5 min of oxygen deprivation. Structural alterations are clearly observed in proximal tubules within 15 min of clamp-induced ischemia (Glaumann et al., 1977; Kahng et al., 1978), as are mitochondrial dysfunctions (Mergner et al., 1977). These malfunctions progress steadily with the duration of ischemia or anoxia. If the anoxic time is limited to ~20 or 30 min, these changes are reversible. Longer anoxia elicits a series of irreversible changes, seen most dramatically as morphological alterations, but also produces functional changes such as damage to mitochondrial and transport functions (Arendshorst et al., 1975; Venkatachalam et al., 1978; Wilson et al., 1984).

Although the irreversible changes have been described in considerable detail, the underlying causes remain unknown. Ca has been implicated by numerous investigators as one of the main mediators of irreversible cell injury (Farber et al., 1981; Trump et al., 1982; Wilson et al., 1984; Burke et al., 1984). The two main mechanisms suggested for this action of Ca are: (*a*) through an increase in cytosolic free Ca and consequent activation of phospholipases and/or proteases that cause membrane damage and (*b*) through Ca accumulation in the mitochondria inducing mitochondrial damage.

This chapter deals with the study of both of these possible mechanisms in a suspension of proximal tubules obtained from the rabbit kidney. In this preparation, the aforementioned conditions of reversibility and irreversibility can be duplicated in vitro (Takano et al., 1985; Weinberg, 1985). The tubules were placed in a specially designed thermostatted chamber, where the P_{O_2} was continuously monitored with an O_2 electrode (Balaban et al., 1980*b*). Anoxia was achieved by closing the chamber and allowing the tubules to consume all the oxygen. Varying degrees of hypoxia were achieved by equilibrating the surface of the suspension with mixtures containing various O_2 concentrations and 5% CO_2 concentrations. The tubules were subjected to anoxia or hypoxia for 10, 20, 30, or 40 min and then reoxygenated for 20 or 40 min. As described in other publications, these tubules rapidly lose ATP and K, and their respiratory rate is inhibited with time of oxygen deprivation (Takano et al., 1985).

Short-Term Ca Accumulation and Tubular Damage

The first question that was investigated concerned the possible deleterious effects of short-term cellular Ca accumulation on proximal tubular functions (Takano et al., 1985). To investigate this question, the relationship between total Ca accumulation and respiratory dysfunction was studied. Preliminary experiments suggested that renal tubules subjected to hypoxia, rather than anoxia, accumulated Ca. These results prompted an investigation into the relationship between Ca accumulation and the degree of hypoxia. The latter was measured spectrophotometrically by the percent oxidation of cytochrome oxidase in the tubular mitochondria. During normoxia, cytochrome oxidase has a high affinity for O_2 and is therefore almost

100% oxidized (Balaban et al., 1980*b*), whereas anoxia leads to complete reduction (0% oxidation) of this cytochrome. Hypoxia is present when the oxygen pressure within the cells is insufficient to keep cytochrome oxidase fully oxidized, leading to partial reduction and impairment of electron flow (Jones and Kennedy, 1982). In the present experiments, this was achieved by decreasing the P_{O_2} of the bathing medium below 1%. Under normoxic conditions, total cellular Ca content averaged 16.6 ± 0.5 nmol/mg of protein ($n = 20$). As shown in Fig. 1, no change in Ca content occurred until the P_{O_2} was decreased sufficiently to create severe hypoxia, as measured by a partial reduction of cytochrome oxidase to the 10–30% oxidation level. This led to a large accumulation of Ca. In contrast, a slight Ca release was actually observed during anoxia.

The Ca accumulation observed during severe hypoxia was studied as a function of the duration of hypoxia, as shown in Fig. 2. Ca content was seen to increase

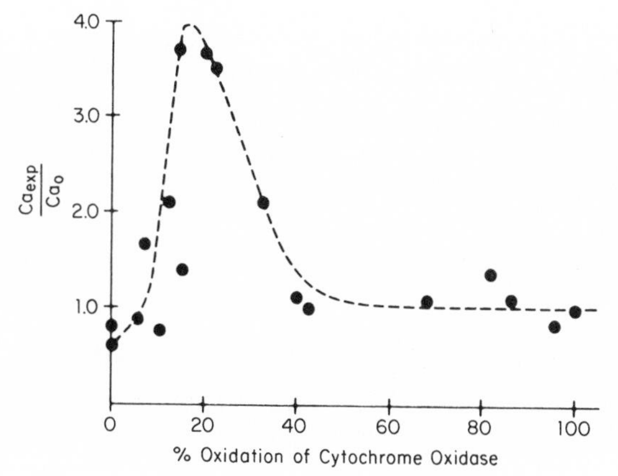

Figure 1. Cellular Ca content of the renal tubules as a function of the severity of hypoxia. Variable degrees of hypoxia were monitored spectrophotometrically as the percent oxidation of cytochrome oxidase. Ca content is expressed as the ratio of experimental to control, in which the latter was measured in a paired experiment under normoxia. Each experimental point represents a separate paired experiment in which the tubules were subjected to 30 min of hypoxia under the stated conditions. The dashed line was drawn by eye. (Reproduced with permission from Takano et al., 1985.)

approximately linearly with time, reaching a maximum level after 30 min of hypoxia. This process was rapidly reversible in all cases, since the tubules regained their original Ca content within 10 min of reoxygenation.

The differences between anoxia and hypoxia were used to determine whether Ca accumulation affected the observed respiratory dysfunction. The latter was assessed through a "stress test" that involved the addition of the ionophore nystatin to increase the intracellular Na concentration, thereby stimulating Na,K-ATPase activity and respiration (Harris et al., 1981). As seen in Table I, the inhibition of nystatin-stimulated respiration was identical whether the tissue was subjected to severe hypoxia (30 min), increasing its total Ca content by 250%, or to anoxia (30 min), with unchanged total Ca. Furthermore, the hypoxia-induced Ca accumulation was abolished by the addition of 3 μM ruthenium red (Table I), an inhibitor of mitochondrial Ca uptake, with no significant ameliorative effect on nystatin-

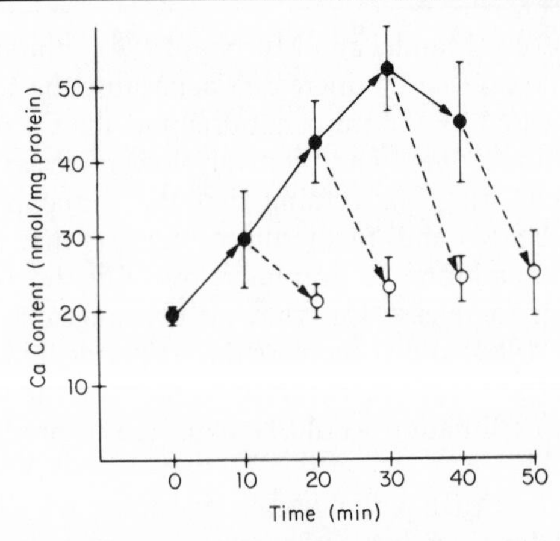

Figure 2. Content of the tubular cells subjected to severe hypoxia for the indicated time, followed by 10 min of reoxygenation. Filled circles: values after hypoxia; open circles: values after reoxygenation (n = 5). (Reproduced with permission from Takano et al., 1985.)

stimulated respiration after hypoxia. Two conclusions can be drawn from these results: (*a*) Ca accumulates in the mitochondria during severe hypoxia but not during anoxia, and (*b*) this accumulation is reversible and appears to be unrelated to the respiratory dysfunction elicited by the short-term hypoxia or anoxia.

Cytosolic Free Ca During Short-Term Anoxia

Next, we investigated whether free Ca (Ca_f) was altered in proximal tubular cells during short-term anoxia. Ca_f was measured in the tubule suspension using fura-2, with some modifications of methods published elsewhere (Jacobs and Mandel, 1987). The effects of substances with known action on Ca transport pathways were first assessed to understand better the regulation of Ca_f in the proximal tubular cells. As shown in Table II, the addition of a mitochondrial uncoupler increased Ca_f from 100 to only 190 nM, whereas Ca ionophores increased Ca_f to ~450 nM. The former agents are known to release Ca from mitochondria and to rapidly

Table I
Cellular Ca Content and Respiratory Dysfunction after Hypoxia or Anoxia

	Control	Severe hypoxia	Anoxia	Severe hypoxia + ruthenium red
Ca content (nmol/mg protein)	16±0.5	40.6±6.8*	15.5±1.6	16.1±1.2
Nystatin-stimulated respiration (nmol O_2/min·mg protein)	32.0±1.5	13.0±61.5	13.8±0.7	16.3±1.7
n	20	8	7	5

The tubules were subjected to hypoxia or anoxia as described in the Methods. The duration of hypoxia or anoxia was 30 min. (Reproduced with permission from Takano et al., 1985.)
* $P < 0.01$.

reduce cellular ATP levels (Mandel and Murphy, 1984), inhibiting the Ca-ATPase. This situation would be expected to increase Ca_f because the active extrusion from the cells of the Ca released by the mitochondria and the Ca passive influx would now occur at an inhibited rate. The relatively low increase in Ca_f under these conditions is consistent with our finding that the mitochondria of these cells contain little Ca (see below) and, furthermore, suggests that the Ca permeability through the plasma membrane is normally low. On the other hand, the Ca ionophores are known to release Ca from all exchangeable intracellular stores (Mandel and Murphy, 1984) and to increase the Ca permeability across the plasma membrane. Both of these would be expected to increase the availability of Ca to the Ca-ATPase, and this situation would be expected to produce a large increase in Ca_f.

These results suggest that pathological conditions whose initial effect is to inhibit mitochondrial function (e.g., ischemia), leading to a decrease in cellular ATP levels, may not by themselves alter Ca_f by much. The low Ca permeability of these cells may ensure that the Ca-ATPase activity be normally low, with a minimal requirement for ATP that could be supplied even in a metabolic inhibited state.

Table II
Effect of a Mitochondrial Uncoupler and Ca Ionophores on Cytosolic Free Ca

	Control	Experimental
10 μM 1799 ($n = 6$)	113±6	192±17
6 μM 1799 ($n = 4$)	109±17	160±13
10 μM Br-A23187 ($n = 6$)	132±13	452±36
10 μM ionomycin ($n = 4$)	113±21	466±18

Values for Ca_f in the presence of 1799, Br-A23187, and ionomycin were calculated 2 min after the addition of each agent.

Large increases in Ca_f could only be obtained through specific alterations that either caused sustained intracellular releases of Ca or increased the Ca permeability across the plasma membrane.

In a series of preliminary experiments, Ca_f was measured as a function of time in anoxia. Under normoxic conditions, Ca_f averaged 100 ± 14 nM ($n = 6$) and was unchanged after 5, 15, or 40 min of anoxia ($n = 3$). This is a preliminary result that needs to be explored in further detail, particularly because it differs from a recent result of Borle et al. (1986). These authors, using aequorin in rabbit proximal tubules suspended in a fiberglass matrix, found that Ca_f increased after ~30 min of anoxia. They found that glucose removal was essential to observe the increase in Ca_f. It will be extremely interesting to determine in future experiments whether damage to the plasma membrane preceded this increase in Ca_f under their experimental conditions. Our own preliminary experiments suggest that no increase in Ca_f occurs under conditions that cause demonstrable damage to the renal cell. Therefore, changes in cytosolic Ca may not be necessary to initiate the cascade of events that lead to the damage. Conversely, when Ca_f does increase, it may already be as a result of plasma membrane disruption. These possibilities and others clearly require further experimentation.

Compartmentation of Ca in the Intracellular Environment of Proximal Tubular Cells

How Does It Change During Anoxia?

As shown in Fig. 1, there are changes in total cellular Ca that occur during anoxia and hypoxia in renal cells. To understand the role of Ca in the cell injury that accompanies these states, it is important to know which intracellular compartments show changes in Ca content. This was approached by use of electron probe x-ray microanalysis (EPMA). This technique allows the measurement of Ca content within subcellular compartments in thin cryosections obtained from biological samples (Kitazawa et al., 1983). The application of these techniques to proximal renal tubules has been described in detail elsewhere (LeFurgey et al., 1986). A typical spectrum obtained from a mitochondrion of a proximal renal tubule is shown in Fig. 3. It shows low Na and Cl contents and high K, which are typical of

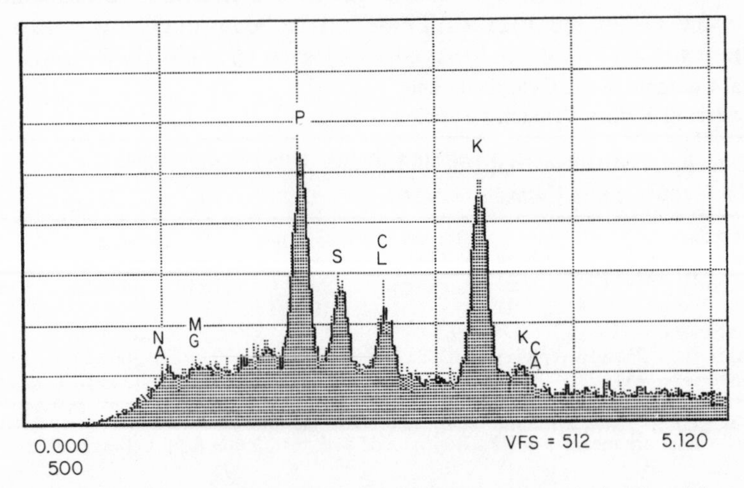

Figure 3. Representative energy dispersive x-ray spectrum from a mitochondrion of a renal proximal tubule. For experimental details, see LeFurgey et al. (1986).

a healthy cell. Its Ca content is extremely low and has to be calculated by a deconvolution of the K-K_β peak from the overlapped Ca-K_α peak (Kitazawa et al., 1983). Average values for the elemental contents in the cytoplasm and the mitochondria are shown in Table II. The cells considered viable display a K/Na ratio of 3–4 and Ca contents of ~3–4 mmol/g dry wt (4–5 nmol/mg protein), and they constitute the great majority of cells in the suspension. Table III, part *B*, shows the elemental contents of the cells labeled "nonviable." These show a K/Na ratio of ~1, large accumulations of Ca in the mitochondria, and a high cytoplasmic Ca content. The low mitochondrial Ca content of the viable cells is consistent with the possibility that mitochondrial Ca controls matrix dehydrogenase activity in these cells, as also discussed in two other chapters in this volume (Somlyo et al., 1987; Hansford and Staddon, 1987). It is unclear how the other cells became nonviable; however, this state seems to be accompanied by an accumulation of Ca, as previously described by other investigators (Trump et al., 1982; Wilson et al., 1984). The percentage of nonviable cells in the population can be calculated from the total Ca content of the suspension. The viable cells contain an average of 4–5 nmol/mg of protein of Ca (Table III, part *A*) and therefore cannot account for the observed Ca content of ~17 nmol/mg protein for the suspension (Table I). A simple calculation shows that if nonviable cells constituted 5% of the population, they would account for this total value in the suspension. This would also mean that the nonviable cells contained ~75% of the total Ca in the suspension. Therefore, a sharp heterogeneity in Ca compartmentation appears to be normally present in the tubules, and this needs to be taken into consideration in interpreting results obtained from experiments in which total Ca content or Ca fluxes are measured in a cellular preparation (LeFurgey et al., 1986).

Only very preliminary data are presently available concerning the effects of anoxia on Ca compartmentation. The suggestion is emerging that after 40 min of anoxia, the Ca content of the majority of cells may be unchanged. No data are yet available on hypoxia. Future work will determine whether most of the changes in Ca content shown in Figs. 1 and 2 occur in the nonviable cells, which would

Table III
Elemental Contents in the Cytoplasm and Mitochondria of Kidney Proximal Tubules

	Elemental content (mmol/kg dry weight)							n
	Na	Mg	P	S	Cl	K	Ca	
(*A*) Viable cells								
Cytoplasm	125±12	27±4	356±18	124±8	141±11	348±23	4.1±1.4	23
Mitochondria	95±7	27±3	349±22	158±6	131±12	349±22	3.1±1.1	23
(*B*) Nonviable cells								
Cytoplasm	261±16	26±5	288±16	147±14	70±10	206±11	15±2	9
Mitochondria	154±29	37±16	612±85	112±17	187±23	124±16	685±139	10

n = number of 500-s raster probes obtained from each region. The tubules were preincubated at 37°C for 20 min. Values are means ± SEM. (Reproduced with permission from LeFurgey et al., 1986.)

demonstrate a strong Ca homeostasis for the viable cells. Conversely, if the Ca content of the viable cells changes, it may provide an opportunity to study the conditions necessary to obtain these changes and their effect on cell viability.

Effects of Extracellular Ca Removal on Anoxia-induced Damage

Experiments in which Ca is removed from the extracellular medium provided the clearest indication that Ca selectively affects aspects of cellular function during anoxia. Decreasing extracellular Ca from 1 mM to 2.5 μM drastically reduced the percent of lactate dehydrogenase (LDH) loss from the tubules during 30 min of anoxia (Takano et al., 1985), which suggests that less membrane damage had occurred in the low-Ca medium. However, the low-Ca medium appeared to be harmful to the tubules during reoxygenation, since the percent of LDH release increased significantly in this medium, whereas no significant change was observed after reoxygenation in 1 mM Ca (Takano et al., 1985). Therefore, it appears that low extracellular Ca may play a mildly protective role during anoxia but may be somewhat deleterious during reoxygenation.

Conclusions

Proximal renal tubules subjected to short-term anoxia or hypoxia show the following. (*a*) Mitochondrial Ca accumulation appears to be reversible and not associated with hypoxic-induced damage. (*b*) Cytosolic free Ca does not seem to increase during anoxia. Since numerous cellular alterations occur under these conditions, it seems unlikely that changes in cytosolic free Ca are required to initiate these events in the proximal kidney.

References

Arendshorst, W. J., W. F. Finn, and C. W. Gottschalk. 1975. Pathogenesis of acute renal failure following temporary renal ischemia in the rat. *Circulation Research.* 37:558–568.

Balaban, R. S., L. J. Mandel, S. Soltoff, and J. M. Storey. 1980*a*. Coupling of active ion transport and aerobic respiratory rate in isolated renal tubules. *Proceedings of the National Academy of Sciences.* 77:447–451.

Balaban, R. S., S. Soltoff, J. M. Storey, and L. J. Mandel. 1980*b*. Improved renal cortical tubule suspension: spectrophotometric study of O_2 delivery. *American Journal of Physiology.* 258:F50–F59.

Borle, A. B., C. C. Freudenrich, and K. W. Snowdowne. 1986. A simple method for incorporating aequorin into mammalian cells. *American Journal of Physiology.* 251:C323–C326.

Burke, T. J., P. E. Arnold, J. A. Gordon, R. E. Bulger, D. C. Dobyan, and R. W. Schrier. 1984. Protective effect of intrarenal calcium membrane blockers before or after renal ischemia: functional, morphological, and mitochondrial studies. *Journal of Clinical Investigation.* 74:1830–1841.

Farber, J. L., K. R. Chien, and S. Mittnacht, Jr. 1981. The pathogenesis of irreversible cell injury in ischemia. *American Journal of Pathology.* 102:271–281.

Gerlach, E. B. Deuticke, and R. H. Dreisbach. 1963. Zum Verhalten nucleotiden und ihren dephosphorylierten abbauproduckten in der niere bei ischamie und kurzzeitiger postischamischer wiederdurchblutung. *Pflügers Archiv.* 278:296–315.

Glaumann, B., H. Glaumann, I. K. Berezesky, and B. F. Trump. 1977. Studies on cellular recovery from injury. II. Ultrastructural studies on the recovery of the pars convoluta of the proximal tubule of the rat kidney from temporary ischemia. *Virchows' Archives of Cell Pathology, Series B.* 24:1–18.

Hansford, R. G., and J. M. Staddon. 1987. The relationship between the cytosolic free calcium ion concentration and the control of pyruvate dehydrogenase. *In* Cell Calcium and the Control of Membrane Transport. L. J. Mandel and D. C. Eaton, editors. The Rockefeller University Press, New York. *Society of General Physiologists Series.* 42:241–257.

Harris, S. I., R. S. Balaban, L. Barrett, and L. J. Mandel. 1981. Mitochondrial calcium accumulation and respiration in ischemic acute renal failure in the rat. *Journal of Biological Chemistry.* 256:10319–10328.

Jacobs, W. R., and L. J. Mandel. 1987. Fluorescent measurements of intracellular free calcium in isolated toad urinary bladder epithelial cells. *Journal of Membrane Biology.*, In press.

Jones, D. P., and F. G. Kennedy. 1982. Intracellular oxygen supply during hypoxia. *American Journal of Physiology.* 243:C247–C253.

Kahng, M. R., I. K. Berezesky, and B. F. Trump. 1978. Metabolic and ultrastructural response of rat kidney cortex to *in vitro* ischemia. *Experimental and Molecular Pathology.* 29:183–198.

Kitazawa, T., H. Shuman, and A. P. Somlyo. 1983. Quantitative electron probe analysis: problems and solutions. *Ultramicroscopy.* 11:251–262.

LeFurgey, A., P. Ingram, and L. J. Mandel. 1986. Heterogeneity of calcium compartmentation: electron probe analysis of renal tubules. *Journal of Membrane Biology.* 94:165–170.

Mandel, L. J., and E. Murphy. 1984. Regulation of cytosolic free calcium in rabbit proximal renal tubules. *Journal of Biological Chemistry.* 259:11188–11196.

Mason, J., F. Beck, A. Dorge, R. Rick, and K. Thurau. 1981. Intracellular electrolyte composition following renal ischemia. *Kidney International.* 20:61–70.

Mergner, W. J., L. Marzella, C. Mergner, M. W. Kahng, M. W. Smith, and B. F. Trump. 1977. Studies on the pathogenesis of ischemic cell injury. VII. Proton gradient and respiration of renal tissue cubes, renal mitochondrial and submitochondrial particles following ischemic cell injury. *Beitrage Pathologie.* 161:230–243.

Soltoff, S. P., and L. J. Mandel. 1986. Potassium transport in the rabbit renal proximal tubule. Effects of barium, ouabain, valinomycin, and other ionophores. *Journal of Membrane Biology.* 94:153–161.

Somlyo, A. P., A. V. Somlyo, M. Bond, R. Broderick, Y. E. Goldman, H. Shuman, J. W. Walker, and D. R. Trentham. 1987. Calcium and magnesium movements in cells and the role of inositol trisphosphate in muscle. *In* Cell Calcium and the Control of Membrane Transport. L. J. Mandel and D. C. Eaton, editors. The Rockefeller University Press, New York. *Society of General Physiologists Series.* 42:77–92.

Takano, T., S. P. Soltoff, S. Murdaugh, and L. J. Mandel. 1985. Intracellular respiratory dysfunction and cell injury in short-term anoxia of rabbit renal proximal tubules. *Journal of Clinical Investigation.* 76:2377–2384.

Trump, B. F., I. K. Berezesky, and R. A. Cowley. 1982. The cellular and subcellular characteristics of acute and chronic injury with emphasis on the role of calcium. *In* Pathophysiology of Shock, Anoxia, and Ischemia. R. A. Cowley and B. F. Trump, editors. Williams & Wilkins, Baltimore, MD. 6–46.

Venkatachalam, M. A., D. B. Bernard, J. F. Donohoe, and N. G. Levinsky. 1978. Ischemic damage and repair in the rat proximal tubule: differences among the S_1, S_2, and S_3 segments. *Kidney International.* 14:31–49.

Weinberg, J. M. 1985. Oxygen deprivation-induced injury to isolated rabbit kidney tubules. *Journal of Clinical Investigation.* 76:1193–1208.

Wilson, D. R., P. E. Arnold, T. J. Burke, and R. W. Schrier. 1984. Mitochondrial calcium accumulation and respiration in ischemic acute renal failure in the rat. *Kidney International.* 25:519–526.

Chapter 19

Calmodulin as a Regulator of Cell Growth and Gene Expression

Colin D. Rasmussen and Anthony R. Means

Department of Cell Biology, Baylor College of Medicine, Houston, Texas

Introduction

Calmodulin (CaM) is a multifunctional intracellular calcium receptor involved in the regulation of such essential cellular processes as cyclic nucleotide and glycogen metabolism, calcium-dependent phosphorylation, cell division, and cell motility and contractility (Means et al., 1982). The overall effect exerted by a multifunctional protein such as CaM must be the result of a critical balance between the regulated systems involved. As a consequence, the perturbation of intracellular CaM levels should be expressed through alterations in those systems that are regulated at the level of intracellular calcium concentration.

To date, only two circumstances have been reported in which alterations in CaM levels can occur. First, the CaM concentration abruptly doubles as cells traverse the G1/S boundary of the cell cycle (Chafouleas et al., 1982; Sasaki and Hidaka, 1982). Second, in cells (or tissues) transformed by oncogenic viruses, chemical carcinogens, or hormone treatment, the level of CaM is consistently increased (Watterson et al., 1976; LaPorte et al., 1980; Chafouleas et al., 1981; Connor et al., 1983; Veigel et al., 1984; Zendequi et al., 1984).

Progress through the cell cycle and cell transformation are accompanied by alterations in cell morphology, cyclic nucleotide metabolism, metabolic rate, and intracellular calcium levels. In addition, transformed cells lose both the requirement for anchorage to a substrate and the high levels of extracellular calcium required for proliferation by their nontransformed counterparts.

An increase in the intracellular levels of CaM would be expected to result in all the abnormal properties observed in transformed cells. Our previous studies (Chafouleas et al., 1982) with the anti-CaM drug W-13 suggest that the elevation of CaM levels at the G1/S boundary is required for cells to progress through the S-phase. W-13 also synergizes with bleomycin in preventing recovery from DNA damage introduced in response to the latter drug, which indicates that CaM may play a role in some aspect of DNA repair (Chafouleas et al., 1984). Finally, it has also been shown in vitro that the effect of calcium on microtubule stability is mediated through CaM (Marcum et al., 1978). This suggests that CaM might play a pivotal role in cytoskeletal regulation.

The investigation of the consequences of changes in CaM levels on cell growth and gene expression have recently become feasible as a result of the isolation and characterization of the chicken CaM gene in our laboratory (Simmen et al., 1985). Knowledge of the sequence of the promoter region and the structural organization of the gene has made possible the construction of eukaryotic expression vectors with which CaM levels can be altered through increases in the CaM gene dosage upon transfection of tissue culture cells. We have chosen to express the CaM gene in a bovine papilloma virus (BPV)-based vector in mouse C127 cells (Sarver et al., 1981). C127 cells were originally derived from a mammary tumor (Lowy et al., 1978). They are an excellent host for BPV and, though they are of tumor origin, are not considered by the usual criteria (e.g., inability to grow in soft agar) to be transformed. Therefore, cells transformed by BPV-based vectors can be identified by the appearance of transformed foci in a manner that is indistinguishable from those identified using a selectable marker.

In addition, the transformed cells will still grow to saturation density and will re-enter the cell cycle in response to several mitogenic stimuli (Lowy et al., 1978;

Sarver et al., 1981; Lusky and Botchan, 1984). We selected BPV because it is maintained in mouse C127 cells as a nuclear episome. The result is that the gene of interest is maintained in a consistent chromatin environment, and allows for cell lines with a wide range of copy numbers to be selected. Finally, the lack of random integration into the host cell genome precludes the possibility of altering (or extinguishing) a gene required for normal cell function.

Construction of Expression Vectors

We have constructed two vectors that are each comprised of a promoter, pieces of CaM DNA, and BPV and pBR322 sequences. The vector BPV-CM contains a CaM mini-gene placed 3′ to the CaM promoter isolated from a chicken genomic clone by our group (Simmen et al., 1985). A mini-gene construction was used because it allowed the removal of nearly 8 kb of genomic DNA, which comprise the first two introns of the CaM gene. The CaM promoter was used because one of our interests is in the role that 5′ flanking sequences play in the regulation of CaM gene expression.

A second vector, BPV-hMCM, contains the same CaM mini-gene ligated to the promoter from the human metallothionein IIa gene (hMTIIa) (Karin and Richards, 1982). The hMTIIa promoter was selected since it is known to be regulated both by divalent cations and glucocorticoids and thus allows CaM mRNA and protein levels to be increased in an inducible fashion. Both BPV-CM and BPV-hMCM contain the 69% transforming region of the BPV genome.

Transfection of C127 Cells and Selection of Stably Transformed Cell Lines

Mouse C127 cells at 70–80% confluency were transfected with vector DNA by the calcium phosphate precipitate method, including a brief glycerol shock to improve transformation efficiency. After a 24-h recovery from the transfection procedure, the cells were split 1:2 into 100-mm culture dishes. The cultures were fed every 2 d, and after 10–14 d, foci of transformed cells were observed. Individual foci were selected and subcultured into 60-mm dishes and grown to confluency. At this point, cells were passaged to larger dishes, and samples were retained for initial characterization of the state of the transfected DNA.

Stable cell lines containing episomal and unrearranged copies of the input DNA were selected and characterized in further detail. The cell lines currently under study are designated as shown in Table I.

Results and Discussion

From each transfection experiment, the cell lines that contained episomal copies of the input plasmid were examined to determine the level of vector-derived and total CaM mRNA. It is possible to make this distinction since, under stringent conditions, a probe homologous to the 3′ untranslated region of the chicken CaM cDNA hybridizes only to vector-derived mRNA, whereas a probe derived from the coding region of the same cDNA hybridizes to both vector-derived and mouse CaM mRNA.

Several BPV-CM–transformed cell lines showed small increases in CaM mRNA levels. However, in one cell line (CM-1), CaM mRNA levels are constitutively increased nearly 100-fold over the level observed in cells transformed by BPV alone (BPV-1 cells) or untransformed C127 cells. Northern analysis of cytoplasmic RNA showed that the increase was due solely to transcription from the BPV-CM vector. Expression is stable and has continued, unabated, during >1 yr of continual passage of the CM-1 cell line (>600 cell divisions).

The BPV-CM vector produces a single mature cytoplasmic mRNA in CM-1 cells that is ~100–200 nucleotides shorter than the corresponding mRNA observed in chicken, from which the CaM mini-gene present in the expression vector was derived. Evidence suggests that the difference in mRNA length must be due entirely to differences in the length of the 3′ untranslated region, since the 5′ cap site in mRNA isolated from CM-1 cells is identical to that of mRNA isolated from chicken gizzard. Interestingly, the sequence GATAAAA occurs in the vicinity in which we would predict the 3′ end of the BPV-CM mRNA to be. This sequence has been previously found to be an alternative poly-A addition site in the rat parvalbumin gene (Epstein et al., 1986).

In cells transformed with BPV-hMCM, cell lines were initially screened for cells that showed an increase in vector-derived CaM mRNA when cultured in the

Table I
Relative BPV Copy Numbers of Characterized Cell Lines

Cell Line	Vector	Number of episomes per cell
BPV-1	pd-BPV1	150
CM-1	BPV-CM	90
MCM-4	BPV-hMCM	50

The cell line BPV-1 serves as a control cell line for the effects of transformation by BPV alone as compared with effects of increased CaM in the CM-1 cell line. For the MCM-4 cell line, the control is MCM-4 cells cultured in the absence of an inducer.

presence of 50 μM $ZnSO_4$ for 4 h. One cell line, MCM-4, had a low basal level of expression, which increased nearly sevenfold in response to Zn^{2+} induction. Expression was found to be concentration dependent and peaked at an extracellular Zn^{2+} concentration of 50 μM. Zn^{2+} concentrations greater than this were found to inhibit cell growth. Time course studies show that expression is maximal at 4–6 h, with a subsequent decline to a level twofold greater than the basal level of expression when Zn^{2+} is present in the culture medium for >6 h. Preliminary results also indicate that expression of the BPV-hMCM vector is inducible by glucocorticoids. The addition of dexamethasone (150 nM) resulted in a slightly more than twofold increase in CaM mRNA after 4 h. As is the case in CM-1 cells, the CaM mRNA produced by the BPV-hMCM vector is shorter than the corresponding mRNA in chicken.

Consequences of Increased CaM Gene Expression

CaM protein levels were measured in each of the cell lines under investigation by radioimmunoassay (RIA). BPV-1 cells have CaM levels ~50% higher than untransformed C127 cells. In CM-1 cells, however, CaM levels are constitutively fivefold higher than can be accounted for by transformation with BPV alone. Intact CaM can be isolated from CM-1 cells that co-migrates with both CaM from C127 cells and CaM produced by a bacterial expression vector (Putkey et al., 1985), as determined by sodium dodecyl sulfate-polyacrylamide gel electrophoresis (SDS-PAGE). In addition, CaM isolated from CM-1 cells shows a shift in migration on SDS-PAGE identical to purified CaM when samples are electrophoresed in the presence of Ca^{2+}, which suggests that Ca^{2+} binding of the vector-produced protein is normal. Finally, to confirm the results obtained by RIA, relative CaM levels were determined by densitometric scanning of Coomassie-stained polyacrylamide gels and found to agree with the results obtained by RIA.

The observation that CaM protein levels only are increased 5-fold in CM-1 cells despite a 50–100-fold increase in mature CaM mRNA suggests that intracellular CaM levels are elevated at the translational or post-translational level. Preliminary results indicate that all the vector-derived mRNA is associated with polysomes, which indicates that the latter mechanism may be operative. Detailed studies are in progress to determine the precise manner in which CaM protein levels are modulated in CM-1 cells.

So far, we have observed four differences between CM-1 and BPV-1 cells that can be specifically attributed to elevated CaM in CM-1 cells. First, CM-1 cells plateau at a saturation density that is significantly higher than either BPV-1 cells or untransformed C127 cells. This must be a consequence of increased intracellular CaM concentration since the same effect is observed in MCM-4 cells when CaM levels are increased by Zn-induced expression of the BPV-hMCM vector. In these cells, the saturation density is increased >50% when cells are cultured in the presence of 50 μM Zn^{2+}. BPV-1 cells showed no difference in saturation density as a result of Zn^{2+} treatment. Second, the generation time of CM-1 cells (12 h) is 2 h shorter than in BPV-1 cells (14 h). Detailed analysis of the cell cycle has revealed that all of the difference in generation time is due to a reduction in the length of the G1 period in CM-1 cells.

We have examined the levels of three different mRNAs during the cell cycle in both BPV-1 and CM-1 cells. The mRNA for histone H4 is regulated in a similar fashion in both cell lines, and the pattern of expression is consistent with previously reported results. In addition, even though CM-1 cells have 100-fold more CaM mRNA than BPV-1 cells, the relative changes in CaM mRNA levels during the cell cycle show a nearly identical pattern in both cell lines. This suggests that the chicken CaM promoter present in the BPV-CM vector is capable of responding to the signals that are responsible for cell-cycle–dependent regulation of CaM mRNA levels. However, while the regulation of both CaM and histone mRNA levels appears to be unaltered by increased CaM, we have observed one mRNA whose pattern of expression is changed during the cell cycle as a consequence of increased CaM levels. The mRNA for myosin light-chain kinase (MLCK), a CaM-dependent enzyme, increases 5–10-fold, as compared with early G1 levels, at both the G1/S and S/G2 boundaries of the cell cycle in BPV-1 cells. In CM-1 cells, MLCK mRNA levels do not show periodic increases during the cell cycle and remain at a level characteristic of G1 cells. Thus, it would appear that CaM may potentially regulate at least one CaM-dependent enzyme at the level of its mRNA.

Finally, we have observed a decrease in the levels of both α- and β-tubulin, but not β-actin mRNAs in CM-1 cells relative to BPV-1 cells. Cleveland et al. (1981) have suggested that tubulin mRNA levels are decreased in response to an increase in the tubulin monomer/polymer ratio. Therefore, these data are consistent with our earlier contention that CaM may be the mediator of the Ca^{2+}-dependent depolymerization of microtubules (Marcum et al., 1978).

Conclusions

This study is designed to answer how changes in the intracellular concentration of CaM affect cell growth and gene expression. We have found that an increase in CaM as seen in CM-1 cells alters the rate of cell proliferation through a reduction in the cell-cycle duration. The fact that this effect is mediated through an alteration in only the length of the G1 period would suggest that CaM may play a role in regulating some event that determines the timing of the onset of DNA synthesis in these cells. These studies also support earlier work in which CaM was shown to be elevated in transformed cells and required for progression into and through the S-phase. The observed decrease in both α- and β-tubulin mRNA levels indicates that CaM may affect the rate at which tubulin monomers cycle on and off microtubules. The data also support the contention that the Ca^{2+} sensitivity of microtubule depolymerization might be mediated by CaM. Finally, increased CaM levels appear to ablate the normal cell-cycle regulation of MLCK mRNA levels. Studies are under way using nucleic acid probes for other CaM-dependent enzymes to determine whether this phenomenon is specific for MLCK or is a general feature of proteins whose functions are modulated by CaM.

The availability of the cell line that we have described allows us to question directly how altered CaM levels affect: (*a*) cells in exponential growth; (*b*) cells that are grown to the plateau phase; and (*c*) cells in the plateau phase that are induced to re-enter the cell cycle by mitogenic stimulation. Criteria that can be examined include: (*a*) cell-cycle kinetics; (*b*) expression of genes reported to be important in the cell cycle or cell structure; (*c*) possible changes in CaM-binding proteins; (*d*) the requirements for the extracellular Ca^{2+} concentration required for growth, as well as potential changes in the intracellular Ca^{2+} concentration and localization.

Acknowledgments

The authors would like to thank Elizabeth MacDougall and Charles Mena for their excellent technical assistance, and Lisa Gamble for preparation of the manuscript.

This work was supported by American Cancer Society grant BC-326M (320 G13047) to A.R.M.

References

Chafouleas, J. G., W. E. Bolton, H. Hidaka, A. E. Boyd, and A. R. Means. 1982. Calmodulin and the cell cycle: involvement in regulation of cell progression. *Cell.* 28:41–50.

Chafouleas, J. G., W. E. Bolton, and A. R. Means. 1984. Potentiation of bleomycin lethality by anticalmodulin drugs: a role for calmodulin in DNA repair. *Science.* 224:1346–1348.

Chafouleas, J. G., R. L. Pardue, B. R. Brinkley, and A. R. Means. 1981. Regulation of intracellular levels of calmodulin and tubulin in normal and transformed cells. *Proceedings of the National Academy of Sciences.* 78:996–1000.

Cleveland, D. W., M. A. Lopata, P. Sherline, and M. W. Kirschner. 1981. Unpolymerized tubulin modulates the level of tubulin mRNA's. *Cell.* 25:537–546.

Connor, C. G., P. B. Moore, R. C. Brady, J. P. Horn, R. B. Arlinghaus, and J. R. Dedman. 1983. The role of calmodulin in cell transformation. *Biochemical and Biophysical Research Communications.* 112:647–654.

DiMaio, D., R. Treisman, and T. Maniatis. 1982. A bovine papilloma virus vector that propagates as a plasmid in both mouse and bacterial cells. *Proceedings of the National Academy of Sciences.* 79:4030–4034.

Epstein, P., A. R. Means, and M. Berchtold. 1986. Isolation of a rat parvalbumin gene and full length cDNA. *Journal of Biological Chemistry.* 261:5886–5891.

Karin, M., and R. I. Richards. 1982. Human metallothionein genes: primary structure of the metallothionein-II gene and a related processed gene. *Nature.* 299:797–802.

LaPorte, D. C., S. Gidwitz, M. J. Weber, and D. R. Storm. 1980. Relationship between changes in the calcium dependent regulatory protein and adenylate cyclase during viral transformation. *Biochemical and Biophysical Research Communications.* 86:1169–1177.

Lowy, D. R., E. Rand, and E. M. Scolnik. 1978. Helper independent transformation by unintegrated Harvey sarcoma virus DNA. *Journal of Virology.* 26:291–298.

Lusky, M., and M. R. Botchan. 1984. Characterization of the bovine papilloma virus plasmid maintenance sequences. *Cell.* 36:391–401.

Marcum, J. M., J. R. Dedman, B. R. Brinkley, and A. R. Means. 1978. Control of microtubule assembly-disassembly by calcium-dependent regulator protein. *Proceedings of the National Academy of Sciences.* 75:3771–3775.

Means, A. R., J. G. Chafouleas, and J. S. Tash. 1982. Physiological implications of the presence, distribution and regulation of calmodulin in eukaryotic cells. *Physiological Reviews.* 62:1–39.

Putkey, J. A., G. R. Slaughter, and A. R. Means. 1985. Bacterial expression and characterization of proteins derived from the chicken calmodulin cDNA and a calmodulin processed gene. *Journal of Biological Chemistry.* 260:4704–4712.

Sarver, N., P. Gruss, M. F. Law, G. Khoury, and P. M. Howley. 1981. Bovine papilloma virus deoxyribonucleic acid: a novel cloning vector. *Molecular and Cellular Biology.* 1:486–496.

Sasaki, Y., and H. Hidaka. 1982. Calmodulin and cell proliferation. *Biochemical and Biophysical Research Communications.* 104:451–456.

Simmen, R. C. M., T. Tanaka, K. F. Ts'ui, J. A. Putkey, M. J. Scott, E. C. Lai, and A. R. Means. 1985. The structural organization of the chicken calmodulin gene. *Journal of Biological Chemistry.* 260:907–912.

Tobey, R. A., and J. Seagrave. 1984. Inducibility of metallothionein throughout the cell cycle. *Molecular and Cellular Biology.* 4:2243–2245.

Veigel, M. L., T. C. Vanaman, M. E. Branch, and W. E. Sedwick. 1984. Calcium and calmodulin in cell growth and transformation. *Biochimica et Biophysica Acta.* 738:21–48.

Watterson, D. M., L. J. Van Eldik, R. E. Smith, and T. C. Vanaman. 1976. Calcium-dependent regulator protein of cyclic nucleotide metabolism in normal and transformed chicken embryo fibroblasts. *Proceedings of the National Academy of Sciences.* 73:2711–2715.

Zendequi, J. G., R. E. Zielinski, D. M. Watterson, and L. J. Van Eldik. 1984. Biosynthesis of calmodulin in normal and virus-transformed chicken embryo fibroblasts. *Molecular and Cellular Biology.* 4:883–889.

List of Contributors

Pennsylvania School of Medicine, Philadelphia, Pennsylvania.

List of Contributors

David Armstrong, Department of Biology, University of California, Los Angeles, California

Kamlesh Asotra, Department of Physiology, University of California, Los Angeles, California

Dennis A. Ausiello, Renal Unit, Massachusetts General Hospital, and Harvard Medical School, Boston, Massachusetts

P. F. Baker, Department of Physiology, MRC Secretory Mechanisms Group, King's College London, London, England

H. Banfić, Max-Planck-Institut für Biophysik, Frankfurt/Main, Federal Republic of Germany

Robert M. Bell, Department of Biochemistry, Duke University Medical Center, Durham, North Carolina

Christopher D. Benham, Department of Physiology, Yale University Medical School, New Haven, Connecticut

J. R. Berlin, Department of Physiology, University of Maryland School of Medicine, Baltimore, Maryland

M. Bond, Cleveland Clinic Foundation, Research Institute, Cleveland, Ohio

R. Broderick, Pennsylvania Muscle Institute and Department of Physiology, University of Pennsylvania School of Medicine, Philadelphia, Pennsylvania

Dennis Brown, Renal Unit, Massachusetts General Hospital, and Harvard Medical School, Boston, Massachusetts

M. B. Cannell, Department of Pharmacology, University of Miami School of Medicine, Miami, Florida

Ernesto Carafoli, Laboratory of Biochemistry, Swiss Federal Institute of Technology (ETH), Zurich, Switzerland

John Chad, Department of Neurophysiology, University of Southampton, Southampton, England

Sara Cohen, Departments of Cell Biology and Immunology, Hospital for Sick Children, and Department of Biochemistry, University of Toronto, Toronto, Ontario, Canada

Kathleen E. Coll, Department of Biochemistry and Biophysics, University of Pennsylvania School of Medicine, Philadelphia, Pennsylvania

Norman Davidson, Divisions of Biology and Chemistry, California Institute of Technology, Pasadena, California

Michael Delay, Department of Medical Physiology, University of Calgary, Calgary, Alberta, Canada

L. Eckhardt, Max-Planck-Institut für Biophysik, Frankfurt/Main, Federal Republic of Germany

Charles Filburn, Gerontology Research Center, Laboratory of Molecular Aging, National Institute on Aging, Baltimore, Maryland

L. Fink, Departments of Pharmacology and Physiology, Yale University, New Haven, Connecticut

A. Fox, Departments of Pharmacology and Physiology, Yale University, New Haven, Connecticut

Julia D. Goetz-Smith, Departments of Cell Biology and Immunology, Hospital for Sick Children, and Department of Biochemistry, University of Toronto, Toronto, Ontario, Canada

Y. E. Goldman, Pennsylvania Muscle Institute and Department of Physiology, University of Pennsylvania School of Medicine, Philadelphia, Pennsylvania

Sergio Grinstein, Departments of Cell Biology and Immunology, Hospital for Sick Children, and Department of Biochemistry, University of Toronto, Toronto, Ontario, Canada

Yusuf A. Hannun, Department of Medicine, Duke University Medical Center, Durham, North Carolina

Carl A. Hansen, Department of Biochemistry and Biophysics, University of Pennsylvania School of Medicine, Philadelphia, Pennsylvania

Richard G. Hansford, Energy Metabolism and Bioenergetics Section, Laboratory of Cardiovascular Science, Gerontology Research Center, National Institute on Aging, Baltimore, Maryland

John Hartwig, Hematology-Oncology Unit, Massachusetts General Hospital, and Harvard Medical School, Boston, Massachusetts

Peter Ingram, Research Triangle Institute, Research Triangle Park, North Carolina

William R. Jacobs, Department of Physiology, Duke University Medical Center, Durham, North Carolina

Roy Johanson, Department of Biochemistry and Biophysics, University of Pennsylvania School of Medicine, Philadelphia, Pennsylvania

L. K. Kaczmarek, Departments of Pharmacology and Physiology, Yale University, New Haven, Connecticut

Daniel Kalman, Department of Biology, University of California, Los Angeles, California

T. Kemmer, Max-Planck-Institut für Biophysik, Frankfurt/Main, Federal Republic of Germany

D. E. Knight, Department of Physiology, MRC Secretory Mechanisms Group, King's College London, London, England

W. J. Lederer, Department of Physiology, University of Maryland School of Medicine, Baltimore, Maryland

Ann LeFurgey, Department of Physiology, Duke University Medical Center, Durham, North Carolina

John P. Leonard, Divisions of Biology and Chemistry, California Institute of Technology, Pasadena, California

Henry A. Lester, Divisions of Biology and Chemistry, California Institute of Technology, Pasadena, California

Stefano Longoni, Cardiovascular Research Laboratory, University of California at Los Angeles, School of Medicine, Los Angeles, California

Hermann Lübbert, Divisions of Biology and Chemistry, California Institute of Technology, Pasadena, California

Lazaro J. Mandel, Department of Physiology, Duke University Medical Center, Durham, North Carolina

Anthony R. Means, Department of Cell Biology, Baylor College of Medicine, Houston, Texas

L. J. Mullins, Department of Biophysics, University of Maryland School of Medicine, Baltimore, Maryland

Joël Nargeot, Cèntre de Recherches de Biochemie Macromoléculaire, C.N.R.S., Montpellier, France

David G. Nicholls, Department of Biochemistry, University of Dundee, Dundee, Scotland

Colin D. Rasmussen, Department of Cell Biology, Baylor College of Medicine, Houston, Texas

J. Requena, Instituto Internacional de Estudios Avanzados, Centro de Biociencias, Caracas, Venezuela

Jose Sanchez-Prieto, Department of Biochemistry, University of Dundee, Dundee, Scotland

S. Schnefel, Max-Planck-Institut für Biophysik, Frankfurt/Main, Federal Republic of Germany

I. Schulz, Max-Planck-Institut für Biophysik, Frankfurt/Main, Federal Republic of Germany

H. Shuman, Pennsylvania Muscle Institute and Department of Physiology, University of Pennsylvania School of Medicine, Philadelphia, Pennsylvania

Talvinder S. Sihra, Department of Biochemistry, University of Dundee, Dundee, Scotland

Terry P. Snutch, Divisions of Biology and Chemistry, California Institute of Technology, Pasadena, California

Stephen P. Soltoff, Tufts University School of Medicine, Boston, Massachusetts

A. P. Somlyo, Pennsylvania Muscle Institute and Departments of Physiology and Pathology, University of Pennsylvania School of Medicine, Philadelphia, Pennsylvania

A. V. Somlyo, Pennsylvania Muscle Institute and Department of Physiology, University of Pennsylvania School of Medicine, Philadelphia, Pennsylvania

James M. Staddon, Energy Metabolism and Bioenergetics Section, Laboratory of Cardiovascular Science, Gerontology Research Center, National Institute on Aging, Baltimore, Maryland

J. A. Strong, Departments of Pharmacology and Physiology, Yale University, New Haven, Connecticut

Takehito Takano, Tokyo Medical and Dental University, Tokyo, Japan

F. Thévenod, Max-Planck-Institut für Biophysik, Frankfurt/Main, Federal Republic of Germany

D. R. Trentham, Medical Research Council, National Institute for Medical Research, London, England

Richard W. Tsien, Department of Physiology, Yale University Medical School, New Haven, Connecticut

Julio Vergara, Department of Physiology, University of California, Los Angeles, California

Arthur Verhoeven, Central Laboratory of the Blood Transfusion Service, Amsterdam, The Netherlands

J. W. Walker, Medical Research Council, National Institute for Medical Research, London, England

John R. Williamson, Department of Biochemistry and Biophysics, University of Pennsylvania School of Medicine, Philadelphia, Pennsylvania

Michael T. Williamson, Department of Biochemistry and Biophysics, University of Pennsylvania School of Medicine, Philadelphia, Pennsylvania

Subject Index